P9-DTA-234

NATIONAL GEOGRAPHIC

SCIENCE OF EVERYTHING

HOW THINGS WORK IN OUR WORLD

FROM CELL PHONES, SOAP BUBBLES & VACCINES TO GPS, X-RAYS & SUBMARINES

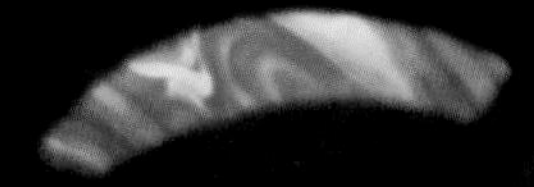

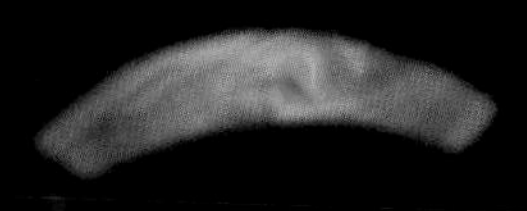

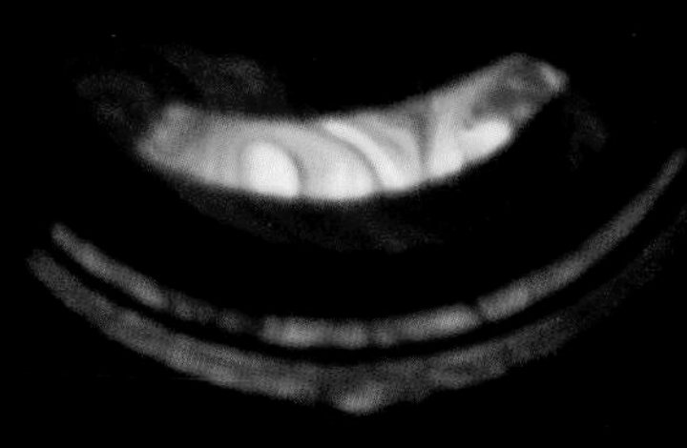

NATIONAL GEOGRAPHIC
Washington, D.C.

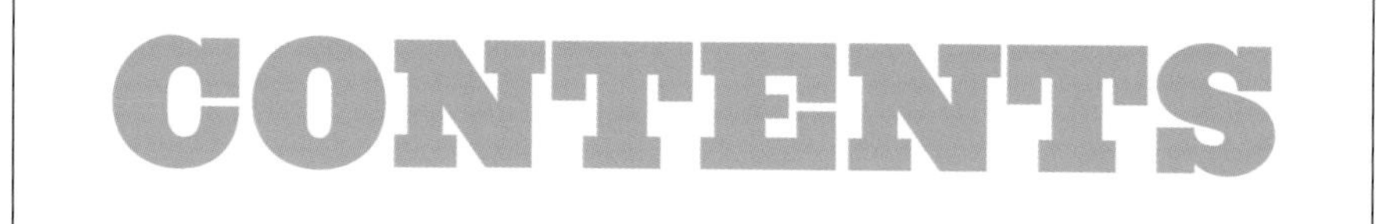

CONTENTS

PART 2
NATURAL FORCES

PART 3
MATERIALS AND CHEMISTRY

PART 4
BIOLOGY AND MEDICINE

The silhouetted figure is standing behind 21 square feet of human skin substitute.

FOREWORD

DAVID POGUE

As universes go, ours is a pretty cool one. It's bursting with brain-frying miracles: rainbows, volcanoes, music, squids, tides, electricity, gravity, fireworks, earthquakes, cellos. And those are just the ones you can see.

Furthermore, the natural world is only the beginning. Consider some of human technology's greatest hits, like microwave ovens. Sunscreen. Superglue. Ziploc bags. Noise-canceling headphones. Probiotic yogurt. If the ancients saw our modern inventions in action, they'd fall down and worship us as gods.

We developed the skills for invention, of course, only because we're a curious species. From the beginning, we've tried to ferret out the secrets behind the mind-blowers all around us. How do they work? How can they be possible? How can the miraculous be explained?

The Greeks dreamed up an archipelago of gods. Medieval scientists imagined mood-altering fluids in our veins. To this day, plenty of people credit unexplained phenomena to special human powers, like witch doctors, faith healers, and late-night TV psychics.

Well, with all due respect to the gullible and the ancient, it's all just science.

Actually, the phrase "just science" doesn't do the point justice. Knowing that science explains the boggling spectacles of our universe doesn't make them any less miraculous. In fact, knowing that every one of nature's magic tricks actually boils down to a few scientific principles makes them even more amazing.

Lately, it seems as though our society has entered a weird, anti-science phase. When school-system budgets get tight, science programs often get the axe. No matter how much proof they're shown, skeptics still doubt the reality of solidly established scientific theories like global warming and evolution. When it comes to funding, congressional lawmakers still make science a low priority.

And unless you want to be depressed, don't look at the rankings of American students' science scores compared with other countries. We come in 25th, just behind Latvia.

But it turns out that some of the greatest challenges ahead—sustainable energy, global warming, clean water, health care, food supply—will require scientific and technical solutions. And in the next five years, there will be twice as many science and technology jobs as other kinds of jobs.

Bottom line: This is the wrong time for the world to go anti-science.

Besides, it's easy to make the case that spirituality isn't at odds with science. Many scientists are filled with a deep, reverent awe at the astonishingly complex, effective systems that nature has come up with all by itself.

But never mind all that. You didn't pick up this book for a lecture or a bummer. The point is, science doesn't have to be intimidating, or scary, or dry; presented well, the stories of science are incredibly interesting—especially when they peel back the operations of everyday things.

And that, of course, is the joy of this book. The number of "aha" moments per page is stunning.

Shellac comes from a bug's back. Air conditioners don't add coolness to a room; they actually subtract heat. Color-boosting detergents work by leaving a fluorescent film on your clothes. All the gold ever mined would fit into a 60-foot cube.

You'll also find that the science of everyday future things is also well represented here. Sustained nuclear fusion. 3-D printing. Hybrid cars. Geothermal energy. Earthquake-proof buildings. Magnetically levitating ships. Cloning and gene therapy. Meat grown in a lab, without killing animals.

Even if you're not especially interested in the science of the world around you, it's thrilling to find out the secrets of how everything works. The world teems with technologies that seem like magic—and we just take them for granted. How do lasers work? GPS, cellphones, WiFi, and Bluetooth? How do the TSA's airport body scanners work? What about holograms?

We assume that somebody, somewhere knows how it was invented and how it works. Why not you?

Whether you read cover to cover or just dip in for occasional micro-enlightenments, it's a pretty good bet that this book will leave you subtly changed. You'll find yourself both wiser and more amazed. You'll be a better conversationalist.

Above all, you'll have a new appreciation for the magic of science, the genius of human ideas, and the coolness of the universe you've been handed.

HOW TO USE THIS BOOK

THERE ARE FOUR TYPES OF PRESENTATIONS:

CHAPTER 11

Elements

We are surrounded by elements. They're part of the air we breathe, the food we eat, the clothes we wear, and the sidewalks we walk on. If it weren't for the centuries of study that led to the discovery of these elements and their properties, the world would be a very different place.

The elements are the most basic building blocks of chemistry, and the science of chemistry examines their properties and ability to be combined to create compounds and materials.

Most of the materials, tools, and conveniences we use can be traced to the discovery of an element or its combination with others. Each new element on the periodic table (a table that lists the chemical elements according to the number of protons each has in its nucleus) has meant another step forward for humankind, from the discovery of neon to gold and silver and their hundreds of uses.

The discovery of neon forever changed the world at night, thanks to the creation of neon signs that can be seen miles away and drive customers to a business's door. Thanks to the element lithium, all our electronics stay charged longer, including what has become one of the most vital electronic instruments to humankind, the cell phone.

Without iodine, humans would suffer from thyroid disease, and aluminum's production led to lighter-weight aircraft and space exploration.

Nations' economies have been based on gold since 700 B.C. (beginning in the ancient civilization of Lydia), and our eating utensils and jewelry would be lusterless without silver. Chlorine keeps our drinking water safe. Without silicon the world wouldn't have sand, clay, or electrical steel. Argon keeps the filaments in light bulbs from corroding, and titanium creates brilliantly white paint.

Without understanding the properties of these and the other elements, life would be more difficult, dimmer, less productive, and, most likely, less enjoyable.

CHAPTER INTRODUCTION Each chapter groups related scientific principles and applications and begins with an exploration of the context.

PRINCIPLE FEATURE Clear, concise introductions provide an appreciation of the science behind the inventions and phenomena in the world around us.

LIGHT

PRINCIPLE

UNSEEN LIGHT

Ultraviolet Light

The frequencies of light occupy a very wide spectrum, and we can see only part of it—the part called visible light. However, there are electromagnetic waves that we can't see, and ultraviolet (UV) is one example. Discovered in 1801, it sits past the violet end of the spectrum and has shorter wavelengths than visible light. Although our human eyes can't detect it, the eyes of some insects can, such as bumblebees. Ultraviolet light comes from the Sun and is classified according to its intensity with A being the least harmful and C being the most harmful.

Ultraviolet Light

UV LIGHT AND EYES

Just because the human eye can't see ultraviolet light doesn't mean that UV isn't dangerous. In particular, UV-A and UV-B rays pose the greatest threat because these are the ones that tend to get through Earth's protective ozone layer. UV light damages DNA in cells by causing a reaction among molecules of one of the bases in DNA (thymine). Prolonged exposure to the eyes can lead to all sorts of problems, ranging from cataracts to macular degeneration. So does this mean you should spend your time indoors and only come out at night? Hardly. Easy ways to protect your eyes from UV light include wearing sunglasses and broad-brimmed hats when outdoors and seeing an eye doctor on a regular basis to detect problems early.

»»See Also

RELATED PRINCIPLES: Radiation, 239 · Health and Medicine, 354

RELATED APPLICATIONS: Spectrum of Wavelengths, 248 · UV-Blocking Fabric, 308

215

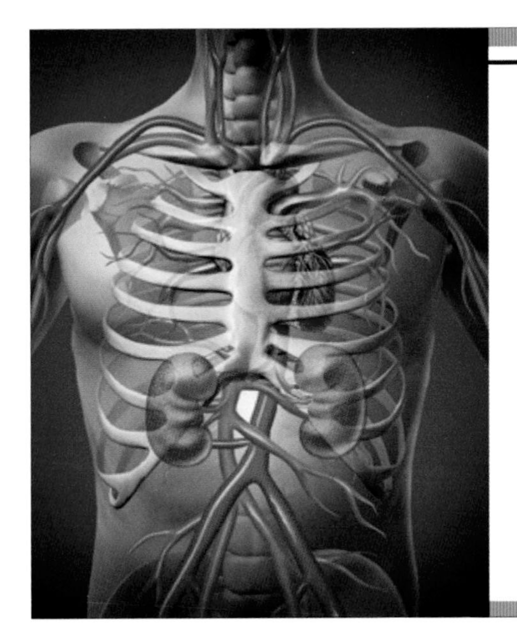

HEALTH AND MEDICINE

Bioengineering

KIDNEY DIALYSIS

Kidney function can be knocked out suddenly in the case of injury or overdose, or it can decline gradually as a result of diabetes or high blood pressure. Either way, the failure of these bean-shaped organs in the lower back to rid the blood of wastes and excess water can be deadly. The blood becomes overloaded with urea, a metabolic waste product, and minerals such as sodium (which can increase blood pressure), potassium (which can lead to irregular heartbeat), and phosphorus (which causes a loss of calcium in the bones).

In dire cases of kidney failure, patients can turn to a treatment called dialysis, which takes over the kidneys' most critical roles. It can be used during recovery from an acute problem or on an ongoing basis. The machine works by drawing the patient's blood through tubing into a bath of purifying liquid, then returning it to the body. The tubing is made of semipermeable membrane. When exposed to such a membrane, particles in solution always move from an area of higher concentration to one of lower concentration, according to the principle of osmosis. So in dialysis, the toxic waste particles in the blood cross the membrane into the dilute water; the larger particles, such as red blood cells, are too big to pass through the membrane's pores.

It was the anguish of watching a young man die slowly of kidney failure that moved the Dutch physician Willem Kolff to design the first dialysis machine in the early 1940s. He had to scrounge parts for its construction. His initial test of the prototype came in 1943; the first several patients died, but in 1945, the machine revived a comatose patient. By keeping patients with kidney failure alive, dialysis paved the way for the first successful kidney transplants. Although kidney dialysis is most often administered in a dialysis center, home dialysis has been growing as an option.

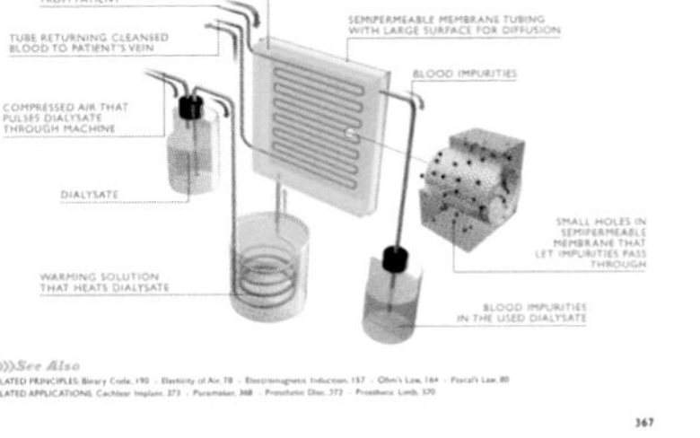

See Also

367

SIDEBAR FEATURE Related principles and applications as well as spotlights on inventors are featured throughout.

APPLICATION FEATURE Articles explain how things work—from the everyday to the extraordinary—with accompanying photos and diagrams.

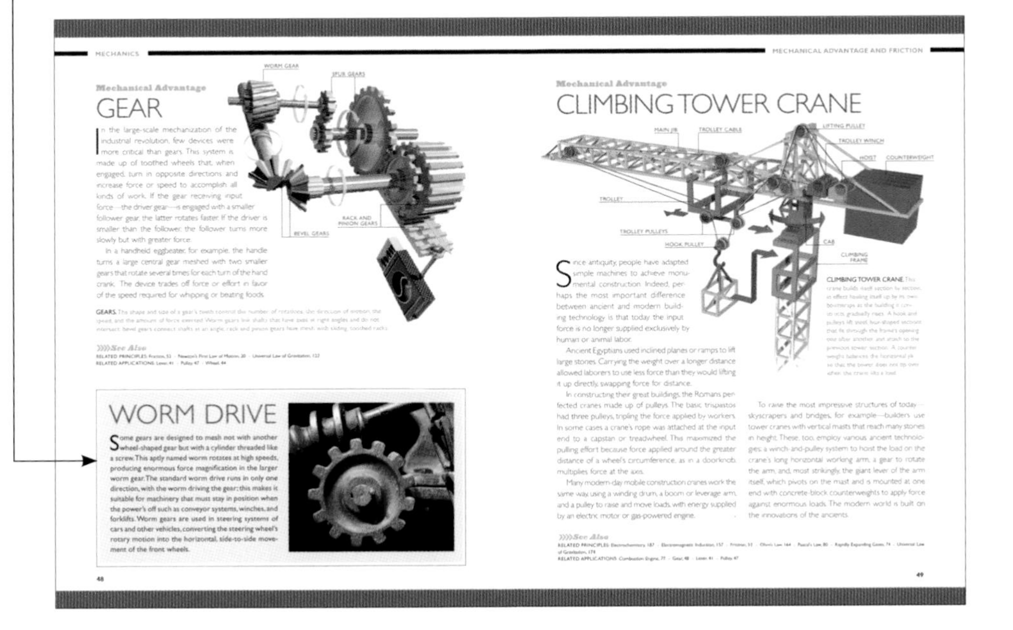
MECHANICS

MECHANICAL ADVANTAGE AND FRICTION

Mechanical Advantage

GEAR

In the large-scale mechanization of the industrial revolution, few devices were more critical than gears. This system is made up of toothed wheels that, when engaged, turn in opposite directions and increase force or speed to accomplish all kinds of work. If the gear receiving input force—the driver gear—is engaged with a smaller follower gear, the latter rotates faster. If the driver is smaller than the follower, the follower turns more slowly but with greater force.

In a handheld eggbeater, for example, the handle turns a large central gear meshed with two smaller gears that rotate several times for each turn of the hand crank. The device trades off force or effort in favor of the speed required for whipping or beating foods.

See Also

WORM DRIVE

Some gears are designed to mesh not with another wheel-shaped gear but with a cylinder threaded like a screw. This aptly named worm rotates at high speeds, producing enormous force magnification in the larger worm gear. The standard worm drive runs in only one direction, with the worm driving the gear; this makes it suitable for machinery that must stay in position when the power's off such as conveyor systems, winches, and forklifts. Worm gears are used in steering systems of cars and other vehicles, converting the steering wheel's rotary motion into the horizontal, side-to-side movement of the front wheels.

48

Mechanical Advantage

CLIMBING TOWER CRANE

Since antiquity, people have adapted simple machines to achieve monumental construction. Indeed, perhaps the most important difference between ancient and modern building technology is that today the input force is no longer supplied exclusively by human or animal labor.

Ancient Egyptians used inclined planes or ramps to lift large stones. Carrying the weight over a longer distance allowed laborers to use less force than they would lifting it up directly, swapping force for distance.

In constructing their great buildings, the Romans perfected cranes made up of pulleys. The basic trispastos had three pulleys, tripling the force applied by workers. In some cases a crane's rope was attached at the input end to a capstan or treadwheel. This maximized the pulling effort because force applied around the greater distance of a wheel's circumference, as in a doorknob, multiplies force at the axis.

Many modern-day mobile construction cranes work the same way, using a winding drum, a boom or leverage arm, and a pulley to raise and move loads, with energy supplied by an electric motor or gas-powered engine.

To raise the most impressive structures of today—skyscrapers and bridges, for example—builders use tower cranes with vertical masts that reach many stories in height. These, too, employ various ancient technologies: a winch-and-pulley system to hoist the load on the crane's long horizontal working arm, a gear to rotate the arm, and, most strikingly, the giant lever of the arm itself, which pivots on the mast and is mounted at one end with concrete-block counterweights to apply force against enormous loads. The modern world is built on the innovations of the ancients.

See Also

49

Performance by a member of the Canadian national parachute team

A wooden kitchen match bursts into flame.

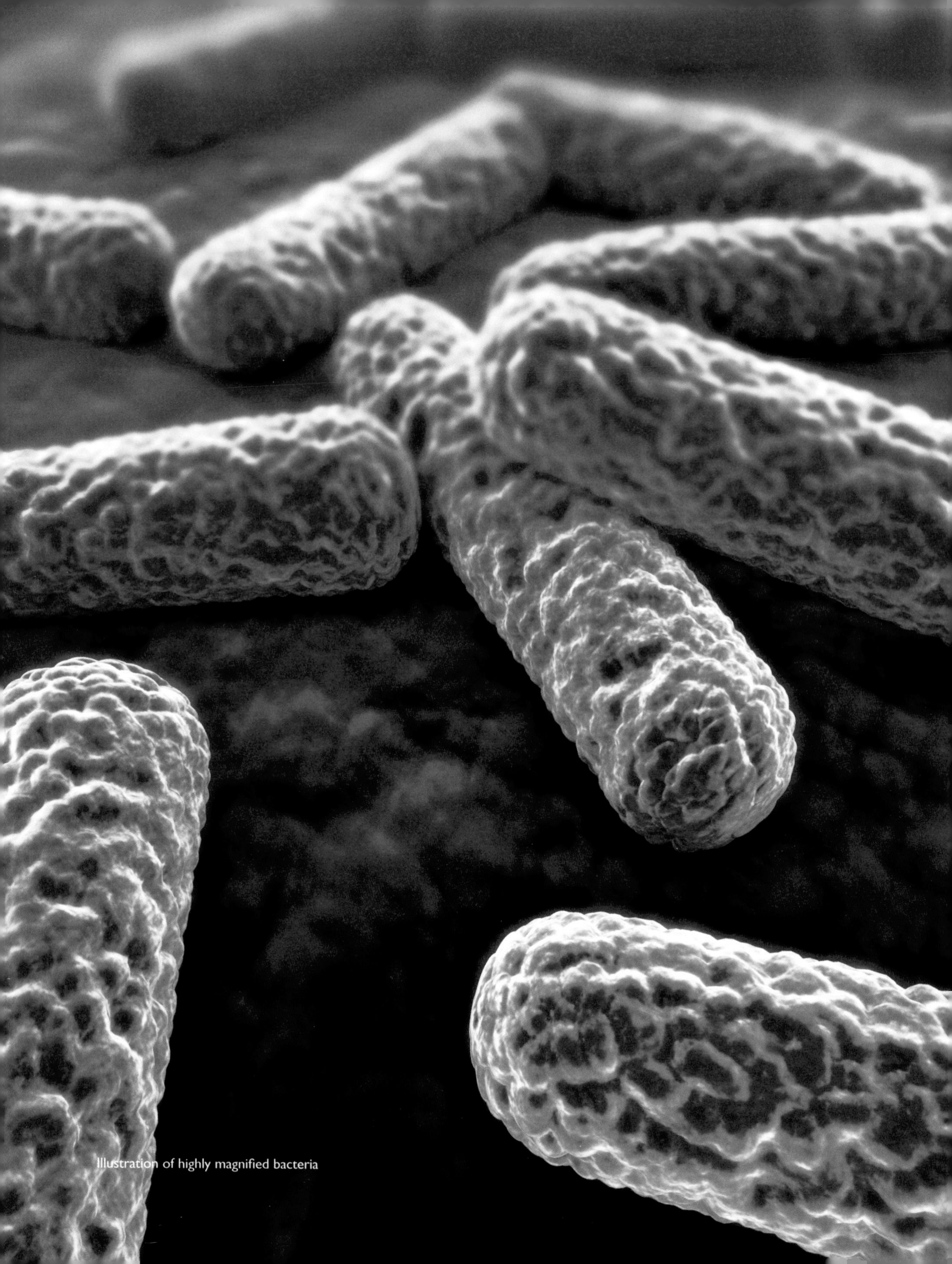
Illustration of highly magnified bacteria

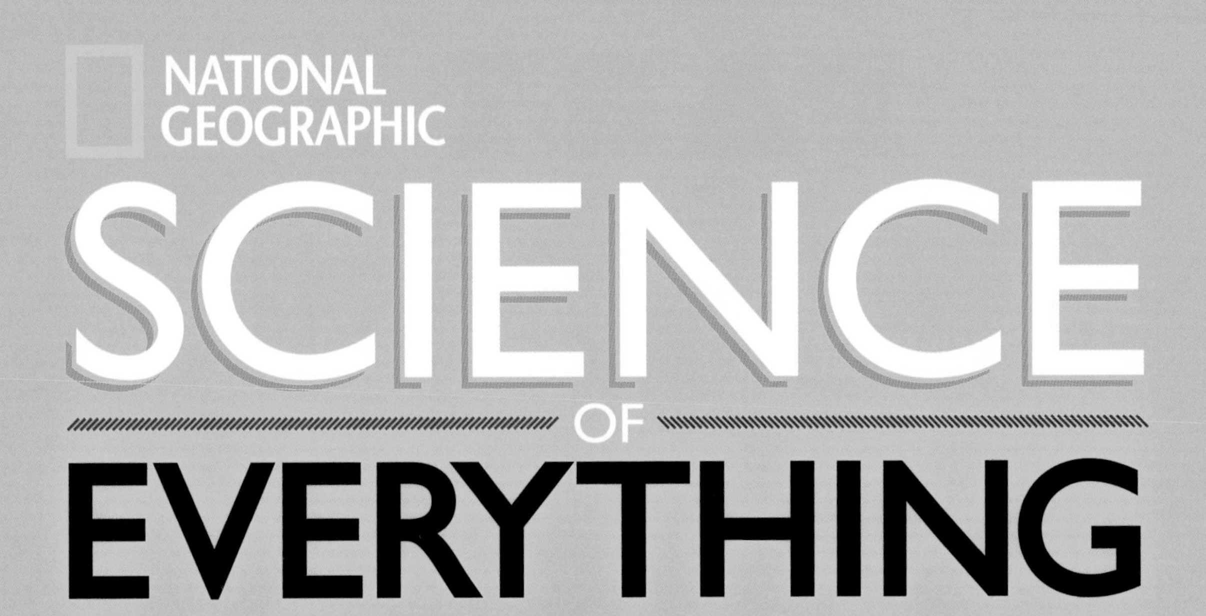
NATIONAL
GEOGRAPHIC
SCIENCE
OF
EVERYTHING

PART 1
MECHANICS

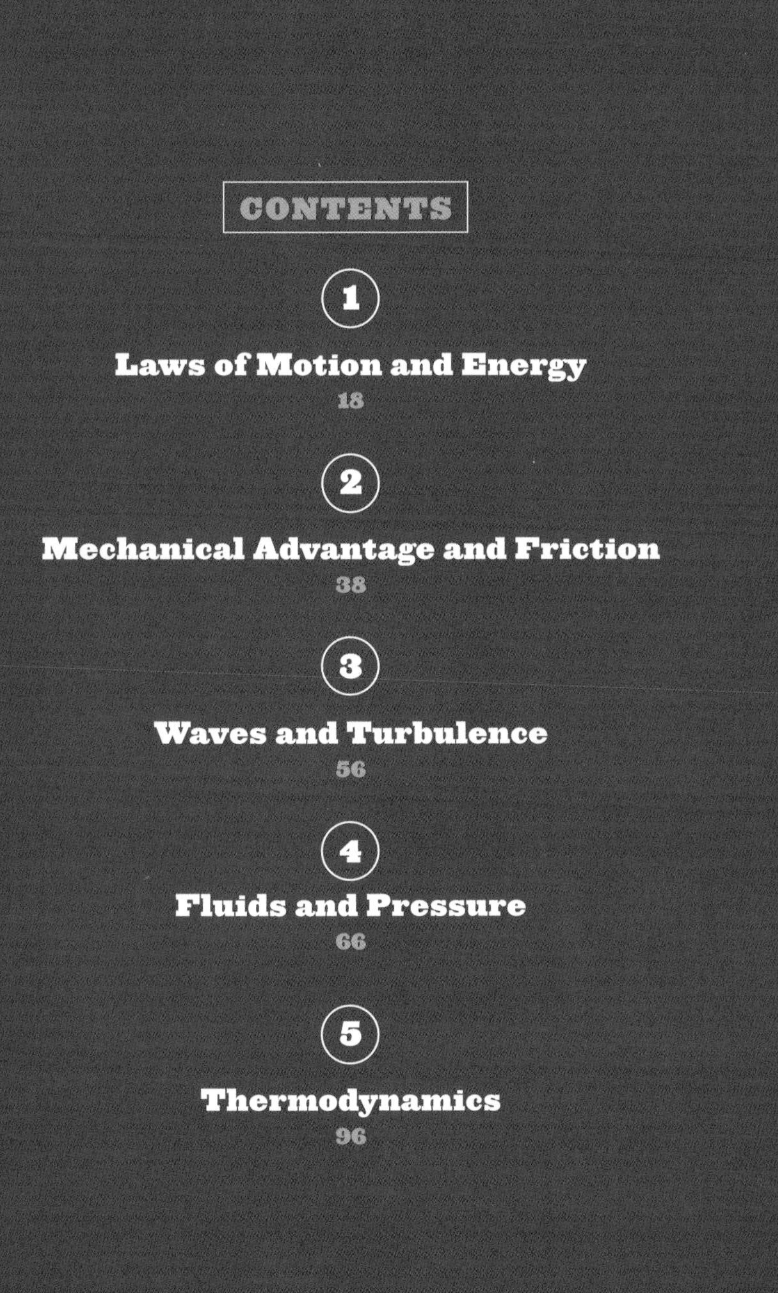

CONTENTS

CHAPTER 1

Laws of

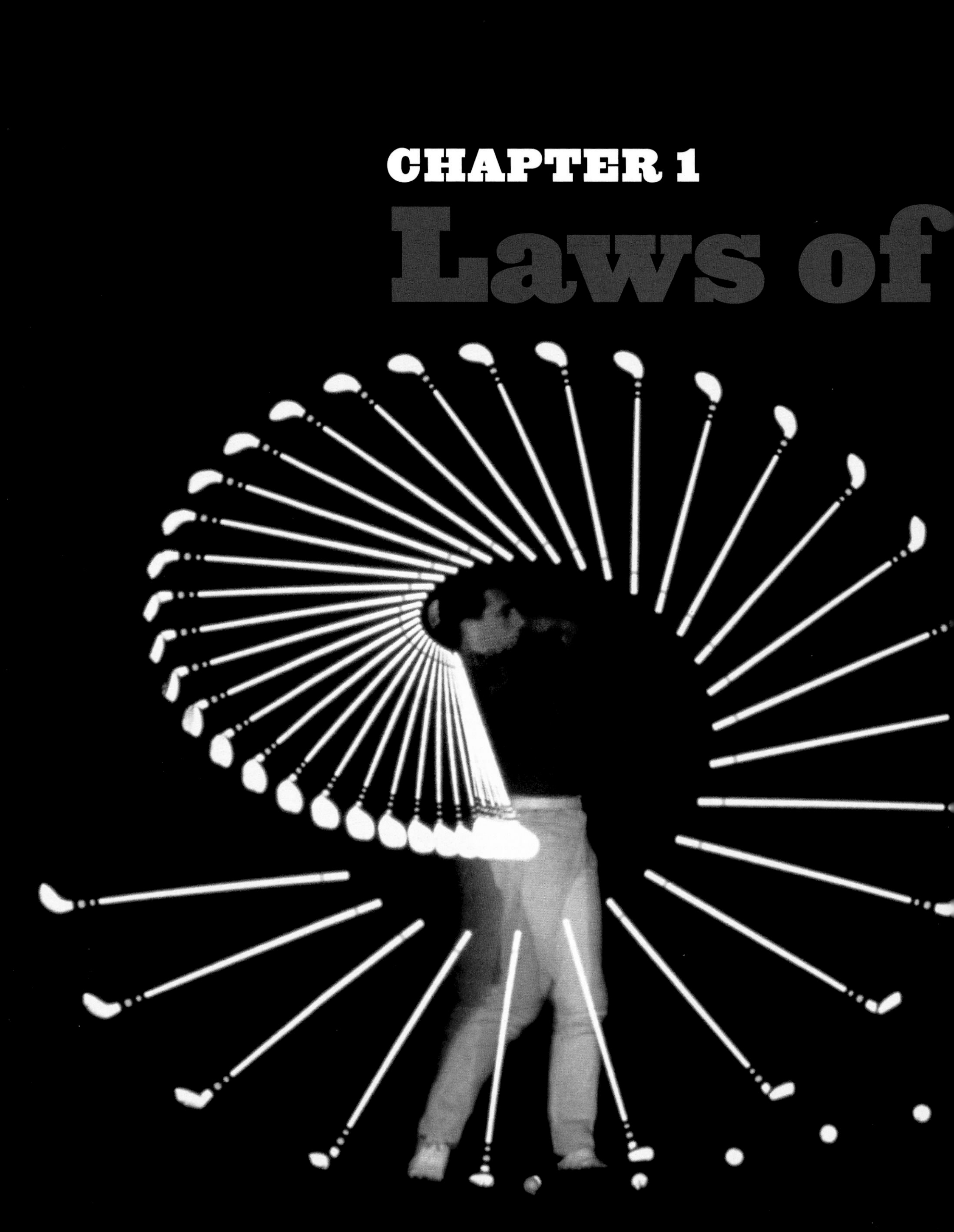

Motion and Energy

Sir Isaac Newton in his 1687 *Philosophiae Naturalis Principia Mathematica* published observations that would form the foundation of classical mechanics. Perhaps most famously, Newton's book lays out the law of universal gravitation. But just as important, the *Principia* puts forth Newton's three laws of motion and energy.

The first law says that an object will retain its state of motion unless some external force acts on it. If it's moving, it will keep moving; if at rest, it will remain so.

The second law describes how much force is required to achieve a given change in the velocity of a particular mass (Force = Mass × Acceleration). Put another way, it predicts the acceleration of a mass when a given force is applied. Newton's third law asserts that for every action, there is a reaction of equal magnitude in the opposite direction.

"If I have seen further," Newton wrote, "it is by standing on the shoulders of giants." Newton did draw deeply on the work of earlier philosophers, physicists, and mathematicians. But it was he who showed that the physical world is governed by universal, mathematical laws—and who began to set down those laws in elegant formulations borne out by centuries of observation. Newton's laws advanced the scientific revolution of the seventeenth century and laid a foundation for the industrial revolution of the nineteenth century.

These laws describe the behavior of idealized particles or points. Objects can often be treated as points when studying motion, but their movement involves more complex dynamics. A half-century after Newton, Swiss physicist Leonhard Euler extended Newton's work in two laws that describe the movement of bodies made up of an assemblage of points.

PRINCIPLE

RESISTANT TO CHANGE

Newton's First Law of Motion

Absent some force acting on it, a ball sitting on the floor will just sit there. A ball rolling along a perfectly frictionless and even surface will likewise continue rolling, in the same direction and at the same speed, indefinitely.

Indeed, according to Newton's first law of motion—more familiarly known as the law of inertia—all objects inherently resist change in their state of motion. A thing at rest remains at rest, and an object in motion keeps moving in the same direction and at the same speed, unless some outside force disrupts it.

Newton's First Law of Motion

SEGWAY

Unveiled in 2001, the Segway PT (Personal Transporter) is a two-wheeled, self-balancing, electric-powered vehicle designed to mimic the naturalness of walking and turning—and maybe even pausing to smell the roses. Its movement responds to subtle shifts in the user's body weight forward, backward, and to the side while remaining upright. Riders steer by leaning in the direction they want to go. This balancing act involves a high-tech elaboration on Newton's law of inertia. It's based on five miniature silicon gyroscopic devices coupled with electronic sensors that monitor change in the pitch of the vehicle's platform in comparison with the inertial movement of the gyroscopes.

»»See Also

RELATED PRINCIPLES: Binary Code, 190 • Electrochemistry, 187 • Electromagnetic Induction, 157 • Law of Conservation of Energy, 30 • Mechanical Advantage, 38 • Ohm's Law, 164 • Universal Law of Gravitation, 122

RELATED APPLICATIONS: Electric Motor, 158 • Flight Stabilizer, 23 • Gyroscope, 21 • Ship Stabilizer, 22

Newton's First Law of Motion

GYROSCOPE

Just about everyone is familiar with one gyroscopic instrument—the child's spinning top, developed independently in ancient civilizations around the world. Why does it go on spinning? Inertia. A moving object will persist in its motion, along the same axis, unless a force acts to change or stop it.

In 1852 French physicist Léon Foucault created a precision gyroscope and gave the instrument its name, made up of Greek words meaning essentially rotation watcher. While holding his gyroscope cage steady, Foucault showed that the angle of the inner disk's rotation seemed to change very gradually on its own. In fact it was the frame moving; this was an early demonstration of Earth's own rotation on its axis.

Once in motion, the gyroscope's inner wheel maintains the orientation of its rotation even as its outer frame is tilted. Because of this, the gyroscope is a tremendously useful tool in setting or maintaining direction in a number of applications, from aircraft navigation to missile guidance to tunnel mining.

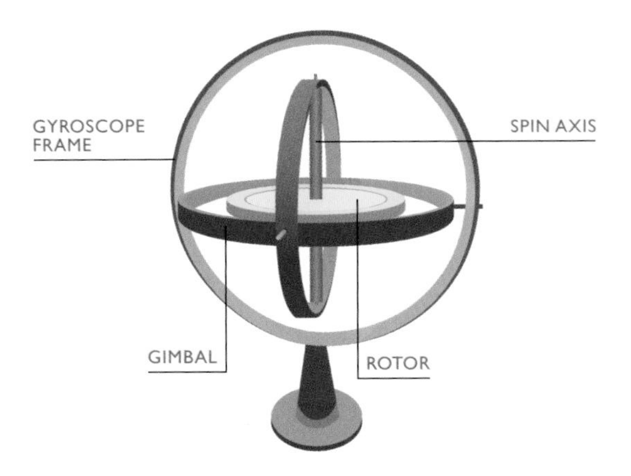

See Also

RELATED PRINCIPLES: Law of Conservation of Energy, 30 • Universal Law of Gravitation, 122
RELATED APPLICATIONS: Flight Stabilizer, 23 • Segway, 20 • Ship Stabilizer, 22

Newton's First Law of Motion

SHIP STABILIZER

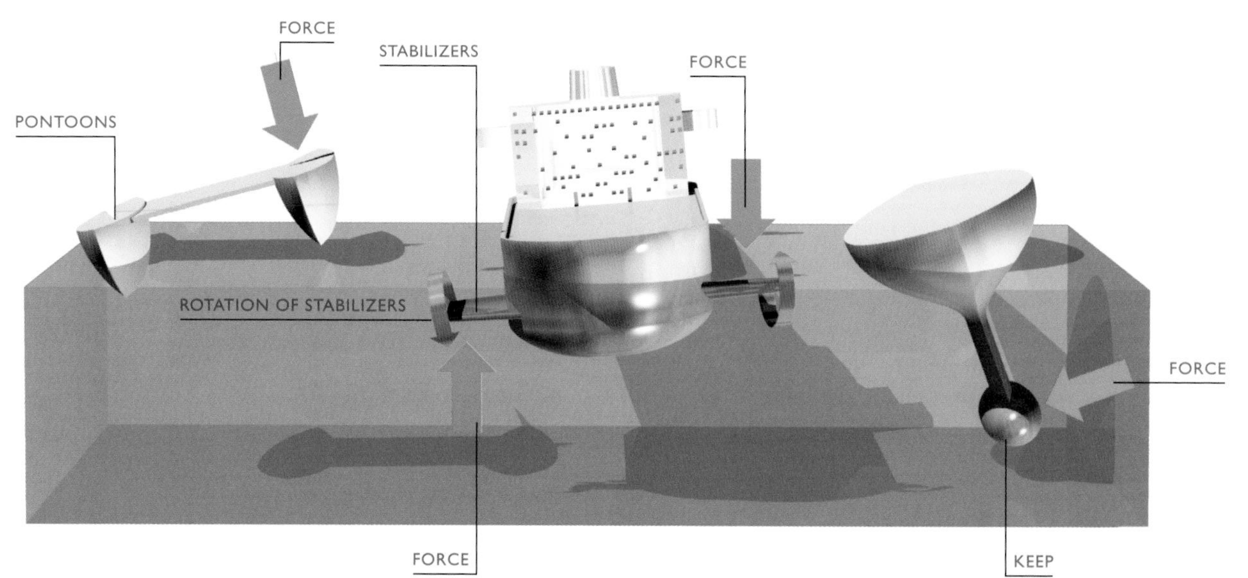

STABILITY. The steadiness of a ship depends on the vessel's center of gravity. Keels and pontoons help maintain stability and minimize the risk of capsizing, but very large vessels may employ stabilizers, winglike protuberances that provide lift to counteract the sea's roll.

Hold the hub of a swiftly spinning bicycle wheel, and try to change its angle of rotation; you'll feel it resist. Its tendency to maintain orientation makes the wheel seem to push back in the other direction.

As we shall see on page 36, angular momentum refers to the tendency of rotating objects to continue rotating unless acted on by a torque. This natural persistence of angular momentum is the basis for gyrostabilizers in ships. As a craft encounters the kind of wave that makes ocean travel hopelessly nauseating for some, the spinning rotor of the stabilizer resists, exerting a counterforce that keeps the vehicle from rolling.

The first gyroscopic ship stabilizers were used in United States naval ships and large ocean liners in the early twentieth century. They relied on the mass of the main stabilizing gyro itself to right the vessel. The Navy's first gyrostabilizer was placed in the hold of a small destroyer and weighed five tons. Later, designers adapted this method by using smaller gyroscopes whose rotational force controls stabilizing fins that extend from the ship's hull.

»»See Also

RELATED PRINCIPLES: Archimedes' Principle, 92 • Binary Code, 190 • Electrochemistry, 187 • Electromagnetic Induction, 157 • Law of Conservation of Energy, 30 • Mechanical Advantage, 38 • Ohm's Law, 164 • Pascal's Law, 80 • Universal Law of Gravitation, 122

RELATED APPLICATIONS: Electric Motor, 158 • Flight Stabilizer, 23 • Gyroscope, 21 • Segway, 20

Newton's First Law of Motion

FLIGHT STABILIZER

Aim a gyroscope's axis at true north, and it will continue to aim there, no matter what orientation is taken by its outer frame. This is the basis for gyrostabilizers used to keep a plane flying level. Inventor Lawrence B. Sperry, who developed the first such stabilizers for ships, first wowed the public with his three-way flight gyrostabilizer at the International Airplane Safety Competition in Paris in 1914. This gyroscope's inertial rotation controlled movement along the plane's three axes of movement—yaw (nose right or left), pitch (nose up or down), and wing-to-wing roll. Sperry thrust his arms in the air, releasing the controls; the gyroscope worked like an automatic pilot.

Similar technology helps point the Hubble Space Telescope, launched in 1990. A wheel inside each of six gyroscopes spins at a rate of 19,200 revolutions per minute on gas bearings. Electronic sensors in the gyros relay information about even slight movements of the scope to Hubble's central computer.

GIMBAL ROTATION
MAIN DRIVE GEAR
COMPASS CARD GEAR
GIMBAL
ADJUSTMENT KNOB
GYRO
ADJUSTMENT GEARS

US Navy Demonstration Squadron Blue Angels flying Boeing F/A-18s

See Also

RELATED PRINCIPLES: Bernoulli's Principle, 68 • Binary Code, 190 • Electrochemistry, 187 • Electromagnetic Induction, 157 • Law of Conservation of Energy, 30 • Mechanical Advantage, 38 • Ohm's Law, 164 • Pascal's Law, 80 • Universal Law of Gravitation, 122
RELATED APPLICATIONS: Electric Motor, 158 • Gyroscope, 21 • Segway, 20 • Ship Stabilizer, 22

Newton's First Law of Motion

WASHING MACHINE

This everyday household appliance ingeniously applies Newton's insights to the wringing of soaked laundry, a chore roundly dreaded in the preindustrial age as the most strenuous part of washday.

According to Newton's first law of motion, matter will move in a straight line unless forced to do otherwise. In the washing machine's spin cycle, the drum rotates rapidly, its inner walls continually pushing the clothes into a circular path. This directional change by definition means the clothes are accelerating and thus subject to external force. The rapid mechanical spinning can exert quite substantial force, as anyone who has witnessed the banging of an unevenly loaded washer can attest.

But holes in the walls of the drum mean no such inward force acts on the water. The wash water is permitted to fly off along the linear path predicted by Newton, effectively wringing the clothes against the spinning drum.

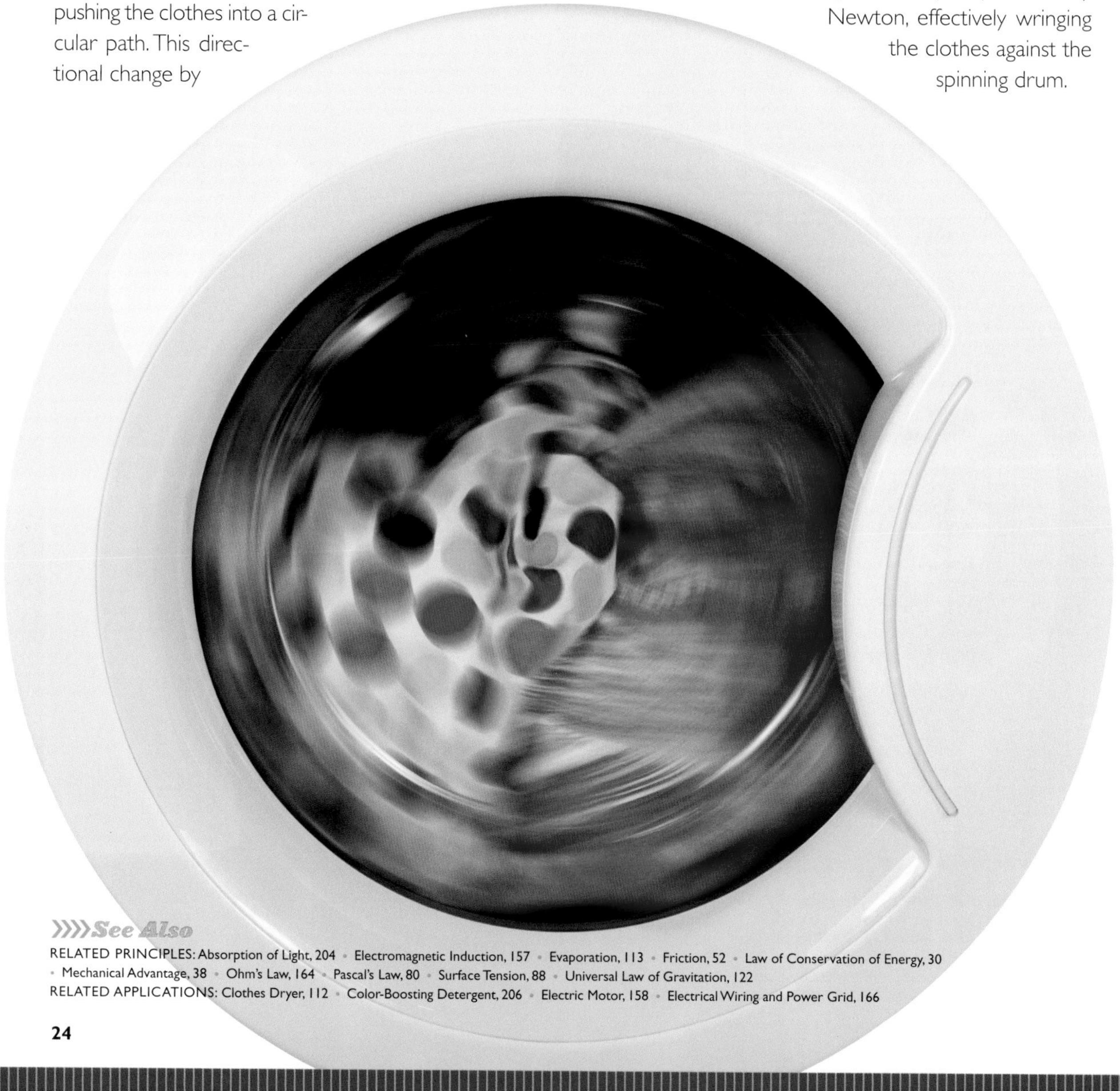

»»See Also

RELATED PRINCIPLES: Absorption of Light, 204 • Electromagnetic Induction, 157 • Evaporation, 113 • Friction, 52 • Law of Conservation of Energy, 30 • Mechanical Advantage, 38 • Ohm's Law, 164 • Pascal's Law, 80 • Surface Tension, 88 • Universal Law of Gravitation, 122

RELATED APPLICATIONS: Clothes Dryer, 112 • Color-Boosting Detergent, 206 • Electric Motor, 158 • Electrical Wiring and Power Grid, 166

FROM THE TOP. Cone-shaped, motor-driven agitator in a top-loading washer moves clothes back and forth in a tub filled with soapy water. When the wash cycle ends, a timer directs the motor to free the agitator and spin a perforated inner basket. Centrifugal force spins water through the holes in the outer tub and presses the clothes against the sides of the basket.

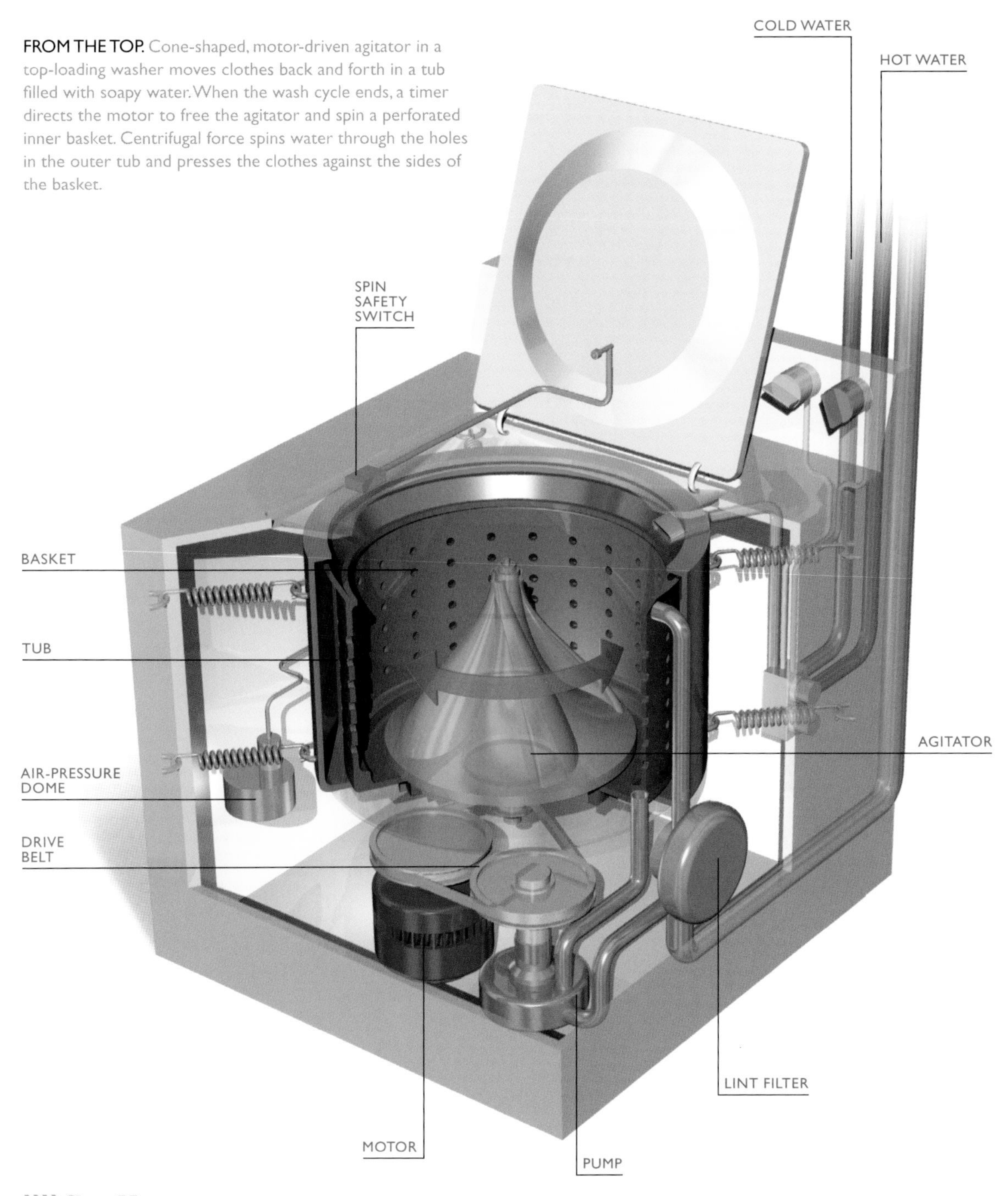

»»See Also
RELATED PRINCIPLES: Friction, 52 • Law of Conservation of Energy, 30 • Newton's First Law of Motion, 20 • Rapidly Expanding Gases, 74 • Universal Law of Gravitation, 122
RELATED APPLICATIONS: Bullet Trajectory, 123 • Firearm, 76 • Fireworks, 74 • Jet Engine, 28 • Rocket Engine, 29 • Squid and Octopus, 27

GUN RECOIL

As a bullet whizzes from the muzzle of a gun with explosive force, the shooter experiences the rearward motion of the firearm itself. According to Newton's third law, the force pushing the gun back—called recoil or kick—is equal to that propelling the projectile forward. Some weapons, such as cannon and certain machine guns, are mounted to transfer this force to the ground; others capture the recoil force in a spring that in turn empties a spent shell and reloads.

PRINCIPLE

PUSH AND PUSH BACK

Newton's Third Law of Motion

For every action, there is an equal and opposite reaction. Elegant in its simplicity, Newton's third law of motion describes such far-ranging phenomena as how squid propel through water and rockets launch into space.

All kinds of interactions create force, from simple pushing, pulling, friction, or tension to the more obscure action of gravitation. But it's always a two-way street, so to speak. Forces come in pairs.

The equal strength of paired forces can be counterintuitive. A classic example is that of a gnat that collides with the windshield of a moving car. Of course the tiny gnat loses in the interaction, but the force it exerts on the car is as great as that which the vehicle exerts on it. One way to think about it: The bug squashes itself on the windshield as much as the windshield squashes the bug.

Newton's Third Law of Motion

SQUID AND OCTOPUS

Evolution, that ingeniously subtle engineer, designed cephalopods such as squid and octopi to exploit the principle of jet propulsion for movement. These creatures propel through water at surprisingly high speeds—in the case of squid, more than 25 miles per hour.

They do it by squirting fluid out of their bodies, which, in keeping with Newton's third law of motion, exerts an equal and opposite force on the cephalopod itself. First, the animal expands its muscular body cavity (known as the mantle), pulling seawater in through an opening near its head. To build pressure in the mantle, the marine creature closes off all orifices except its tube-like funnel, which it can aim. Then, with a vigorous contraction of the mantle, the cephalopod forces water out through the funnel, shooting its own mass in the opposite direction—whether away from a threat or toward its next meal.

»»*See Also*

RELATED PRINCIPLES: Friction, 52 • Law of Conservation of Energy, 30 • Newton's First Law of Motion, 20
RELATED APPLICATIONS: Jet Engine, 28 • Rocket Engine, 29

Newton's Third Law of Motion

JET ENGINE

In the jet engine, the meeting of force and counterforce described by Newton's third law of motion has a special name: thrust. At its simplest, thrust can be observed in the way a balloon when it releases air is propelled in the opposite direction.

A jet sucks large quantities of air into the front of its engine. The air then enters a compressor made of many spinning blades, which, as the name implies, compresses the air into a smaller space, increasing its pressure and potential energy. In the engine's combuster, the pressurized air is sprayed with fuel and ignited. Oxygen and fuel burn, producing very hot, expanding gases. These enter the turbine, rotating its blades (which in turn help power the compressor blades). Finally the hot air, together with cooler air that has flowed around the engine core, blasts through the engine's rear nozzle with enormous energy—enough to push the plane forward.

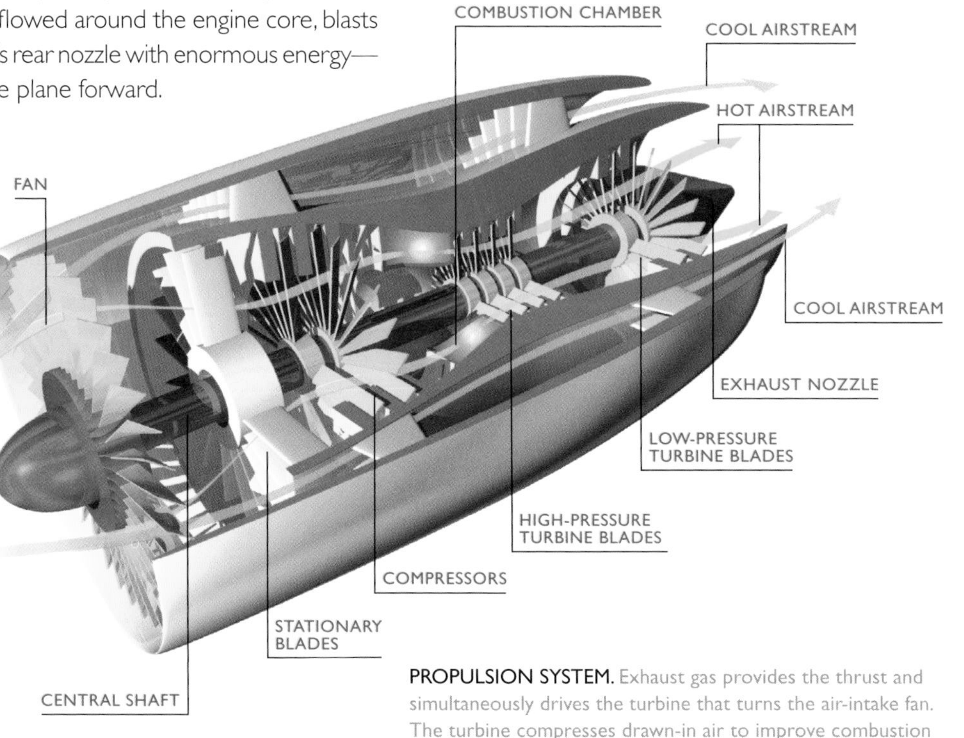

PROPULSION SYSTEM. Exhaust gas provides the thrust and simultaneously drives the turbine that turns the air-intake fan. The turbine compresses drawn-in air to improve combustion and make the exhaust work better.

»»See Also

RELATED PRINCIPLES: Bernoulli's Principle, 68 • Binary Code, 190 • Electrochemistry, 187 • Electromagnetic Induction, 157 • Friction, 52 • Law of Conservation of Energy, 30 • Mechanical Advantage, 38 • Ohm's Law, 68 • Pascal's Law, 80 • Rapidly Expanding Gases, 74 • Second Law of Thermodynamics, 98 • Universal Law of Gravitation, 122

RELATED APPLICATIONS: Combustion Engine, 77 • Electric Motor, 158 • Firearm, 76 • Fireworks, 74 • Flight Stabilizer, 23 • Gyroscope, 21 • Rocket Engine, 29 • Squid and Octopus, 25

Newton's Third Law of Motion

ROCKET ENGINE

The liftoff and flight of massive spacecraft like Saturn 5 or the space shuttle seem so unlikely as to be almost miraculous. But the apparent miracle of rocket flight results from a careful harnessing of natural forces. Like a jet, a space rocket pushes off from its own exhaust, in keeping with Newton's third law of motion. Since the rocket's exhaust at liftoff is directed toward the ground, launch pads are designed to divert potentially damaging blasts sideways. The original launch pads built for Saturn 5 sat atop deep horizontal trenches made of thousands of fireproof bricks.

Of course, once the rocket reaches outer space, it faces a challenge not placed on the jet engine: no oxygen. The rocket must carry its own supply. The airlessness of space, however, does have an advantage. Spared the energy needed to push the air aside, the rocket requires less effort to thrust itself forward once it leaves Earth's atmosphere.

»»See Also

RELATED PRINCIPLES: Bernoulli's Principle, 68 • Binary Code, 190 • Electrochemistry, 187 • Electromagnetic Induction, 157 • Friction, 52 • Law of Conservation of Energy, 52 • Mechanical Advantage, 30 • Pascal's Law, 80 • Rapidly Expanding Gates, 74 • Second Law of Thermodynamics, 98 • Universal Law of Gravitation, 122

RELATED APPLICATIONS: Electric Motor, 258 • Firearm, 76 • Fireworks, 74 • Flight Stabilizer, 23 • Gyroscope, 21 • Jet Engine, 28 • Squid and Octopus, 27

PRINCIPLE

NOTHING NEW
Conservation of Energy

Energy can be neither created nor destroyed; the total amount of energy, in the universe or within any closed system, remains constant over time. So states the law of conservation of energy, which governs everything from roller coasters to hybrid electric cars.

What we think of as the generation of energy is really its conversion from one form to another. Manifestations of energy include heat, light, sound, and motion. There is potential energy, by virtue of position, in a child poised at the top of a slide or a cherry hanging on a tree. There is chemical energy in the glucose the human body harvests, through cellular respiration, to power itself.

Conservation of Energy

BILLIARDS

The game of pool is all about the careful transfer of kinetic energy and linear momentum from one ball to another. A skillful player will strike the sweet spot just above the cue ball's center of gravity so that it rolls rather than slides across the felt, minimizing diversion of kinetic energy into heat due to friction. Thus, when the cue ball strikes a second ball, it passes all or nearly all its kinetic energy to the target ball. This is known as an elastic collision.

»»*See Also*

RELATED PRINCIPLES: Friction, 52 • Mechanical Advantage, 38 • Newton's First Law of Motion, 20 • Newton's Third Law of Motion, 27

RELATED APPLICATIONS: Downhill Skiing, 33 • Lever, 41 • Regenerative Braking, 31 • Roller Coaster, 35 • Wheel, 44

Conservation of Energy

REGENERATIVE BRAKING

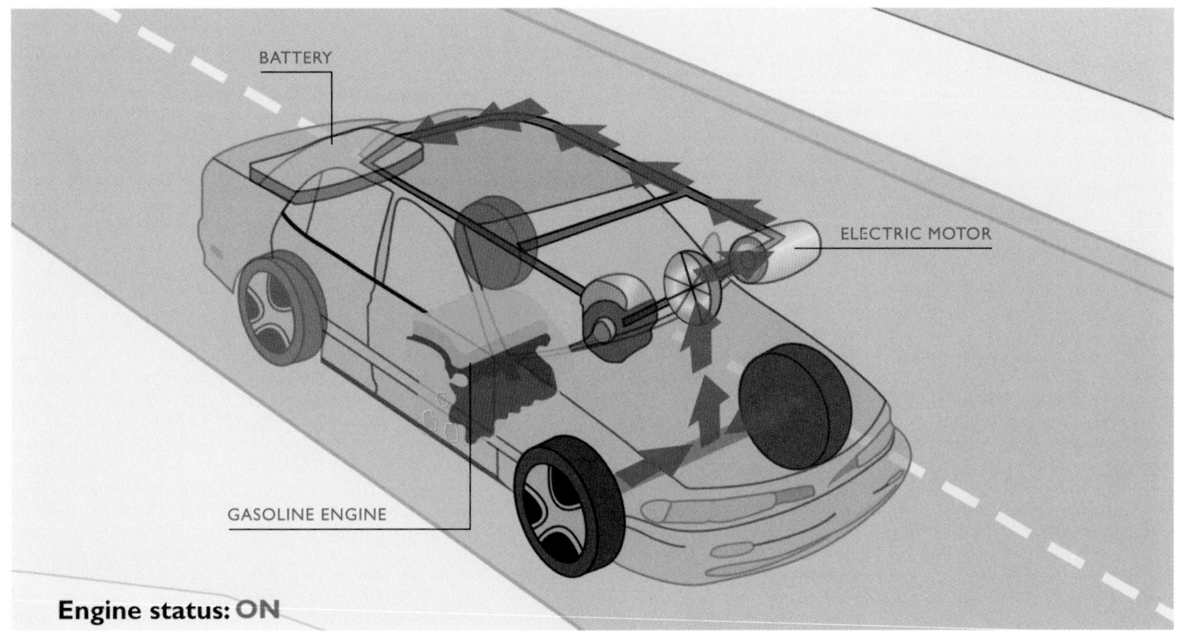

CONVERTING ENERGY. Regenerative braking converts otherwise wasted energy from braking into electricity and stores it in the battery.

An automobile at full throttle doesn't actually destroy or burn energy; it transforms energy. An efficient vehicle is one that captures as much energy as possible in usable forms. Traditional fossil fuel–burning cars are not so successful at this. It is estimated that a mere 15 percent of the energy from fuel goes to move the car or run features like heat and air conditioning.

Take braking, for example. In most cars, the driver presses on a brake pedal that, through a hydraulic system, applies brake pads to the wheels. This slows or stops the car by friction, which converts the kinetic energy of its forward motion into heat energy. But heat doesn't help the car get down the road; it is wasted energy.

Hybrid cars that combine the usual internal combustion engine with an electric, battery-powered motor employ regenerative braking. This system uses the car's electric motor as a generator that captures some of the kinetic energy lost during deceleration and converts it to electrical energy. This energy is stored in the car's battery for later use to power the vehicle.

There are two basic approaches to combining an internal combustion engine with an electric motor in the hybrid vehicle. In one, the gas-fueled engine powers an electrical generator that in turn runs the wheels; this is called a series hybrid. In a parallel hybrid, both the gas motor and the electric motor are linked independently to the drive wheels; it can use one or both power sources at any given time. The parallel arrangement is more costly, but it is more efficient, retaining energy that could be wasted as heat during conversion from chemical (gas) to electrical and finally to mechanical energy.

»»See Also

RELATED PRINCIPLES: Binary Code, 190 • Electrochemistry, 187 • Electromagnetic Induction, 157 • Friction, 52 • Newton's First Law of Motion, 20 • Newton's Third Law of Motion, 27 • Mechanical Advantage, 38 • Rapidly Expanding Gases, 74

RELATED APPLICATIONS: Automobile Transmission, 51 • Billiards, 30 • Downhill Skiing, 33 • Electric Car, 160 • Gear, 48 • Roller Coaster, 35 • Wheel, 44

Conservation of Energy

DOWNHILL SKIING

The goal of a downhill ski racer is speed. Put another way, every competitor wants to convert as much potential energy as possible into kinetic energy and, along with it, velocity. That means minimizing the action of external forces whose work on the skier would leach energy from this conversion: namely, the grab of surface friction and pushback of air resistance. Skiers wax their skis so they glide over the snow. They wear tight-fitting suits and assume a crouching, arms-forward position to reduce the body area pressing against the air. They harness gravitational force, essentially falling down the hill.

As the skier reaches the bottom of the slope, nearly all potential energy has been converted to kinetic energy; now, the skier turns the skis sideways to the slope in a skidding motion. Friction converts motional energy into heat energy.

See Also

RELATED PRINCIPLES: Friction, 52 • Mechanical Advantage, 38 • Newton's First Law of Motion, 20

RELATED APPLICATIONS: Billiards, 30 • Electric Car, 160 • Gear, 48 • Lever, 41 • Regenerative Braking, 31 • Roller Coaster, 35 • Wheel, 44

Conservation of Energy

ROLLER COASTER

As a chain hauls a roller coaster car to its first and highest peak, the riders sense something in the pit of their stomachs: The car is gaining potential energy. The heavier the loaded car and the higher it goes, the more potential energy it takes on. When the car reaches the peak and the stored energy is at its maximum, the chain releases, and it falls.

As the car plunges, what it loses in potential energy it gains in kinetic energy—the energy of motion. Pushed upward once again by the track, the car reverses that transformation, converting motion energy back to the potential energy that comes with elevated position. Indeed, the whole ride can be seen as a thrilling exercise in energy conversion. In theory at least (assuming negligible friction and air resistance), the total mechanical energy of the ride remains constant throughout. Just as the applied force of mechanical lifting invested the system with potential energy at the beginning of the ride, the coaster comes to a stop with the application of brakes—and the final, friction-induced dissipation of motion energy into heat energy.

The first tubular steel track roller coaster opened in Disneyland in 1959, allowing designers to introduce loops for the first time.

»»See Also

RELATED PRINCIPLES: Electromagnetic Induction, 157 • Friction, 52 • Mechanical Advantage, 38 • Newton's First Law of Motion, 20 • Universal Law of Gravitation, 122
RELATED APPLICATIONS: Gear, 48 • Wheel, 44

ROTATING OBJECTS

A spinning figure skater represents a special kind of momentum called angular momentum. And like linear momentum, it is conserved—remains stable—in the absence of external pushes or pulls.

Angular momentum is the product of two quantities: the speed of the object's rotation, and what's called its moment of inertia, which describes both the amount of its mass and where this mass is distributed in the rotation. The closer the mass lies to the axis of rotation, the smaller the moment of inertia.

Thus, as the skater tucks her arms and legs tighter to the vertical axis of her body, reducing her moment of inertia, her spin quickens to a blur.

A platform diver likewise maintains angular momentum during a dive that may involve numerous body positions, from a tight tuck to a more elongated pike to full extension. These different positions vary the diver's moment of inertia and with it the velocity of spin. A tucked position reduces the moment of inertia, increasing spin speed.

It also works in reverse. To slow a spinning object, you increase its moment of inertia. A fast-spinning bicycle wheel will slow down a lot faster if you apply friction (a firm hand, say) near the rim rather than near the axle.

Golf clubs are sometimes designed to offer a high moment of inertia by redistributing weight around the heel and toe of the clubhead. This gives the club resistance to change in its state of rotation; it's less likely to twist and hit the ball in the wrong spot.

See Also

RELATED PRINCIPLES: Friction, 52 • Mechanical Advantage, 38 • Newton's First Law of Motion, 20 • Universal Law of Gravitation, 122

RELATED APPLICATIONS: Bicycle, 43 • Gear, 48 • Gyroscope, 21 • Lever, 41

CHAPTER 2

Mechanical

Advantage and Friction

Since fashioning the first handheld stone chippers, humans have invented devices to amplify the effects of their effort; that is, machines. "Give me a place to stand, and with a lever I shall move the world," the ancient Greek mathematician Archimedes is said to have exclaimed. He was exaggerating of course, but the implication is nonetheless true. Even the simplest machine—a lever—can dramatically increase the force applied to it by the user, giving a mere mortal the strength of a titan. The factor by which a machine multiplies the force put into it is its mechanical advantage.

Because the total amount of work performed remains constant and is always a product of force times distance (Work = Force × Distance), there is a trade-off involved—the greater the distance, the less force required; conversely, the less distance, the greater the force. Returning to Archimedes' boast, (even if he could find a place to stand) he would need a very long lever indeed to provide enough distance to amplify the force exerted with his own muscles to be able to move the world.

Some machines are designed to require greater force than they yield, instead offering advantages of distance or speed. In a tennis racket, for example, the force applied to the handle is greater than the resulting speed and range of motion at the head. A bicyclist pushes harder on the pedals than the output force that rolls the bicycle down the road. The bike moves faster than the pedals, however, covering more distance in the same time. These machines have a mechanical advantage of less than one.

In the real world, mechanical advantage is invariably diminished by friction, a rubbing between objects or fluid layers that opposes motion. Friction results in the loss of energy as heat. Reducing friction in order to make machines more energy efficient is a major concern of engineers.

PRINCIPLE

A PIVOTAL POINT

Mechanical Advantage

In the third century Greek mathematician Archimedes first described the notion of a simple machine—a device that allows an applied force to work against a load force. Such a device can amplify the force or the distance of movement but not both. Classical mechanics identifies six simple machines: the lever, the wheel and axle, the wedge, the screw, the pulley, and the inclined plane. Simple machines, by virtue of mechanical advantage, make it easier to move things. They pervade the modern world, though more often as components of complex machines—everything from cranes to car transmissions.

Mechanical Advantage

LEVER

A lever is a rigid beam resting on a pivot or fulcrum. A person depressing one end of the beam can lift more weight at the other end than he could alone. That's mechanical advantage.

Like other machines, depending on how it's rigged, a lever cannot only magnify force but change its direction as well—you push down on one end, and the other end goes up. A baseball bat, for example, works as a lever where the input force is greater than the output force. The load (the thick

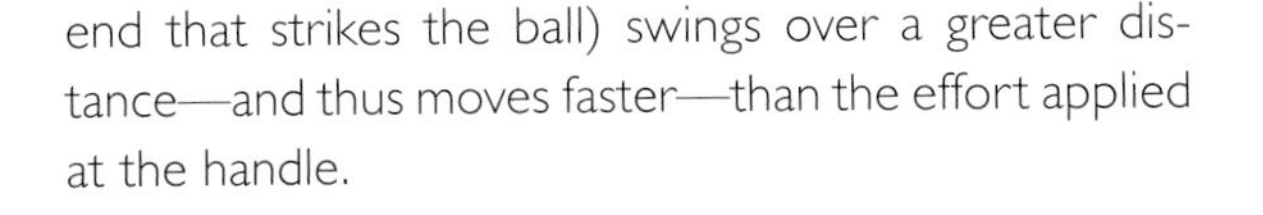
end that strikes the ball) swings over a greater distance—and thus moves faster—than the effort applied at the handle.

See Also

RELATED PRINCIPLES: Friction, 52 • Law of Conservation of Energy, 30 • Universal Law of Gravitation, 122
RELATED APPLICATIONS: Gear, 48 • Pulley, 47 • Wheel, 44

RUDY
champ-sys.com
cyclocross
champion
Champion System
93
126

Mechanical Advantage

BICYCLE

The elegant gearing of a modern multispeed bicycle optimizes the way a rider can apply muscle power to cover various kinds of terrain. A single chain turns around two sets of toothed disks. The chain rings rotate with the pedals; they are driver gears. The smaller sprockets are connected to the rear wheel hub—follower gears. A mechanism called a derailleur moves the chain to achieve different combinations of driver and follower.

As with any gear train, the ratio between the number of teeth in driver and follower gears determines the machine's mechanical advantage. A larger, many-toothed chain ring coupled to a smaller sprocket will make the bike wheel turn faster but demands more force from the legs. Climbing a hill calls for a gear that joins a smaller driver with a larger follower gear. The cyclist won't have to push as hard for each turn of the pedals, but he'll have to pedal faster.

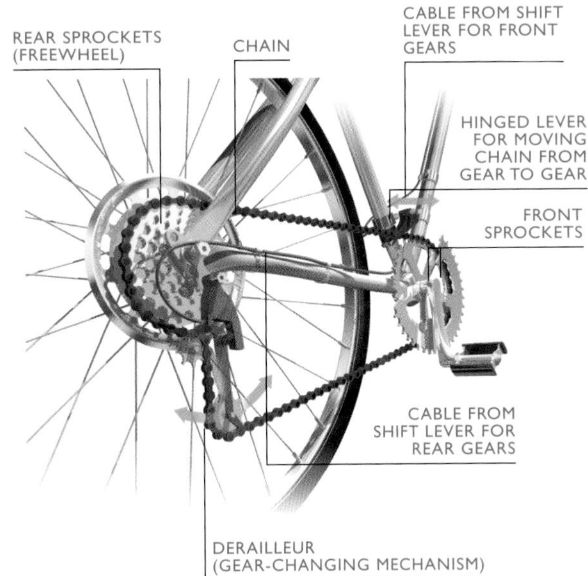

»»See Also

RELATED PRINCIPLES: Friction, 52 • Newton's First Law of Motion, 20 • Universal Law of Gravitation, 122

RELATED APPLICATIONS: Gear, 48 • Lever, 41 • Pulley, 47 • Wheel, 44

Mechanical Advantage

WHEEL

The wheel and axle—a disk with a cylindrical rod attached to the center—is one of the six simple machines that help people move things. As a way to transport heavy objects by rolling instead of dragging them, the wheel is a profoundly significant innovation that emerged more than five thousand years ago in Mesopotamia and elsewhere. The rolling wheels of a loaded cart make much less surface contact with the terrain than would the cart and its contents. This reduces friction, the rubbing between two objects that hinders movement.

A wheel that is bound to a solid axle is actually a round lever. Greater force applied at the axle generates distance and increased speed around the rim. Conversely, as in the case of a doorknob, a relatively small effort exerted on the rim magnifies force at the axle. As with any lever, the mechanical advantage of a wheel.

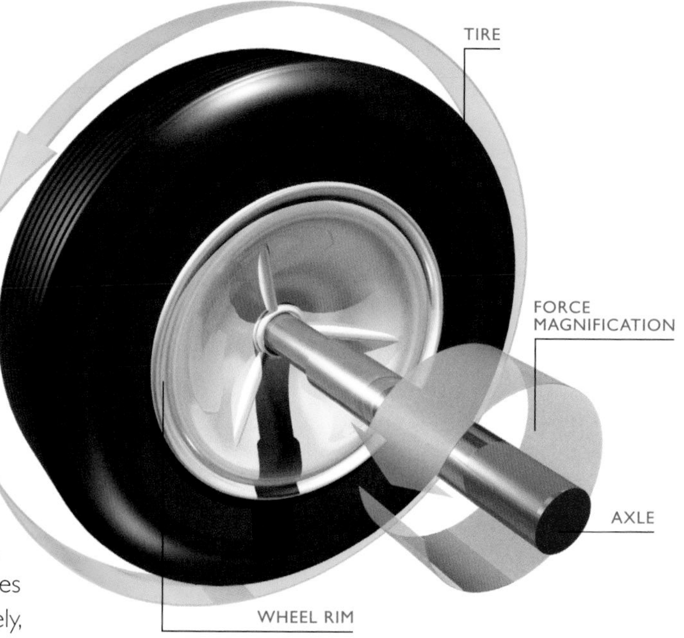

HOW A WHEEL WORKS. With its familiar circular frame, hub, and axle, a wheel effectively transmits power and motion. Its center acts as a fulcrum, making it, in fact, a rotating lever.

»»*See Also*
RELATED PRINCIPLES: Friction, 52 • Newton's First Law of Motion, 20 • Universal Law of Gravitation, 122
RELATED APPLICATIONS: Gear, 48 • Lever, 41 • Pulley, 47

BONES, JOINTS, AND MUSCLES

The human skeleton abounds with levers. Joints act as fulcrums, bones as lifting beams, and muscles as applied force. When you raise a mug of coffee to take a sip, your arm is what's called a class 3 lever. That is, the bicep applies greater force than the output force, but the mug travels a greater distance than this applied force, accomplishing the task quickly. Bite off a piece of Danish along with your coffee, and your lower jaw works as the beam in another class 3 lever.

FERRIS WHEEL

Built for the 1893 World's Columbian Exposition in Chicago, the first Ferris wheel used an axle weighing nearly 90,000 pounds to turn an enormous vertical wheel carrying passenger seats. The ride creates slightly unsettling sensations in riders by ingeniously combining the effects of gravity and centripetal force. Gravity, of course, pushes down toward the center of Earth. Centripetal force, meanwhile, pushes objects into a circular path and always points toward the center of this path.

Mechanical Advantage

PULLEY

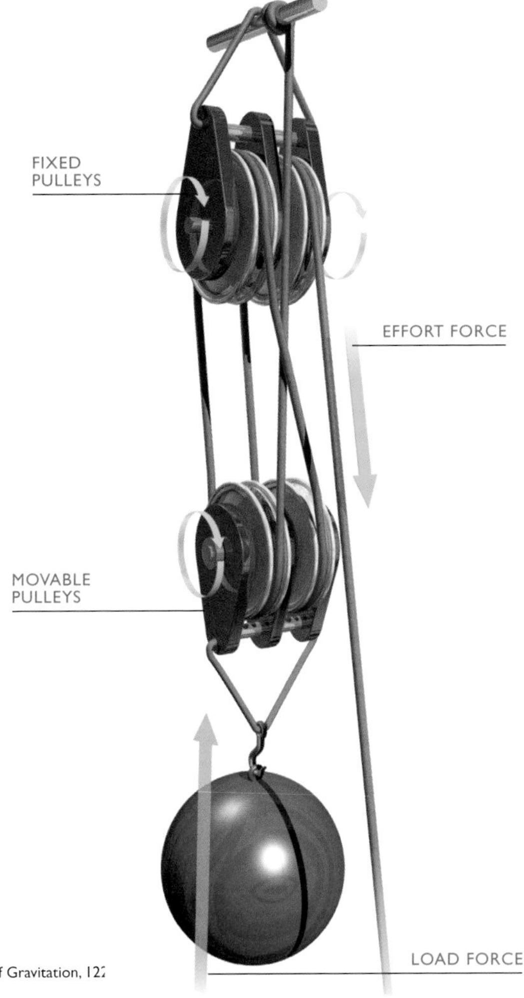

A pulley is a wheel with a rope laid across its rim. It works much like a lever, with an applied force tugging at one end of the rope to lift a load attached to the other end. A single fixed pulley has no mechanical advantage; you must pull with a force equal to the weight you're hoisting. But the device makes lifting easier because you're pulling down, with the force of gravity, and can use your own weight instead of your back muscles. Using two pulleys together spares effort, halving the applied force required to lift a load; the trade-off, once again, is the work is spread over more distance, that of the rope.

»»See Also
RELATED PRINCIPLES: Friction, 52 • Law of Conservation of Energy, 30 • Universal Law of Gravitation, 12
RELATED APPLICATIONS: Lever, 41 • Pulley, 47 • Wheel, 44

CAR SKID

Leverage is what makes a car skid. Slam on the brakes, and the wheels stop spinning, but the car body rocks forward, increasing load on the front wheels and turning them into a pivot or fulcrum. That makes the car a lever beam; it begins to turn on the pivot. How much and how fast depends on the distance between the front and back wheels and how weight is distributed between the front and rear of the car. Antilock braking systems automatically pump the brakes during emergency braking, preventing wheels from locking.

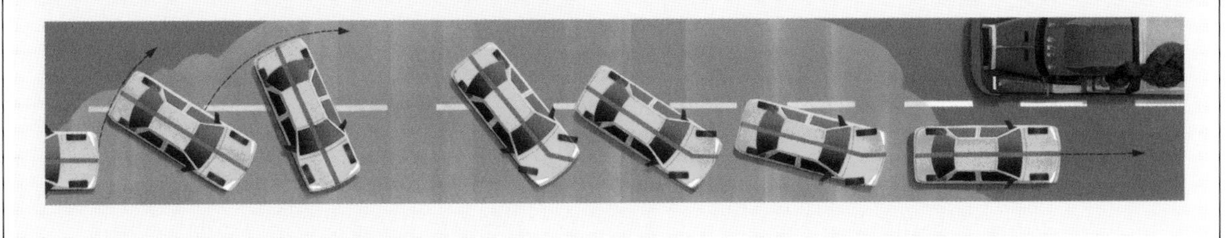

WORM GEAR

SPUR GEARS

RACK AND PINION GEARS

BEVEL GEARS

Mechanical Advantage

GEAR

In the large-scale mechanization of the industrial revolution, few devices were more critical than gears. This system is made up of toothed wheels that, when engaged, turn in opposite directions and increase force or speed to accomplish all kinds of work. If the gear receiving input force—the driver gear—is engaged with a smaller follower gear, the latter rotates faster. If the driver is smaller than the follower, the follower turns more slowly but with greater force.

In a handheld eggbeater, for example, the handle turns a large central gear meshed with two smaller gears that rotate several times for each turn of the hand crank. The device trades off force or effort in favor of the speed required for whipping or beating foods.

GEARS. The shape and size of a gear's teeth control the number of rotations, the direction of motion, the speed, and the amount of force exerted. Worm gears link shafts that have axes at right angles and do not intersect; bevel gears connect shafts at an angle; rack and pinion gears have mesh with sliding, toothed racks.

›››See Also

RELATED PRINCIPLES: Friction, 52 • Newton's First Law of Motion, 20 • Universal Law of Gravitation, 122
RELATED APPLICATIONS: Lever, 41 • Pulley, 47 • Wheel, 44

WORM DRIVE

Some gears are designed to mesh not with another wheel-shaped gear but with a cylinder threaded like a screw. This aptly named worm rotates at high speeds, producing enormous force magnification in the larger worm gear. The standard worm drive runs in only one direction, with the worm driving the gear; this makes it suitable for machinery that must stay in position when the power's off such as conveyor systems, winches, and forklifts. Worm gears are used in steering systems of cars and other vehicles, converting the steering wheel's rotary motion into the horizontal, side-to-side movement of the front wheels.

Mechanical Advantage

CLIMBING TOWER CRANE

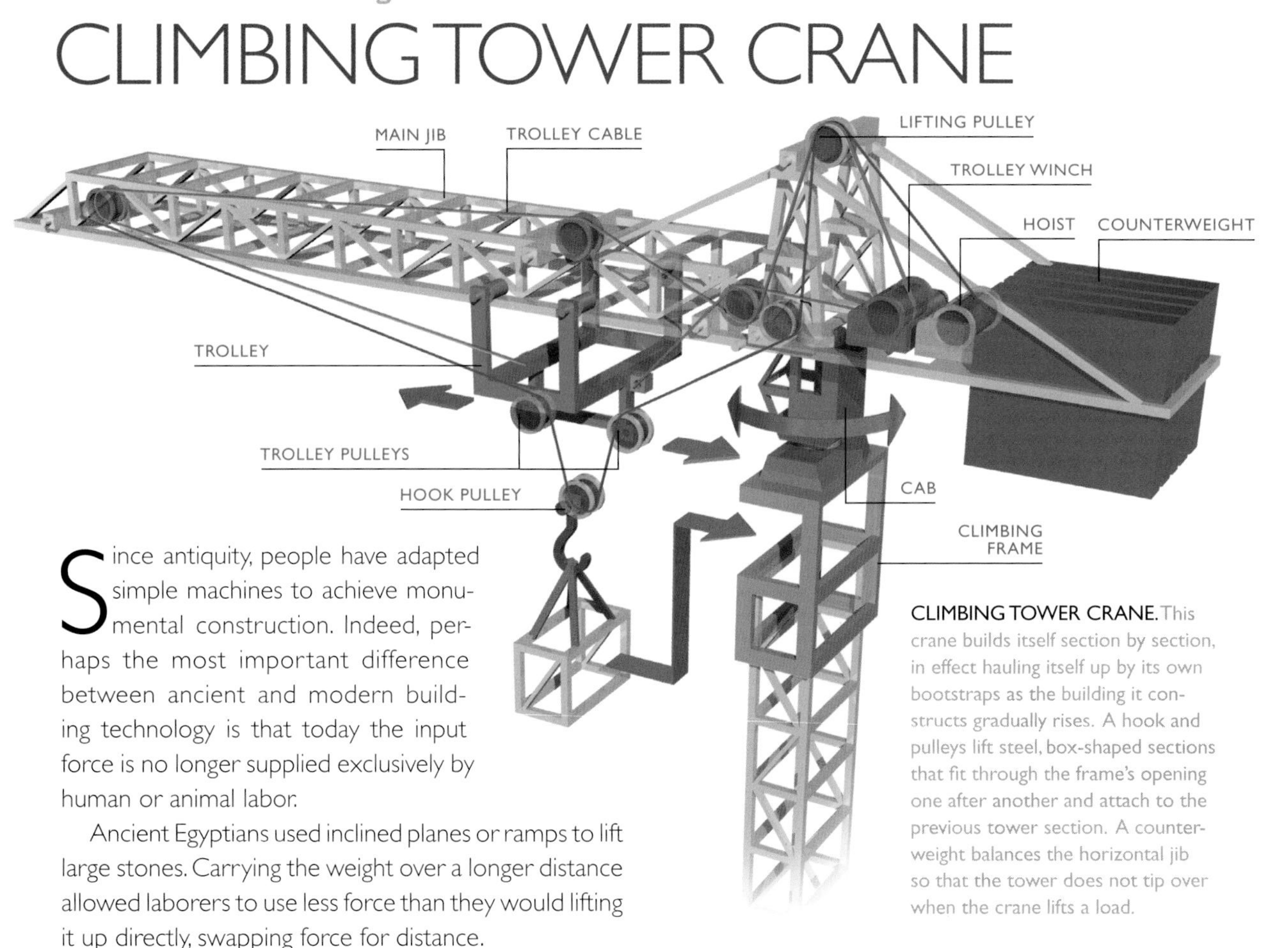

CLIMBING TOWER CRANE. This crane builds itself section by section, in effect hauling itself up by its own bootstraps as the building it constructs gradually rises. A hook and pulleys lift steel, box-shaped sections that fit through the frame's opening one after another and attach to the previous tower section. A counterweight balances the horizontal jib so that the tower does not tip over when the crane lifts a load.

Since antiquity, people have adapted simple machines to achieve monumental construction. Indeed, perhaps the most important difference between ancient and modern building technology is that today the input force is no longer supplied exclusively by human or animal labor.

Ancient Egyptians used inclined planes or ramps to lift large stones. Carrying the weight over a longer distance allowed laborers to use less force than they would lifting it up directly, swapping force for distance.

In constructing their great buildings, the Romans perfected cranes made up of pulleys. The basic trispastos had three pulleys, tripling the force applied by workers. In some cases a crane's rope was attached at the input end to a capstan or treadwheel. This maximized the pulling effort because force applied around the greater distance of a wheel's circumference, as in a doorknob, multiplies force at the axis.

Many modern-day mobile construction cranes work the same way, using a winding drum, a boom or leverage arm, and a pulley to raise and move loads, with energy supplied by an electric motor or gas-powered engine.

To raise the most impressive structures of today—skyscrapers and bridges, for example—builders use tower cranes with vertical masts that reach many stories in height. These, too, employ various ancient technologies: a winch-and-pulley system to hoist the load on the crane's long horizontal working arm, a gear to rotate the arm, and, most strikingly, the giant lever of the arm itself, which pivots on the mast and is mounted at one end with concrete-block counterweights to apply force against enormous loads. The modern world is built on the innovations of the ancients.

»»» *See Also*

RELATED PRINCIPLES: Electrochemistry, 187 • Electromagnetic Induction, 157 • Friction, 52 • Ohm's Law, 164 • Pascal's Law, 80 • Rapidly Expanding Gases, 74 • Universal Law of Gravitation, 174

RELATED APPLICATIONS: Combustion Engine, 77 • Gear, 48 • Lever, 41 • Pulley, 47

Cutaway of an automotive transmission gearbox

Mechanical Advantage

OFFSET PRINTING PRESS

The offset lithographic press used to print enormous runs of newspapers, magazines, and books is made up of a complex system of rolling cylinders, operated by an equally complex gear train.

Three main cylinders of equal size roll against each other at uniform speed: the plate cylinder, the offset or blanket cylinder, and the impression cylinder. The plate cylinder is wrapped with a thin piece of aluminum coated in appropriate places with a material that repels water but attracts the press's sticky ink. When the plate cylinder turns, it rolls against smaller ink and water rollers, which apply ink to the coated surface and wet the nonprinted surface. This inked plate meanwhile rolls against the rubber blanket cylinder, thereby transferring the image to the blanket. The blanket cylinder in turn rolls against the impression cylinder; paper from large reels rolls between these two cylinders, and the image is pressed onto the paper.

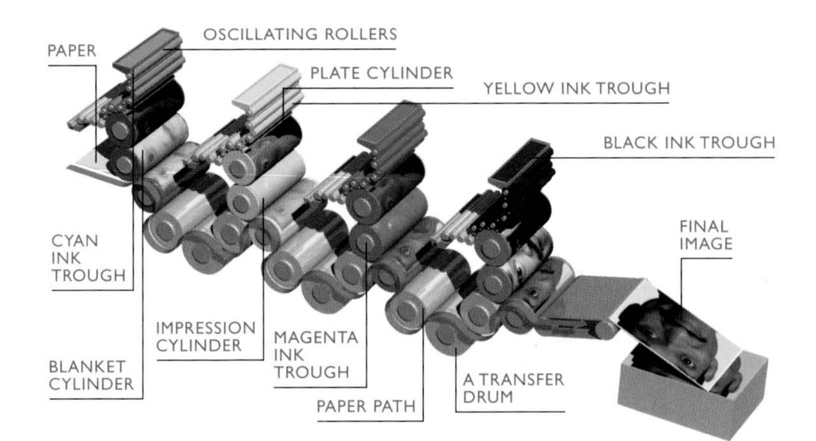

OFFSET PRESS. Color printing using offset technology requires a complex transfer process in which paper runs over and under a meshed system of cylinders, rollers, and drums. First, a photochemical process transfers photographed images and text to a metal plate. Inking rollers apply ink fed from troughs of cyan, yellow, magenta, and black along the line. The ink image then transfers onto the blanket cylinder, where paper picks up the color imprint as it passes between it and the impression cylinder below. The process repeats with each color.

»»*See Also*

RELATED PRINCIPLES: Absorption and Reflection of Light, 18 • Binary Code, 190 • Electromagnetic Induction, 157 • Friction, 52 • Ohm's Law, 164 • Photoelectric Effect, 222

RELATED APPLICATIONS: Gear, 48 • Lever, 41 • Pulley, 47 • Wheel, 44

TRANSMISSION

Automatic automobile transmissions, which include electrical and computer components, are far more sophisticated than bicycle gears. But both vehicles share a basic method for transmitting power to drive wheels. Just as there's an optimal range of speed and force for legs pedaling a bike, the internal combustion engine that runs a car works best at a particular rotational speed.

The centerpiece of the transmission is a set of gears that, in various combinations, adapt the engine's rotational speed to the desired wheel speed. Gear combinations that reduce speed produce more rotational force, or torque, in the drive wheels. One combination actually reverses the car. Car transmissions also have a neutral position that disconnects the engine from the wheels.

PRINCIPLE

RUBBED THE RIGHT WAY

Friction

Friction works against motion and produces heat that represents lost energy. And it happens whenever two objects come into contact or rub against each other.

Because of friction, it's impossible to realize the ideal mechanical advantage of a machine. Rough or malleable surfaces are more apt to produce friction, though even the smoothest surface creates some friction. The key to making a machine run as efficiently as possible is to minimize friction. Just as a greased pig is harder to catch, lubricated machine parts reduce friction and the wear that accompanies it. On the other hand, some devices—brakes, for example—actually use friction to work.

Friction

BALL BEARING

Ball bearings use the rolling action of balls to minimize friction between moving parts of a machine. Balls are mounted between grooved inner and outer rings called races. The inner ring is like an axle, and the outer ring simply rolls on its balls against the axle; it neither slips nor rubs. The spinning of skateboard and car wheels is made smoother by ball bearings. A good yo-yo contains a ball bearing that allows the outer part of the yo-yo to rotate evenly and with a minimum of friction around its axis. Ball bearings reduce the contact area between two metal surfaces to tiny points.

See Also

RELATED PRINCIPLES: Law of Conservation of Energy, 30 • Mechanical Advantage, 38 • Newton's First Law of Motion, 20
RELATED APPLICATIONS: Wheel, 44

Friction

PARACHUTE

Even air creates friction when it rubs or pushes against an object's surface and works against movement. The parachute harnesses this type of friction, known as air resistance.

As a parachutist leaps from an airplane thousands of feet above the ground, he or she will initially plummet with increasing velocity under the force of gravity.

Then the parachute billows open. It is made of lightweight material—originally linen or silk, now nylon—to reduce the force of gravity on it but has a large canopy that maximizes its surface contact with the air. This contact slows the jumper's descent to a safe landing speed.

See Also

RELATED PRINCIPLES: Macromolecular Chemistry, 288 · Universal Law of Gravitation, 122

RELATED APPLICATIONS: Ball Bearing, 52 · Competitive Swimwear, 55 · Nylon, 292 · Stringed Instruments, 54

Friction

STRINGED INSTRUMENTS

Usually, when parts of a machine vibrate, they generate a sound we call noise. But in the case of a violin, our ears interpret string vibrations resonating in the instrument's body as music. This vibration is caused by friction. In fact it results from oscillations in the force of friction called the stick-slip phenomenon. The hairs of a bow drawn across a string alternate between pulling the string with it—sticking, in what's called static friction—and, as the applied force overcomes static friction, slipping, in what's called kinetic friction. The slipping friction is less than the sticking friction; this results in a sudden increase in the bow's speed—a subtle jerking that vibrates the string. Horsehair has long been used in bows because its fibers have a propensity to grip the bow and hold the rosin that enhances stickiness.

»»See Also

RELATED PRINCIPLES: Law of Conservation of Energy, 30 • Mechanical Advantage, 38
RELATED APPLICATIONS: Ball Bearing, 52 • Lever, 41 • Parachute, 53

Friction

COMPETITIVE SWIMWEAR

A good swimmer seems to slip through the water, but water is denser and stickier than air, creating turbulence as it moves around the swimmer's body as well as motion-opposing surface friction against the skin. Competitive swimwear has one overarching purpose: to minimize this drag.

What happens when drag-reducing swimwear begins to detract from the main event—the contest among the athletes themselves? It's a question that came to the fore in 2008 with the introduction of Speedo's high-tech LZR Racer. The suit spawned a host of other shoulder-to-ankle, polyurethane-based suits that compressed and streamlined the body as never before. The LZR Racer's seams were even ultrasonically welded together rather than sewn, to minimize friction. The strategy evidently worked. At the 2008 Beijing Olympics, a great majority of medal-winning and record-breaking swims were performed wearing the LZR. In 2010, the body that oversees official international competition banned the suit and others like it.

»»*See Also*

RELATED PRINCIPLES: Macromolecular Chemistry, 288 • Newton's First Law of Motion, 20

RELATED APPLICATIONS: Ball Bearing, 52 • Parachute, 53 • Stringed Instruments, 54

CHAPTER 3

Waves and Turbulence

What is a wave? It's a disturbance that transfers energy through a medium of interacting particles. This oscillating pattern carries energy from one place to another without moving matter itself. That's why objects bob in waves rather than moving with them. The water is merely the medium through which the energy travels.

There are all kinds of waves—sound waves, waves in water, light waves, and electromagnetic waves that travel at the speed of light. Waves will travel along a string agitated at one end. A stone dropped in a pond will send out waves in concentric circles.

All waves begin with an applied force—a gust of wind on the water, the vibration of a plucked guitar string—that temporarily displaces particles. These displaced particles push or pull on adjacent particles, setting them in motion and transferring energy. As the second wave rises, the first one subsides and comes to rest in its original position. The second wave, meanwhile, pushes or pulls a third into motion and so on.

Waves are orderly. The chaotic movement known as turbulence is much harder for physicists to define or explain. But it, too, is everywhere—a flow of gases or liquids in swirling, unpredictable eddies, a swollen river, a leaping flame, a torrent of exhaust. Apparently random, it may arise from unique conditions in endlessly complex, three-dimensional patterns. Research suggests that smooth, stable (laminar) flows are more apt to occur when the flow is slow, low density, low viscosity, and occurring within a small diameter.

PRINCIPLE

RIDE THE WAVE

Wave Energy

Waves move energy in a ripple effect across a medium. In the ocean, the applied force that starts the process is wind striking the surface, which transfers the air's motion energy to the seawater. A wave's height depends on how fast the wind is moving, the length of time it pushes on the water, and the distance across which it excites the water, known as fetch. A high wave is a high-energy wave. It's not water that travels thousands of miles to crash on a distant shore but energy. The water remains in place, while the energy flows through it in the form of waves. The same principle applies to small ripples in a pond and devastating tsunamis.

Wave Energy

WAVE TURBINE

A typical ten-second, four-foot wave striking a mile of coastline releases enough power to keep a light bulb burning for 30 years. So why not exploit this eminently renewable energy? Engineers have developed devices to do just that including buoy systems that harness the up-and-down motion of waves in an electrical generator, columns near the shoreline in which the pressure of wave peaks and troughs forces air to turn a turbine, and shoreline structures that funnel waves into an elevated reservoir that acts like a hydroelectric dam. Unfortunately, these technologies are costly and, so far, not efficient enough to be used widely.

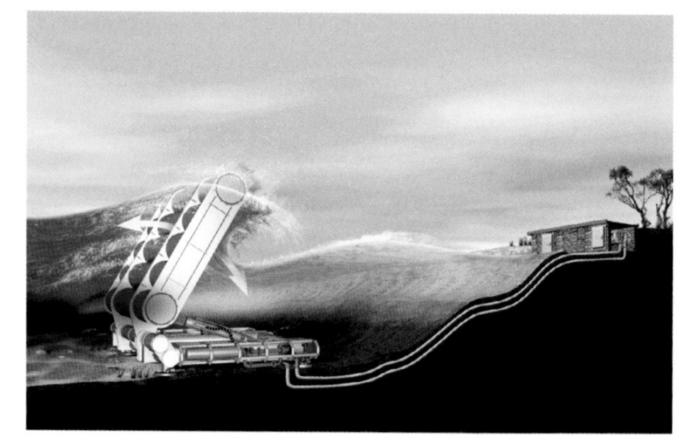

WAVE ENERGY CONVERTER. The waves passing over the upper section cause it to move up and down, and this movement is converted to electrical energy.

»»See Also
RELATED PRINCIPLES: Convection, 130 • Electromagnetic Induction, 157 • Geothermal Energy, 175 • Law of Conservation of Energy, 30 • Mechanical Advantage, 38
RELATED APPLICATIONS: Hydroelectricity, 172 • Tidal Power Generator, 131 • Wind Turbine, 176

Wave Energy

SURFING

Few people study wave energy more avidly than those who hope to catch a ride on it. What they're looking for is a big, powerful wave with a distinct shape that breaks cleanly and slides into shore.

Surfers keep an eye out for high-energy waves by studying forecasts for storms at sea—the strong, steady, far-reaching winds that might create a swell, a series of waves that can travel many miles over deep water. These carry a lot more energy than is produced by local winds lashing coastal waters. Web sites used by surfers predict wave energy as a function of wave height and the time between wave crests (called the period). The greater the period, the faster the wave is moving.

Surfers also get to know the undersea topography of their local surf spots, which determines where and how the waves break. They understand that as a swell approaches shore, it will slow down and dramatically change shape in shallow areas. Eventually the wave falls over—breaks—releasing energy into roaring sound and the movement of sand and water as it returns to sea in undertow and rip currents. Surfers use rip currents to ride back out for the next wave.

»» See Also

RELATED PRINCIPLES: Convection, 103 • Friction, 52 • Geothermal Energy, 175 • Law of Conservation of Energy, 30 • Newton's First Law of Motion, 20 • Specific Heat, 111 • Surface Tension, 88 • Universal Law of Gravitation, 122

RELATED APPLICATIONS: Wave Turbine, 58

PRINCIPLE

A FORCE TO BE RECKONED WITH

Seismic Waves

If waves of energy can move through air and water, they can also move through Earth itself. These are called seismic waves, and they may result from earthquakes, volcanic eruptions, tsunamis, nuclear explosions, or other major impacts on the planet. There are two basic types of seismic waves—body waves, which can vibrate all the way through Earth, and surface waves, which affect only Earth's crust. It's through the study of body waves that scientists have been able to map the unseen interior of Earth.

Seismic Waves

TSUNAMI

A tsunami is fundamentally different from a wind-generated wave. It forms as a result of an undersea earthquake or other sudden seismic disruption that causes enormous quantities of water to shift under the force of gravity. Rather than exhibiting regular peaks and valleys like the usual wave train, a tsunami has a very long distance and time between crests; it plows along horizontally like a great wall of water. The long length of the wave compared to water depth means it travels fast and loses little energy in its epic journey.

When the giant wave approaches shallower shoreline waters, it slows down but gains height, maintaining energy. Finally this energy is loosed on the shore, where it can inflict terrible destruction, moving big swathes of beach, taking out trees and homes, and turning debris of all kinds into water-borne projectiles.

»»» *See Also*
RELATED PRINCIPLES: Friction, 52 • Law of Conservation of Energy, 30 • Newton's First Law of Motion, 20 • Wave Energy, 58
RELATED APPLICATIONS: Earthquake, 61

Seismic Waves

EARTHQUAKE

Along fault lines in Earth's surface, masses of rock move ever so gradually against each other, actually distorting or bending the rock. This strain represents stored elastic energy. When the shifting rock overcomes friction and slips, this elastic energy is released in an earthquake. About a tenth of the energy goes forth in the form of seismic waves, while the rest increases the rupture in Earth's surface or generates heat caused by friction. Several million earthquakes occur in the world every year. Most are of such small magnitude as to go unnoticed. Early detection and better building practices have reduced fatalities.

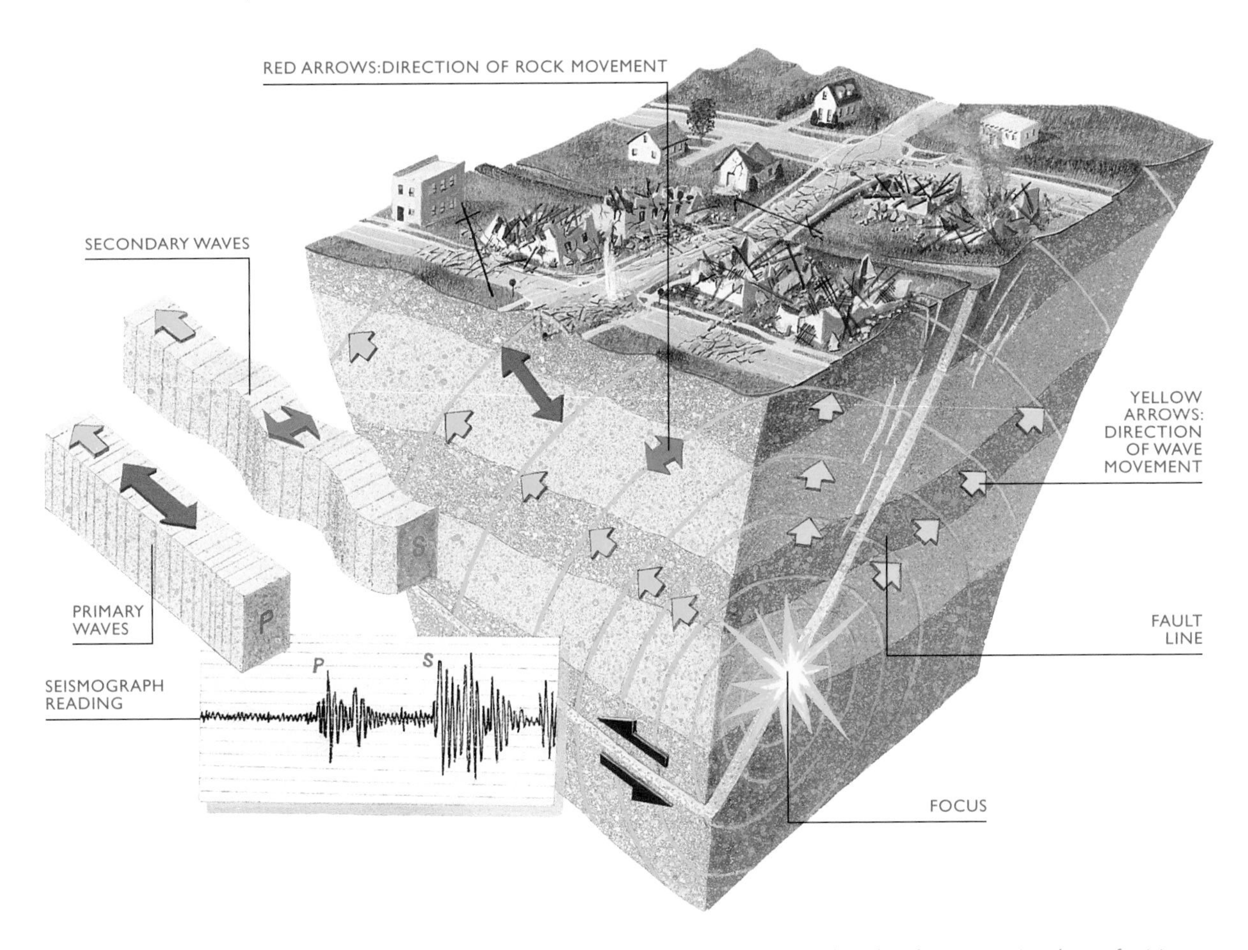

MEASURING QUAKES. Worldwide networks of seismographs enable scientists to track natural earthquakes, pinpointing places of origin and determining relative strengths. Quakes promulgate two kinds of motion at once—compressional primary or P waves and up-and-down secondary or S waves. In general, S waves cause more damage to man-made structures because they shake foundations.

See Also
RELATED PRINCIPLES: Law of Conservation of Energy, 30 • Wave Energy, 58
RELATED APPLICATIONS: Earthquake-Resistant Buildings, 132 • Tsunami, 60

PRINCIPLE

CANCELING EACH OTHER OUT

Destructive Interference

Unlike solid objects, waves—which move energy, not matter—can actually move through each other. When two waves move in the same direction and share the same frequency, the effect will be additive—a wave that combines the amplitude of the first two, still moving in one direction at the original frequency.

If the waves' peaks line up, their combination will result in a higher wave. If, on the other hand, the first wave's peaks line up with the other's troughs, the two wave patterns will cancel each other out. Strange as it seems, in this case, there is no wave. This is called destructive interference.

Destructive Interference

NOISE-CANCELING DEVICE

Noise canceling is sometimes called active noise control to distinguish it from traditional, passive methods of muffling unwanted sound (insulation) or noise-masking approaches such as white noise machines.

Active noise control systems work by generating a sound wave that's the mirror image of the offending sound wave. This wave has the same frequency and direction but opposite crests and valleys; it is out of phase with the first sound wave. The result, theoretically at least, is no sound.

It's a clever technology that's tricky to implement. The waves have to line up in space and time; so the technique works best in simple or enclosed fields where microphones can pick up a sound wave before it reaches the place where noise reduction is required. Active noise control is used in such applications as airplane cabins, air conditioning ducts, and headphones worn by pilots that cancel deafening engine noise but allow the wearer to hear conversation.

»»» *See Also*
RELATED PRINCIPLES: Binary Code, 190 • Electromagnetism, 147 • Ohm's Law, 164 • Wave Energy, 58
RELATED APPLICATIONS: Cell Phone Jammer, 63

Destructive Interference

CELL PHONE JAMMER

Radio waves carry energy through electromagnetic fields. Systematically modified, these waves can send information to an electrical conductor, where they can be translated back to their original form. This is how radio works. It's also how cell phones work, sending and receiving signals from cell phone towers.

When waves of any kind with the same distance between peaks cross each other, their individual heights or amplitudes combine. In the case of radio waves, that collision results in a garbled signal, disabling communications.

Transmitting radio signals can jam radio broadcasts as well as radar guns and cell phones. In the U. S., it's illegal for individuals to do any of the three, although police and security officials have sometimes jammed cell phones, mostly to prevent them from being used to remotely detonate a bomb.

See Also
RELATED PRINCIPLES: Binary Code, 190 • Electromagnetism, 147 • Ohm's Law, 164 • Wave Energy, 58
RELATED APPLICATIONS: Noise-Canceling Device, 62

PRINCIPLE

SOUND BOUNCES

Reflection of Sound Waves

Like ocean waves rolling onto a beach, sound waves that move from one medium to another—hitting a wall after traveling through air, for example—don't just keep on going. Nor do they simply die out. That energy has to go somewhere.

It may be absorbed into softer materials (including air), or it may bounce off harder materials. Sounds reflect off an object or surface at the same angle that they strike the surface; if a sound returns to a listener with enough delay and magnitude, it is perceived as an echo.

Sound waves can also diffract—change direction or bend around a small obstacle in their path. This is especially true of long-wavelength (low-frequency) sounds. That's why a television heard through a doorway will sound like a low murmuring; high-pitched sounds don't diffract well.

Reflection of Sound Waves

SONAR

That pinging sound commonly associated with the submarine is the sound of a century-old technology called sonar, from sound navigation ranging. First extensively used in World War II to detect and attack enemy subs, sonar works much like medical ultrasound, sending out pulses of sound and measuring the distance to various objects by counting the time it takes for echoes to return.

Military applications of sonar now include searching for buried mines on land and at sea. Scientists use sonar to map the ocean floor's mountains, valleys, and plains. Industrial uses include searching for oil and mineral deposits under the seafloor, locating areas safe for fishing nets, and surveying underwater pipelines and other equipment.

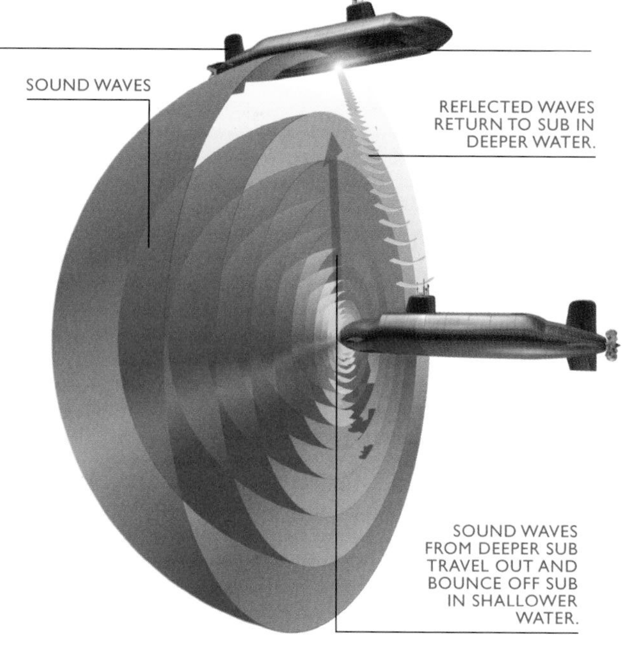

SONAR. Computers tuned to the pinging echoes of electronically generated sound waves can determine the direction from which they come and calculate the time they take to return.

»»See Also

RELATED PRINCIPLES: Binary Code, 190 • Electromagnetic Radiation, 248 • Electromagnetism, 147 • Magnetism, 135 • Ohm's Law, 164 • Radio Waves, 241 • Wave Energy, 58
RELATED APPLICATIONS: Ship, 94 • Submarine, 95

Reflection of Sound Waves

ULTRASOUND

Not unlike the way dolphins navigate by sound echoes, ultrasound is a medical diagnostic technology that turns sounds into a kind of map of the body. Using a probe, the sonographer directs pulses of sound into the body. Each time the sound waves reach a new kind of material with a different density—lung tissue, let's say, or bone—some of the sound energy penetrates the material and goes deeper into the body, while some returns to the probe as an echo. The machine measures the time it takes for the echoes to reach the probe along with their relative intensity. A computer converts these data into an image displayed on a screen.

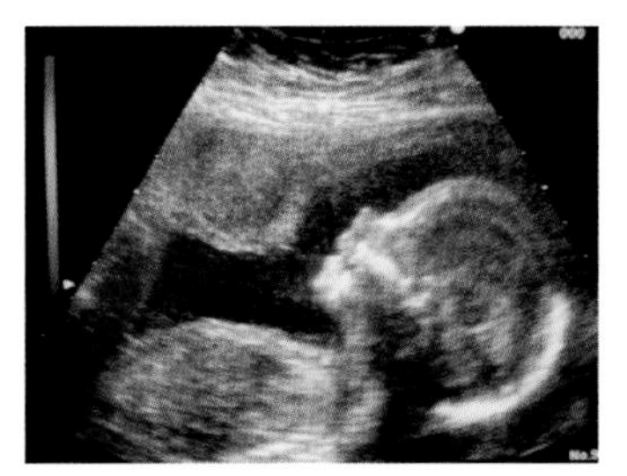

Ultrasound image of a human fetus

First used in the 1960s to image the fetus in the womb, today ultrasound is widely employed to examine many internal organs, including the heart and blood vessels, as well as to guide biopsies.

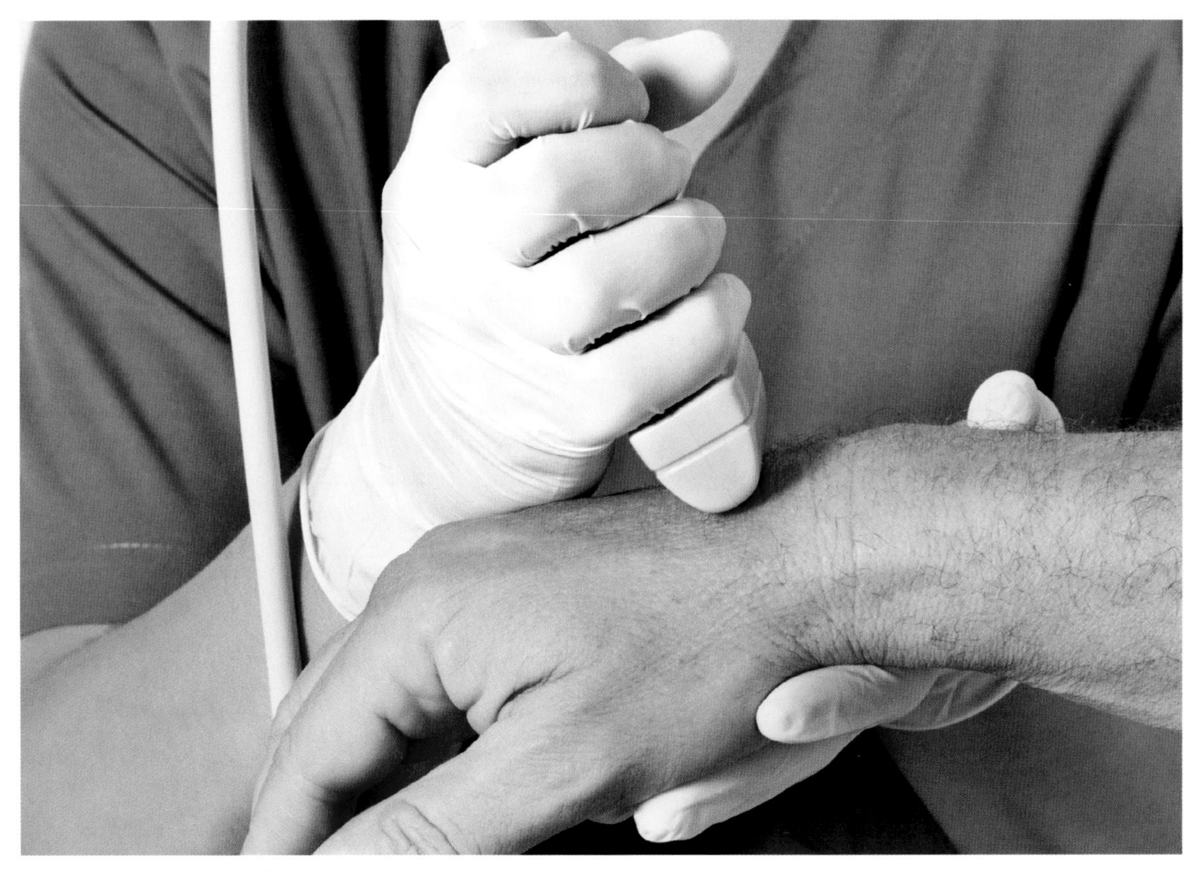

Ultrasound probe used for diagnosis

»»See Also

RELATED PRINCIPLES: Binary Code, 190 • Electromagnetic Radiation, 248 • Electromagnetism, 147 • Magnetism, 135 • Ohm's Law, 164 • Radio Waves, 241 • Wave Energy, 58

RELATED APPLICATIONS: Computerized Axial Tomography, 256 • Magnetic Resonance Imaging, 142 • Sonar, 64 • X-Ray Imaging, 254

CHAPTER 4

Fluids and Pressure

A fluid is a substance whose molecules can move around one another freely. It can flow. This includes liquids and gases, as opposed to solids, whose molecules are fixed in place. The special properties of fluids are responsible for keeping airplanes and helicopters aloft, the sound and light of fireworks, propelling combustion engine automobiles, and even running water in your home.

Fluids do not resist deformation. They take the shape of their container. Unlike, say, a block or a ball, air or water will accommodate itself to a jar. But high-energy gas molecules will spread out to fill the whole container, whereas liquid, under the force of gravity, will settle at the bottom. Because liquids are denser than gases (their molecules are closer together), they are less compressible than gases; they can't be easily squeezed to fit a tighter space.

The study of how fluids behave at rest and two thousand years ago with the Greek scientist Archimedes, who published observations on how things float. The field of fluid mechanics also looks at the stickiness, or viscosity, of a fluid and how it changes with temperature as well as the internal pressure of fluids at rest and in motion. These have had important applications in areas as far ranging as medicine (measuring blood pressure), meteorology (studying air and water currents), and engineering (the use of pressurized fluids to transmit power, as

PRINCIPLE

GO WITH THE FLOW

Bernoulli's Principle

When a flow of fluid (liquid or gas) enters an area of lower pressure, it speeds up. That is, fast-moving fluids exert less pressure than slow-moving ones. This pressure gradient can be exploited to help kites fly, perfume atomizers spray, and boats turn in the water.

This insight published in 1738 by Swiss scientist Daniel Bernoulli neatly conforms to the law of energy conservation. As a fluid gains kinetic energy in faster motion, it loses internal pressure, a form of potential energy.

Bernoulli first studied the phenomenon by observing water flowing through pipes of different diameters. The volume of water passing through a given length of pipe in a given time doesn't vary as the pipe narrows and then widens; through the narrower portions, it simply moves faster, under less internal pressure.

Bernoulli's Principle

SAILBOAT

Skillful sailors know just how to tack into the wind, but the physics of what makes the technique work is complex. A key factor is the pressure differential that forms across the sail. With the boat angled into the wind, the airstream splits around the sail; the outward-arcing downwind side of the sail creates a longer path for the wind, which rushes over it more quickly than it does on the concave side. The faster air forms a low-pressure area, so the higher-pressure air on the sail's concave side pushes into it, essentially changing direction.

The boat would simply be nudged downwind if not for the keel, which hangs down in the water like a wing. Downwind motion of the boat across the water causes water to flow faster around one side of the keel than the other in a pattern opposite that of the air flowing around the sail. The resultant pressure differential causes the water to push against the keel, balancing the direction of the sail force.

»»See Also

RELATED PRINCIPLES: Friction, 52 • Convection, 103 • Mechanical Advantage, 38

RELATED APPLICATIONS: Airplane, 70 • Curveball, 72 • Helicopter, 71 • Hovercraft, 69 • Hydrofoil, 72

Bernoulli's Principle

HOVERCRAFT

Perhaps the most famous hovercraft were those that, until the year 2000 when the Channel Tunnel opened, carried passengers across the English Channel in a half hour. But they're also used in some rescue and military operations and of course for recreation; what could be more fun than zipping around on a cushion of air?

The ultimate all-terrain vehicles, hovercraft can move over slippery, boggy, or bumpy land, not to mention over water. The craft creates lift with a powerful fan that forces air underneath it. This pressurized air is mostly contained within a flexible skirt. Most hovercraft have a propeller at the rear to create thrust, propelling them forward as they push air back. Rudders mounted at the fan's outlet direct the flow of this air. By manipulating the angle of the rudders, the pilot is able to control the craft's direction.

Like the wheel before it, the hovercraft makes travel faster and easier by minimizing surface friction.

»»» ***See Also***

RELATED PRINCIPLES: Electromagnetic Induction, 157 • Electromagnetic Radiation, 248 • Friction, 52 • Mechanical Advantage, 38 • Newton's First Law of Motion, 20 • Newton's Third Law of Motion, 27 • Pascal's Law, 80 • Rapidly Expanding Gases, 74 • Universal Law of Gravitation, 122

RELATED APPLICATIONS: Airplane, 70 • Curveball, 72 • Combustion Engine, 77 • Helicopter, 71 • Hovercraft, 69 • Sailboat, 68

Bernoulli's Principle

AIRPLANE

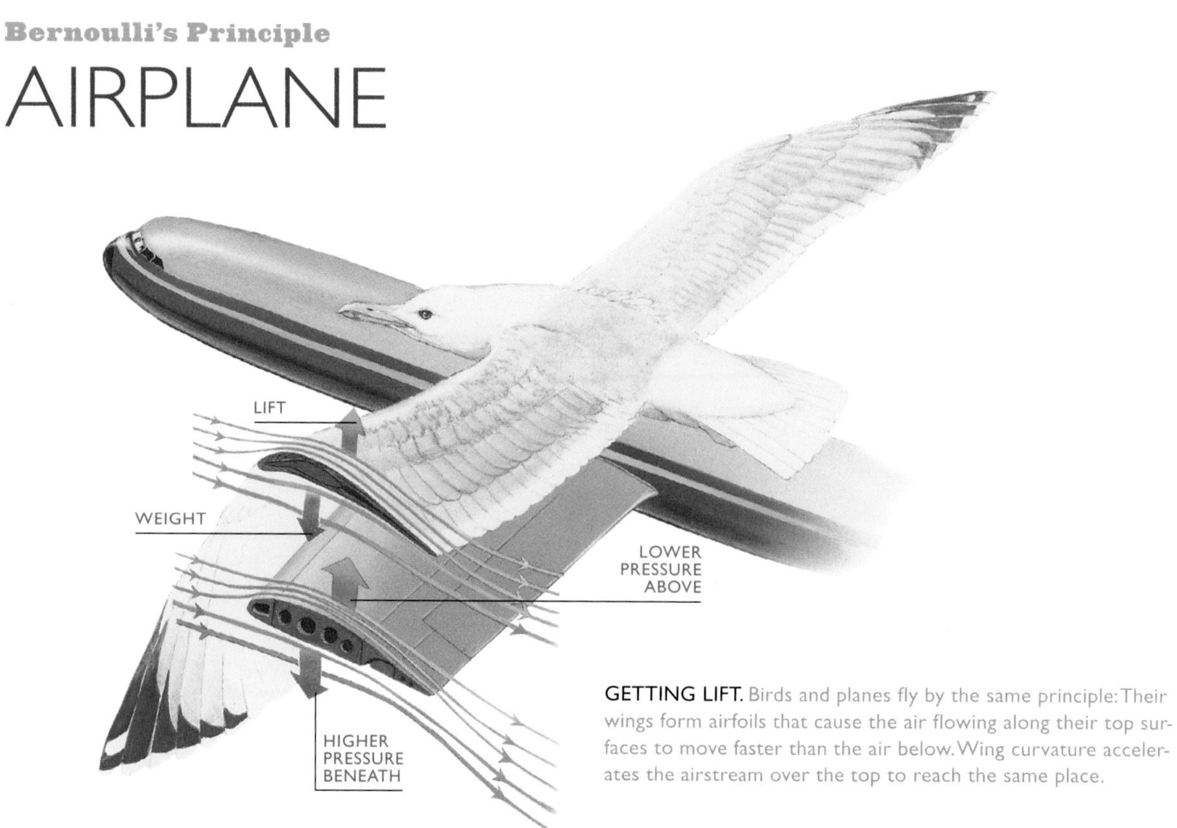

GETTING LIFT. Birds and planes fly by the same principle: Their wings form airfoils that cause the air flowing along their top surfaces to move faster than the air below. Wing curvature accelerates the airstream over the top to reach the same place.

An airplane's forward motion is based on engine thrust—and Newton's law that for every force there is an equal and opposite force. But how does the plane get off the ground and stay aloft? That's all about lift, and wings are the key.

Airplane wings are typically designed with a curved top that lengthens the path of the wind there, creating a fast-flowing, low-pressure zone. The slower-moving air underneath is higher pressure, pushing the plane up. This is Bernoulli's principle at work: Because the air travels longer (and faster) over the upper wing, it creates lower pressure, which is what produces lift.

A wing's angle of attack can be used to create or enhance lift, too. If the pilot pulls up the nose, the wings are angled up against the airstream. The underside of the wing hits the air more directly, and the air, in keeping with Newton's law, pushes on the wing's underside with an equal and opposite force. This force may actually exceed the Bernoulli lift. Using angle of attack for lift, some planes can even fly upside down. Angle of attack is like sticking your hand out of a car window; by tilting your hand, you feel the force upward or downward.

»»*See Also*

RELATED PRINCIPLES: Binary Code, 190 • Electromagnetic Induction, 157 • Electromagnetic Radiation, 248 • Friction, 52 • Mechanical Advantage, 38 • Newton's First Law of Motion, 20 • Newton's Third Law of Motion, 27 • Pascal's Law, 80 • Rapidly Expanding Gases, 74 • Universal Law of Gravitation, 122

RELATED APPLICATIONS: Combustion Engine, 77 • Curveball, 72 • Flight Stabilizer, 23 • Helicopter, 71 • Hovercraft, 69 • Hydrofoil, 72 • Jet Engine, 28 • Sailboat, 68

Bernoulli's Principle

HELICOPTER

Before it became a reality, the idea of vertical flight had engaged the imaginations of engineers for centuries. Leonardo da Vinci sketched what he called an aerial screw in the fifteenth century. In the nineteenth century British engineer Sir George Cayley drew designs for what he dubbed an aerial carriage. But it wasn't until the early twentieth century—shortly after the famous flight of the Wright brothers—that the first "hoppers" got off the ground.

Whereas an airplane relies on forward motion to circulate air against its wings and create lift, a helicopter uses rotating blades for lift. The pilot can control the angle of the blades against the relative current of air in such a way that they push the air down (and the craft up) or, following Bernoulli, create a high-pressure zone under the blades that lifts them up. Tipping the rotor forward produces motion forward and up, while tipping it backward moves the copter in the opposite direction.

CONTROL RODS

ROTOR SHAFT

UPPER SWASH PLATE

LOWER SWASH PLATE

ENGINE DRIVESHAFT

ENGINE DRIVESHAFT

ENGINE DRIVESHAFT

TAIL ROTOR DRIVESHAFT

BLADE

WHIRLING WINGS. A helicopter's rotor blades turn. Control rods connect the blades to a rotating swash plate that tilts, lifts, and lowers. Three engines drive the main shaft via bevel gears.

»»See Also

RELATED PRINCIPLES: Binary Code, 190 • Electromagnetic Induction, 157 • Electromagnetic Radiation, 248 • Friction, 52 • Mechanical Advantage, 38 • Newton's First Law of Motion, 20 • Newton's Third Law of Motion, 27 • Pascal's Law, 80 • Universal Law of Gravitation, 122

RELATED APPLICATIONS: Airplane, 70 • Combustion Engine, 77 • Curveball, 72 • Hovercraft, 69 • Hydrofoil, 72 • Sailboat, 68

Bernoulli's Principle

HYDROFOIL

An airfoil manipulates the flow of air to create and direct movement. A hydrofoil does the same thing, only in the water.

Although the term hydrofoil applies to boat rudders and keels as well as to the flippers, fins, and tails of fish and aquatic mammals, it's most commonly used to describe a ski-like platform attached by a strut to the hull of a boat. When the boat speeds up, this hydrofoil pushes against the water, lifting the boat above the surface much the same way a wing creates lift.

While in ordinary circumstances increasing speed would increase drag on the boat, here, drag is minimized because most of the boat surface is moving against air rather than water, which is much denser.

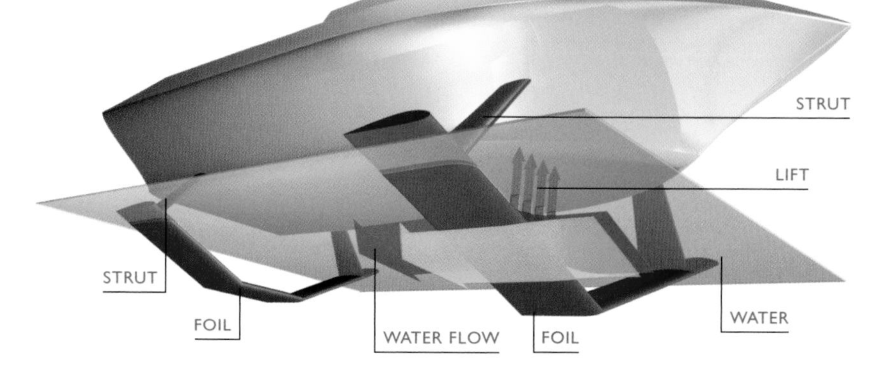

HYDROFOIL. "Water wings" serve as the key element in a hydrofoil's design. As with aircraft wings, higher pressure beneath them than above them creates lift.

»»» *See Also*

RELATED PRINCIPLES: Electromagnetic Induction, 157 • Electromagnetic Radiation, 248 • Friction, 52 • Mechanical Advantage, 38 • Newton's First Law of Motion, 20 • Newton's Third Law of Motion, 27 • Pascal's Law, 25 • Rapidly Expanding Gas, 80 • Universal Law of Gravitation, 122

RELATED APPLICATIONS: Airplane, 70 • Curveball, 72 • Combustion Engine, 77 • Helicopter, 71 • Hydrofoil, 72 • Sailboat, 68

Bernoulli's Principle

CURVEBALL

With a snap of the wrist, the pitcher lets fly the ball. It rises and then, as it nears the plate, sinks into the strike zone.

One of baseball's most thrilling moves, the curveball can be explained by differences in air pressure over and under the ball. Its topspin means the top part of the ball spins in the direction of the pitch—and against the air. This causes greater friction (enhanced by the baseball's traditional stitching) than occurs under the ball, which spins with the airflow. Thus, air flows faster over the bottom of the ball, creating a low-pressure area compared with the air on top. The higher-pressure air pushes the ball down.

HIGHER PRESSURE ON THIS SIDE

LOWER PRESSURE ON THIS SIDE

»»» *See Also*

RELATED PRINCIPLES: Friction, 52 • Mechanical Advantage, 38 • Universal Law of Gravitation, 122

RELATED APPLICATIONS: Airplane, 70 • Helicopter, 71 • Hovercraft, 69 • Hydrofoil, 72 • Parabolic Curve, 122 • Sailboat, 68

PRINCIPLE

UNDER PRESSURE

Downforce

Automobiles, like aircraft, experience lift due to their aerodynamics. Air that takes the shortcut under the car streams by more slowly than does the air passing over the car body. This produces a higher-pressure area underneath that pushes the vehicle upward. In ordinary circumstances it's not enough to lift the car off the ground, but lift can be very dangerous for race cars flying around an oval at well over a hundred miles an hour. That's why they are equipped with extra features to produce the opposite of lift—downforce. These include specially shaped bumpers and protruding wings, all angled to create greater air velocity underneath, increasing pressure on top. This presses the car down, helping it grip the track.

Downforce

ATOMIZER

A spritz of perfume from an old-fashioned atomizer is Bernoulli's principle at work. When you push air from the bulb through a slender nozzle, the airflow speeds up, and its pressure falls. Since the nozzle end is attached to a tube inside the airtight bottle of perfume, this jet of low-pressure air creates a vacuum that sucks the liquid into the airflow, spraying it out in droplets.

Other devices that suck a fluid (liquid or gas) into an accelerated (and thus low-pressure) airstream include a chimney drawing smoke upward and a fuel-injection system spraying vaporized gasoline into a car engine.

»»*See Also*

RELATED PRINCIPLES: Elasticity of Air, 78 • Natural Resins, 283 • Pascal's Law, 80

RELATED APPLICATIONS: Combustion Engine, 77 • Incense and Perfumes, 283 • Siphon, 83 • Vacuum Cleaner, 79

PRINCIPLE

EXPLOSIVE EFFECTS

Rapidly Expanding Gases

When gases are not constrained by some external pressure (say, trapped in a jar), they tend either to expand or contract, depending on the temperature. Gases shrink when the temperature drops; when heated, their molecules get energized and spread apart. The gas expands.

When gases are manipulated to expand extremely rapidly, the result is an explosive release of potential energy. This is the underlying principle behind such applications as fireworks, guns, and even the automobile combustion engine. Some types of solar-energy technology use the sun to heat helium or hydrogen contained in tubes that open onto piston cylinders; as the gas expands, it drives the cylinders. Thunder is an example of rapidly expanding gas in nature. Lightning heats the air, causing it to explode into cooler air at a velocity faster than the speed of sound; that's what causes the boom.

Rapidly Expanding Gases

FIREWORKS

A fireworks display is an explosion designed for maximum visual impact. Before the oohs and ahs, however, the first challenge is to get a packet of chemical reactants into the sky. This is usually accomplished with simple gunpowder, lit by a fuse. The gunpowder is packed so that the resultant expanding gases are forced to escape rearward. This propels the shell upward until the fuse burns into what's called the burst charge at its core. The burst charge chemicals explode, lighting and casting in all directions hundreds of stars, small packages of reactants selected to create color. The reactants are metallic salts: Lithium or strongium produces red; barium nitrates make green; copper compounds result in blue; sodium creates yellow; charcoal and steel produce sparkling gold; and titanium makes white. The chemicals undergo a fast and violent reaction; bonds among molecules of the solid chemicals break apart, yielding hot, concentrated gases that spread out in a flash, transforming energy into sound, movement, and colored light.

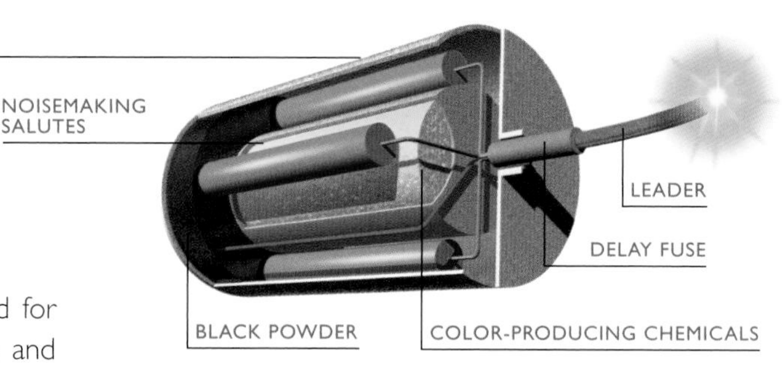

SOUND AND LIGHT. The leader fuse burns down to a lift charge of black powder and propels the shell skyward. Inside, a time-delay fuse causes the color-producing chemicals to ignite and scatter; it also sets off the noisemaking salutes.

»»See Also
RELATED PRINCIPLES: Absorption of Light, 204 • Atomic Structure, 263 • Law of Conservation of Energy, 30 • Newton's Third Law of Motion, 27 • Universal Law of Gravitation, 122
RELATED APPLICATIONS: Combustion Engine, 77 • Firearm, 76 • Lithium, 265 • Rocket Engine, 29

Rapidly Expanding Gases

FIREARM

If you have ever shot a bullet, you have harnessed the power of hot, expanding gases. In every firearm, a propellant (such as the black powder invented by the Chinese as early as the ninth century) is nestled against a projectile in the confined space of the barrel. Once ignited, the powder creates superheated gases whose molecules burst apart against the projectile, blowing it out the firearm's muzzle with great force and velocity.

Early firearms like cannon and muskets were loaded through the muzzle, with powder and bullet or ball inserted separately. In most modern firearms, a cartridge is loaded into a chamber at the back of the barrel. The cartridge contains bullet and powder, as well as a primer that ignites when struck by a trigger-released hammer. The technology has been refined over the centuries but the principle remains the same.

»»» ***See Also***
RELATED PRINCIPLES: Law of Conservation of Energy, 30 • Newton's First Law of Motion, 20 • Newton's Third Law of Motion, 27 • Universal Law of Gravitation, 122
RELATED APPLICATIONS: Bullet Trajectory, 123 • Combustion Engine, 77 • Firearm, 76 • Rocket Engine, 29

Rapidly Expanding Gases

COMBUSTION ENGINE

The four-stroke internal combustion engine seen in today's cars and trucks is an elegant study in fluid mechanics.

Its centerpiece: a piston in a cylinder. Each stroke of the piston accomplishes a different task. First, a downstroke creates a partial vacuum that sucks in vaporized fuel mixed with air. Next, an upward stroke compresses this mixture and, at the top, sets off the spark plug, which ignites the gases. In what's referred to as the power stroke, the hot gases expand, pushing the piston down and transferring power to the crankshaft. An exhaust valve opens, and the piston rises again, expelling the combusted air-and-fuel mixture. A third stroke draws the fuel mixture into the cylinder, and a fourth compresses it to start the cycle again.

This process is sometimes called the Otto cycle, after Nikolaus Otto, the German inventor who built the first working four-stroke internal combustion engine in 1876.

POWER STROKES. Vaporized gasoline ignites and burns inside the cylinders of a four-stroke internal combustion engine. The downward stroke opens an inlet valve, an upward stroke compresses the air and fuel mixture; then a spark plug ignites the fuel, followed by the expulsion of gases through an open valve.

»»**See Also**

RELATED PRINCIPLES: Bernoulli's Principle, 68 · Friction, 52 · Mechanical Advantage, 38
RELATED APPLICATIONS: Atomizer, 73 · Firearm, 76 · Fireworks, 74 · Jet Engine, 28

PRINCIPLE

FULL OF AIR

Elasticity of Air

Air, which is made up of gases—nitrogen, oxygen, and carbon dioxide, among others—is compressible. You can squeeze it into a tighter space by pressing the molecules together. Air is also elastic; once released, its molecules will tend to spread out again to their former volume. In this process, the potential energy (pressure exerted by the condensed air) is transformed to kinetic energy (expansion of the air).

The ability of air to release kinetic energy through compression and decompression has proved a useful source of power in numerous technologies from power tools to jet planes.

Elasticity of Air

VACUUM TUBE

Clear those pesky gas particles out of the way, and negatively charged electrons will leap straight for a positively charged metal plate. In other words, an electrical current will flow more efficiently through a vacuum, a space free of matter.

This is the basis for a key invention of the early twentieth century: the vacuum tube. It consists of a glass bulb emptied of air. On one side is a metal filament, on the other a metal plate. An electrical charge heats the filament, freeing its electrons, which are attracted to the metal plate.

Applying a current to a small grid placed between the filament and the plate makes it possible to not only transmit but also amplify electrical signals along their path. This elaboration on the vacuum tube, known as a triode vacuum tube, revolutionized communications in the early twentieth century, enabling long-distance telephony, radio, and TV.

»»See Also

RELATED PRINCIPLES: Electromagnetic Induction, 157 • Electromagnetism, 147 • Ohm's Law, 164

RELATED APPLICATIONS: Atomizer, 73 • Vacuum Cleaner, 79

Elasticity of Air

VACUUM CLEANER

Electric vacuum cleaners use suction to pick up particles. A motor-driven fan creates a partial vacuum that sucks up dirt loosened by a beater brush and deposits it in a bag or another type of collector. (Vacuum cleaner fan blades spin as fast as 18,000 times a minute; jet engine blades spin only 7,000 to 8,000 times a minute.) An upright vacuum is a single unit with a bag usually attached along the backside of the handle. The canister type has a long, detachable cleaning wand attached to a rolling unit that houses a more powerful motor, the fan, and the dust bag.

The vacuum cleaner became popular in the twentieth century when electricity became widely available. But even before then, there were manual models that used the same idea—low-pressure suction. These sweepers required the household laborer to create the suction by hand, working a bellows or cranking a pulley device.

Electric vacuum cleaners are one of the great labor-saving devices of the modern era. They made a cleaner home possible in a fraction of the time.

DUST BAG

FAN

MOTOR

BEATER BRUSH

DRIVE BELT

CLEANING MACHINE. In the scientific sense, a perfect vacuum means a complete absence of matter within a space—a rare, if not impossible, phenomenon. A motorized fan creates suction and a partial vacuum in the home appliance.

»» *See Also*

RELATED PRINCIPLES: Electromagnetic Induction, 157 • Ohm's Law, 164

RELATED APPLICATIONS: Atomizer, 73 • Electric Motor, 158 • Vacuum Tube, 78

PRINCIPLE

SQUEEZED

Pascal's Law

Unlike gases, liquids such as oil and water cannot be compressed; their molecules can't easily be squeezed closer together, nor do they disperse when not contained. Just as people have developed ingenious ways to harness the power of gases as they expand from a compacted state, they've also used the incompressibility of liquids to change the world.

The keys to tapping liquid power were established by Frenchman Blaise Pascal in the mid-seventeenth century. Pascal's law says that in a contained, incompressible fluid, any external pressure applied at one point, such as pushing on a water-filled balloon, raises pressure equally at every point. Pascal also observed that water pressure increases with depth and conducted experiments showing that the weight of the atmosphere can push liquids up an enclosed tube (as in a barometer).

Pascal's Law

DIVING AND THE BENDS

As a diver descends to an undersea world, the weight of all that water exerts tremendous pressure on his or her body—nearly 15 pounds per square inch for every 33 feet of depth. It would be impossible to inflate the lungs under such force if not for the high-pressure air delivered from a scuba tank, which counteracts surrounding pressure. These inhaled, pressurized gases are pushed in higher-than-usual concentrations into the bloodstream, much the way carbon dioxide is forced into water under high pressure to create soda water.

If the diver rises too quickly to the surface, it's like uncapping a seltzer bottle. A sudden drop in ambient pressure lets gases come out of solution too rapidly, forming bubbles in the diver's bloodstream. These bubbles, particularly of nitrogen gas, can create the painful and potentially lethal condition known as the bends. To prevent this problem, also called decompression sickness, divers ascend gradually and sometimes pause before surfacing in order to give their bodies time to release nitrogen gas slowly.

»»See Also
RELATED PRINCIPLES: Archimedes' Principle, 92 • Universal Law of Gravitation, 122
RELATED APPLICATIONS: Home Plumbing System, 82 • Hydraulic Brakes, 84 • Hydraulic Excavator, 85 • Hydraulic Fracturing, 84 • Oil Drill, 87

Pascal's Law

HOME PLUMBING SYSTEM

Plumbing is one of those marvels of modern living: The more optimally it functions, the less we notice it.

But how does water rise to a home's upper floors to flow at the turn of a tap? In a word, pressure. It's generated by exploiting Pascal's insight that water pressure increases with depth (not overall volume). Water towers achieve this depth through elevation; often sited on hills or on the roofs of tall city buildings, a tank is raised many feet in the air. The weight of the water column drives downward, pressing water through supply lines with enough force to lift it to upper floors. That said, people on the ground floor usually get better water pressure than people on the higher floors.

Wastewater leaves the home via an entirely separate system that depends on gravity; all pipes are angled to let the water drain down.

Among common plumbing fixtures, the humble toilet is probably the most elaborate. It flushes by a siphon effect. Push the flush handle, and a chain lifts open a valve in the toilet tank, dumping water into the bowl. The water enters quickly enough to flood the toilet's drainpipe, which curves up and then down as it leaves the bowl. As in any siphon, the flow downward creates a low-pressure area at the top of the pipe's curve, which sucks the rest of the water from the bowl, emptying it into the waste stream.

WATER AND WASTE. Risers and supply branches (shown in red for hot water, blue for cold) distribute water from a main line. Wastewater flows out through drainpipes (green) to sewers or septic tanks. Vent pipes link drainpipes to rooftop stack vents and rid the system of gases.

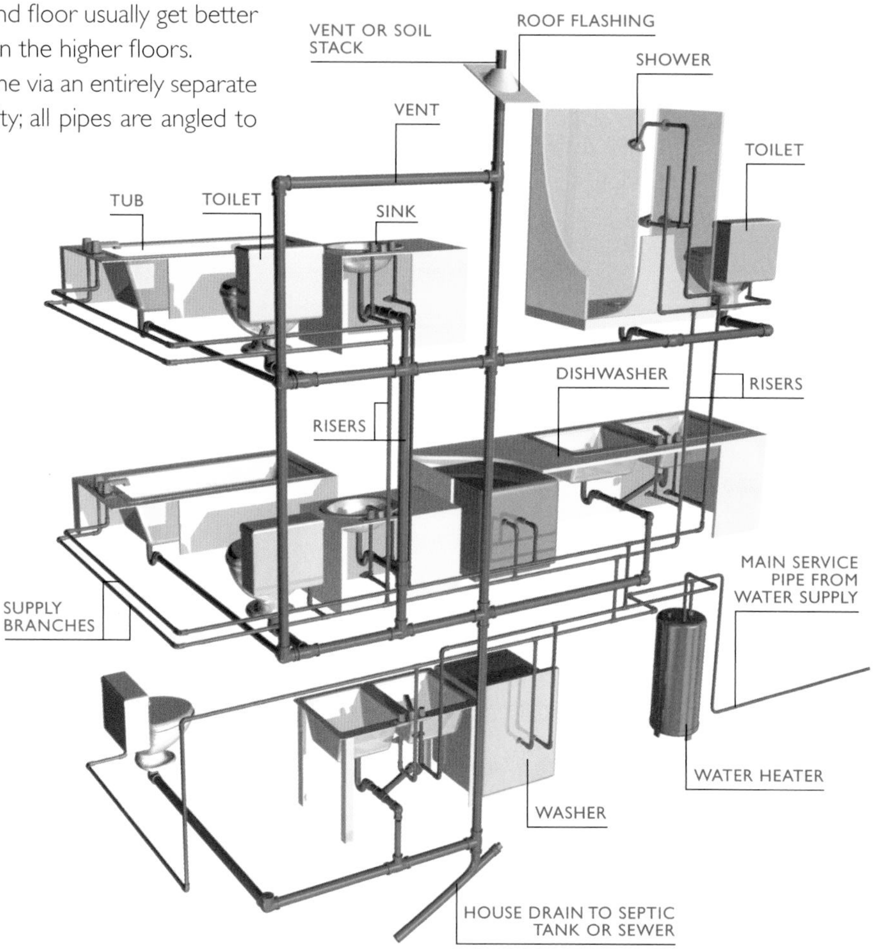

»»See Also

RELATED PRINCIPLES: Universal Law Of Gravitation, 122

RELATED APPLICATIONS: Artesian Well, 83 • Diving and the Bends, 80 • Hydraulic Brakes, 84 • Hydraulic Fracturing, 84 • Oil Drill, 87 • Siphon, 83

Pascal's Law

SIPHON

Gasoline siphoned from a tank flows up the tube on its way out of the tank before descending through the tube's other end into a waiting can.

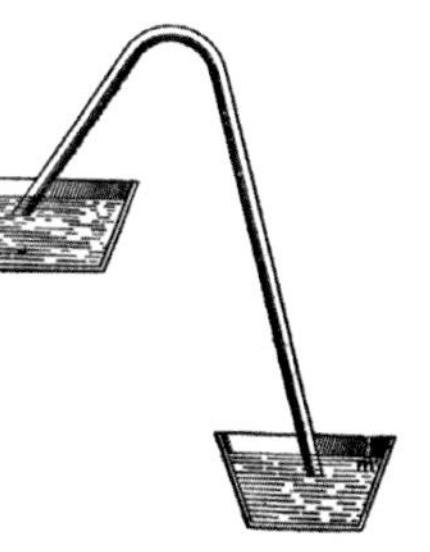

How does it work? First, a few conditions must be met. The gas tank has to be open to the atmosphere. The receiving can must be lower than the gas tank. And the tube has to be filled by suction or some other method to get the flow started. The process begins with gas falling by force of gravity from the filled tube into the can. This creates a lower-pressure zone at the bent top of the tube. The gas at the bottom of the tank, meanwhile, is under the combined pressure of the air and the gas itself. This high pressure pushes the gas up to the lower-pressure area at the top of the tube. Siphons have been employed since ancient times. The Egyptians and Greeks were the first to use them.

»»See Also
RELATED PRINCIPLES: Elasticity of Air, 78
RELATED APPLICATIONS: Artesian Well, 83 • Divers and the Bends, 80 • Hydraulic Excavator, 85 • Hydraulic Fracturing, 84 • Hydraulic Brakes, 84 • Oil Drill, 87 • Vacuum Cleaner, 79

Pascal's Law

ARTESIAN WELL

Named for a French province where they were once common, artesian wells are not unlike water towers in that they use water depth to create pressure, making pumps unnecessary. Engineers punch the well into a deep underground aquifer confined between impermeable rock layers. The aquifer must be one that receives its waters from a higher-level source. This water column pushes down on the trapped aquifer with enough force to press it up through the well.

WATER WORKS. These fountains in Athens, Greece are powered by an artesian well.

»»See Also
RELATED PRINCIPLES: Universal Law of Gravitation, 122
RELATED APPLICATIONS: Diving and the Bends, 80 • Hydraulic Brakes, 000 • Hydraulic Excavator, 84 • Hydraulic Fracturing, 85 • Oil Drill, 84 • Siphon, 83

Pascal's Law

HYDRAULIC BRAKES

In modern vehicles, stepping on the brakes doesn't require much leg strength, thanks to braking systems that magnify pedal power with liquid power.

When you press the brake pedal, a lever activates a pushrod that in turn presses a plunger-like piston through a master cylinder containing brake fluid (typically ethylene glycol). This forces the fluid out of the cylinder through tubing that leads to each of four wheel cylinders. In these smaller cylinders, the fluid depresses pistons whose force applies brake pads to the wheels.

Recall that, according to Pascal, pressure applied to the contained brake fluid at the master cylinder will be distributed undiminished throughout the body of fluid—through the tubing and against the output pistons near the wheels. Pressure is force per unit of area. So by transmitting pressure from a small area (the master cylinder) to a larger one (the output cylinders near the wheels), the fluid multiplies force.

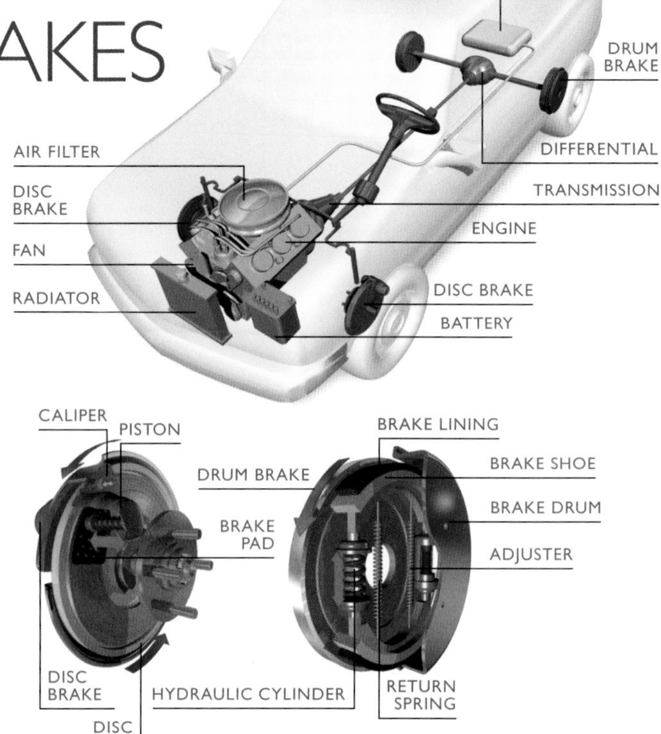

HYDRAULIC BRAKING. In front-wheel disc brakes (left), friction pads inside U-shaped calipers straddle discs that turn with the wheels; brake fluid activates the calipers to press the pads. In rear-wheel drum brakes (right), stationary shoes lie inside cupped drums that turn with the wheels; brake fluid forces the shoes against the drum surfaces, slowing the automobile.

»»See Also
RELATED PRINCIPLES: Friction, 52 • Law of Conservation of Energy, 30 • Mechanical Advantage, 38 • Rapidly Expanding Gases, 74
RELATED APPLICATIONS: Artesian Well, 83 • Combustion Engine, 77 • Diving and the Bends, 80 • Gear, 48 • Home Plumbing System, 82 • Hydraulic Excavator, 85 • Hydraulic Fracturing, 85 • Lever, 41 • Oil Drill, 87 • Siphon, 83 • Wheel, 44

HYDRAULIC FRACTURING

Just as a diver's lungs are compressed by the weight of deep water, rock layers far below the Earth's surface are compacted by the weight of the rock above them. That makes it hard for natural gas to escape from small pores in the shale to be harvested for commercial use.

In the late 1990s energy companies began addressing this challenge with a controversial but effective mining method called hydraulic fracturing. A well is drilled to depths of up to a mile below ground, sometimes with a horizontal extension into the rock layer. The well is lined with concrete. Then workers lower a perforating gun to the location of the trapped gas reservoir; it releases an explosive charge that blasts holes in the well casing. Finally, pumps at the surface push water at extremely high pressure down the well and into the holes, cracking open fissures in the rock. This allows gas to flow out into the well. The controversy involves the potential for contamination of ground water.

Pascal's Law

HYDRAULIC EXCAVATOR

Once upon a time, the steam shovel held sway in construction yards; a coal-fired boiler produced hot, expanding steam whose force ran a lifting winch and bucket. Today you're likely to find an excavator with a diesel engine that powers a mechanical system based on liquid power. The engine runs hydraulic pumps that draw oil from a tank and force it around interlocking gears, pressurizing it. This oil is pushed through a set of valves controlled by the driver-operator and onto tubing that leads to pistons on the arm and basket as well as to hydraulic motors that use the force of pressurized oil to turn the excavator's caterpillar treads and spin it around on its base.

It's all based on Pascal's law; pressure applied to the oil at the pump is transmitted undiminished over considerable distance through narrow tubes to perform a variety of quick, powerful mechanical tasks.

»» See Also

RELATED PRINCIPLES: Friction, 52 · Law of Conservation of Energy, 30 · Mechanical Advantage, 38 · Rapidly Expanding Gases, 74 · Universal Law of Gravitation, 122

RELATED APPLICATIONS: Artesian Well, 83 · Combustion Engine, 77 · Diving and the Bends, 80 · Gear, 48 · Home Plumbing System, 82 · Hydraulic Brakes, 84 · Hydraulic Fracturing, 84 · Lever, 81 · Oil Drill, 87 · Siphon, 83 · Wheel, 44

Pascal's Law

OIL DRILL

In the movies, as soon as drillers strike an underground oil deposit, the black gold bubbles to the surface naturally. Texas tea! That happens when underground pressure is sufficient to push oil up the well. But often, energy companies must use artificial extraction methods, manipulating pressure to drive the petroleum where they want it to go.

The first approach is to place a pump jack over the well; an engine-powered lever pushes a plunger-like rod up and down in the well hole. On the upstroke, the plunger pushes air molecules out of the way, creating a lower-pressure zone that draws the oil up. In some cases, oil workers may drill a secondary well near the producing well. Into this hole they may inject water that presses into rock fissures to sweep oil into and up the primary well. Or they may inject high-pressure steam, whose expansion likewise pushes the oil, and whose heat thins the oil so it flows more readily.

»»See Also

RELATED PRINCIPLES: Combustion Engine, 77 · Mechanical Advantage, 38

RELATED APPLICATIONS: Artesian Well, 83 · Diving and the Bends, 80 · Gear, 48 · Home Plumbing Network, 48 · Hydraulic Brakes, 82 · Hydraulic Excavator, 85 · Hydraulic Fracturing, 84 · Lever, 41 · Siphon, 83

PRINCIPLE

TOUGH EXTERIOR

Surface Tension

When you jump in a pool, it's breaking the surface that requires the most force; once submerged, it's easier to plunge deeper. This effect has to do with the way liquid molecules attract each other. Those on top have neighbors only on one side and cling all the more tenaciously to them, forming a kind of skin. On a waterproof surface, rain will run in droplets, with all the outside molecules drawn inward.

Soap reduces the surface tension of water so it gets into clothes, hair, and other textured things by essentially coating the water molecules and making them more slippery. One end of the soap molecule is drawn to water molecules, while the other end repels water.

Surface Tension

BUBBLE

Because liquid molecules are strongly attracted to each other, the surface of a liquid—the place where it meets air or other gases—will shrink to the smallest possible area. For example, when air is injected into water, the water molecules cling to each other and minimize their contact with the air. The smallest possible surface area required to enclose the air happens to be a sphere: a bubble. The bubble, weighing far less than the water it displaces, rises to the surface.

Competitive high divers sometimes practice new moves using a bubble machine that sends a stream of air up from the bottom of the swimmimg pool; this creates a more forgiving, cushioned landing zone by continually disrupting the water's surface tension.

»»» *See Also*
RELATED PRINCIPLES: Pascal's Law, 80
RELATED APPLICATIONS: Turgor Pressure in Plants, 91

Surface Tension

TURGOR PRESSURE IN PLANTS

Water and various particles travel into and out of cells across a filmy, selectively permeable membrane called the cell membrane. There's a distinct pattern to this movement: Water with a lower concentration of particles will flow across the membrane into water with a higher concentration of solute particles until the concentration is balanced across the membrane. That's called osmosis.

In people (and all animals), cell membranes stretch as the cell takes on water. Rapidly flooding the body with too much dilute water can bloat and even burst the cell membrane in a rare condition known as water poisoning.

Plant cells, however, have a special feature that allows them to use the pressure of water inside the cell as a kind of skeleton. A tough cell wall surrounds the delicate membrane, preventing it from overexpanding. Instead, when the cell is swollen with water, its membrane presses out against the cell wall, stiffening the cell. This is called turgor pressure, and it keeps the plant sturdy and upright. In drought conditions, on the other hand, there's more pure water inside the cell than in the surrounding environment, so water drains from the cells. They lose their turgidity, and the plant droops. But the cell wall helps protect the cells from deformation. If rain comes soon enough, the plant perks up.

»»See Also
RELATED PRINCIPLES: Pascal's Law, 80
RELATED APPLICATIONS: Bubbles, 88

HAJJ TERMINAL

Every year hundreds of thousands of Muslims converge on Mecca to perform the Hajj, the pilgrimage that is a duty of their faith. Designed to host this influx, the Hajj Terminal at Saudi Arabia's King Abdulaziz International Airport is a strikingly modern, open-air tent village covering 120 acres.

Most buildings hold their structure up by the weight of things pushed together—stacked bricks and joists, for example. But the Hajj Terminal relies instead on tension, the force created when things are pulled apart.

In the roofs of the Hajj Terminal, vast swathes of lightweight, flexible, Teflon-coated fiberglass fabric are suspended in cone shapes on taut steel cables.

PRINCIPLE

SINK OR FLOAT

Archimedes' Principle

The ancient Greek mathematician Archimedes of Syracuse is credited with demonstrating a basic principal of buoyancy: An object immersed in a fluid is buoyed up with a force equal to the weight of the fluid the object displaces. A stone, much heavier than the same volume of water, will sink. Anything filled with air—a flotation mattress or rubber ducky, for example—will float on the surface.

Most people float to some extent. Those with large body volumes made up of relatively lightweight tissues like fat are buoyed up with a greater force than wiry, dense-bodied people who displace a smaller area of water.

Archimedes' Principle

FLOATING CONTINENTS

The solid continental plates of Earth's crust float on the endless seas of molten mantle rock beneath them. Archimedes' principle explains why.

Archimedes' principle says that the buoyancy force acting on an object is equal to the weight of the fluid that object displaces. So how pieces of Earth's crust ride in the dense fluid-like material of the mantle depends on both their volume and density. Continental crust, of lower-density rock than seabed rock, floats higher. At the same time, large and heavy features of Earth's crust—Greenland, for example, covered with an enormous volume of ice—may push deeper like the submerged portion of an iceberg. They may not project very high above the mantle, but they do float.

FLOATING CONTINENTS. Archimedes' principle explains how Earth's solid continents ride on the seas of molten mantle rock. Continental rock averages about 2.8 times the density of water while mantle rock averages around 3.3 times the density of water. These land masses do not project very much above the mantle because the difference in densities is not very great.

»»See Also

RELATED PRINCIPLES: Seismic Waves, 60 • Universal Law of Gravitation, 122

RELATED APPLICATIONS: Earthquake, 61 • Ship, 94 • Submarine, 95

Archimedes' Principle

SHIP

Thousands of years ago people fashioned the first small rafts and dugout canoes, harnessing the principle of flotation to travel and move goods. Today, cruise ships with casinos, restaurants, and swimming pools carry millions of pounds through the world's oceans by the same principle.

Lightweight materials are used where possible. Even so, packed tightly together those materials would surely sink. Instead, they're shaped to take up a great deal of space and enclose a lot of air. Thus a giant cruise ship displaces a volume of water equal to its own mass, which lets it hover at the water's surface. In 1912, this balance was tragically disrupted for the 900-foot luxury cruise ship *Titanic*. An iceberg punched holes in several watertight compartments, which flooded. Dense water replaced air, and the ship went down. As the dense water replaced the air, the ship lost its buoyancy and eventually sank to the bottom of the sea.

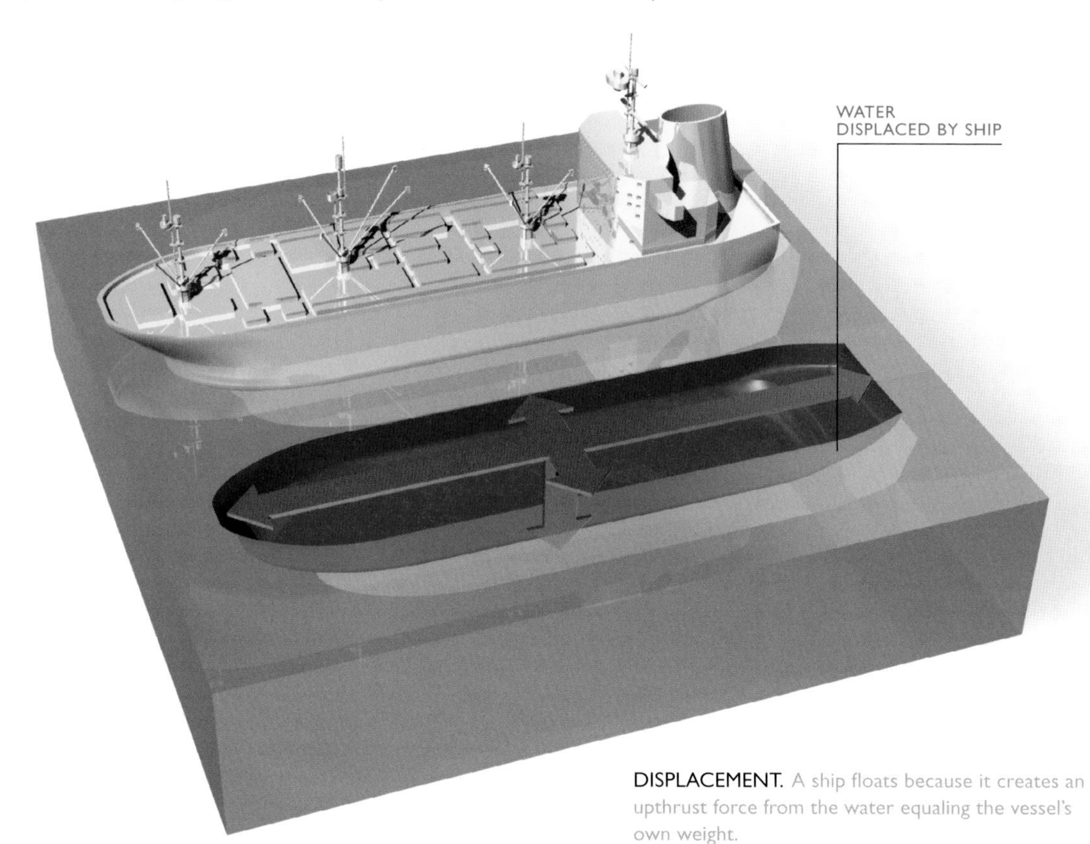

DISPLACEMENT. A ship floats because it creates an upthrust force from the water equaling the vessel's own weight.

»»See Also
RELATED PRINCIPLES: Universal Law of Gravitation, 122
RELATED APPLICATIONS: Floating Continents, 92 • Ship Stabilizer, 22 • Submarine, 95

Archimedes' Principle

SUBMARINE

A ship floats because it displaces a volume of water equal to itself in weight. A submarine dives and surfaces by varying its weight.

It does this by filling ballast tanks with air or water. When diving, the sub takes on water to increase its mass in relation to the surrounding water; it sinks. To surface, the crew discharges water using compressed air stored onboard. The vessel lightens, and buoyancy is restored.

Divers take a similar approach to controlling flotation. They often wear weight belts to help them sink to depths despite wetsuits and other buoyant equipment; when they want to surface, they may use a special vest with bladders that can be filled with compressed gas to increase buoyancy.

GETTING AROUND. A rounded configuration and double-walled hull resist crushing deepwater ocean pressures. A submarine maneuvers by adjusting the volume of water and air in ballast tanks.

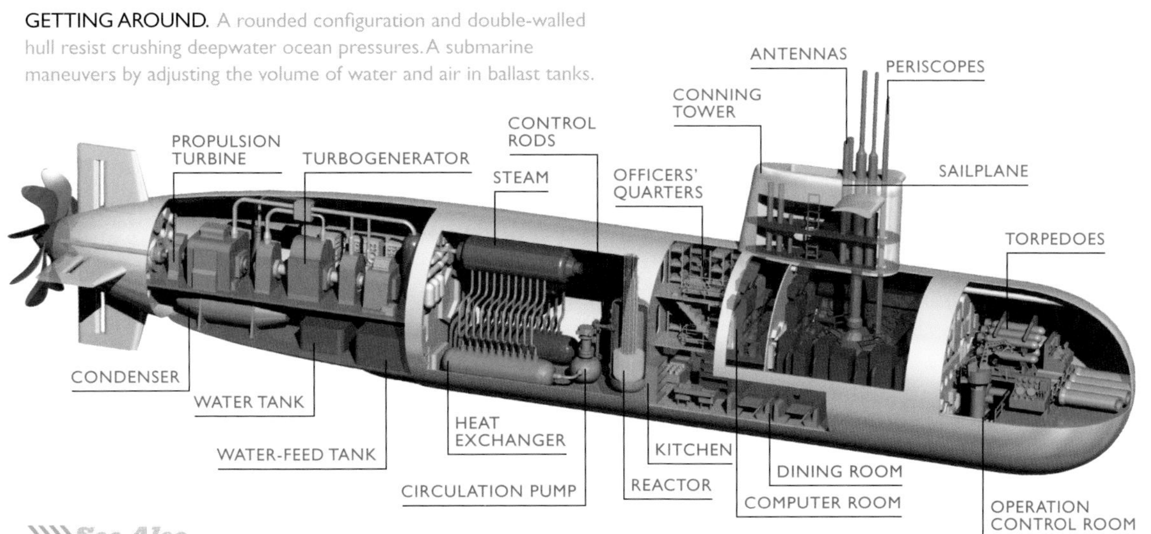

See Also

RELATED PRINCIPLES: Universal Law of Gravitation, 122

RELATED APPLICATIONS: Floating Continents, 92 • Ship, 94 • Ship Stabilizer, 22

CHAPTER 5
Thermod

ynamics

Heat is the primordial form of energy. Humans have sought to understand and harness it since they discovered the power of fire to warm their bodies and cook their food. But the term thermodynamics itself wasn't coined until the nineteenth century, when inventors were exploiting heat energy in the first steam engines. The term combines Greek words meaning heat and power.

As a science, thermodynamics describes the relationship of heat to work—the ability to move things—and to other energy forms. Heat is not something an object contains. It is a flow of energy; temperature simply measures an object's ability to transfer heat energy to another object or system. The hot water in a tub is full of fast-moving, high-energy molecules that can move their energy to the bather's cooler body. Heat always flows from areas of high temperature to areas of lower temperature.

Thermodynamics is concerned especially with the extent to which heat flow can perform useful work. Entropy, a key concept in thermodynamics, is the proportion of thermal energy in a given system or object that is unavailable for useful work due to the random motion of particles. For example, a steam engine derives mechanical energy from the flow of heat; however, because the motion of individual steam molecules is unpredictable, the engine isn't wholly efficient; it can't harvest all the energy of those active molecules to dig and lift. Entropy is also a measure of the disorder of a system.

PRINCIPLE

DISORDERLY CONDUCT

Second Law of Thermodynamics

The second law of thermodynamics recognizes that natural processes have certain stubborn preferences. While energy is neither created nor destroyed, it does change form, and real-world processes invariably end up with less energy that can do work. They use up their potential energy. Entropy increases.

An engine converts chemical energy (gasoline) to motion and heat, and when all the gas is converted, the engine stops. A firework explodes with the force of expanding gases, after which its usable energy is dissipated. The process doesn't run in reverse. Similarly, heat energy moves spontaneously from hot things to cold things, never the other way around. A mug of hot coffee will transfer thermal energy to its cooler surroundings until the coffee and surroundings reach what's called thermodynamic equilibrium—both are at room temperature. The system retains the same amount of energy overall, but its usable energy is spent.

JOULE-KELVIN EFFECT

In the mid-nineteenth century, James Prescott Joule and William Thomson (Lord Kelvin) conducted experiments in which they allowed gases contained in a cylinder to expand through a porous plug into a second, empty cylinder. This set-up insulated the gas against transfers of heat or other energy from outside the cylinders.

Their finding: Generally speaking, gases cool as they expand. Put another way, pressure affects temperature. Such vital modern technologies as refrigeration and air conditioning are based on this insight, manipulating gas pressure to affect temperature.

James Prescott Joule

As gas molecules become more spread out, the fast-moving particles slow down, and their kinetic energy decreases. Temperature, a measure of kinetic energy, also decreases. But energy cannot be destroyed, only transformed. So where does it go? It takes the form of potential energy—the pull among spread-apart gas molecules.

Second Law of Thermodynamics

AIR CONDITIONER

Rather than say an air conditioner cools a room, it would be more accurate to say it removes heat from the space.

To understand air conditioning, you have to know that the state of any substance is related to temperature. Apply heat to a liquid, and it boils; increase the heat, and the liquid vaporizes. Withdraw heat from a gas, and it liquefies; further reduce temperature, and the liquid freezes. Air conditioning accomplishes heat transfer by forcing a chemical called Freon to cycle between liquid and gaseous forms.

Freon has a very low boiling point—that is, the attraction among molecules is weak. In an air conditioner, Freon is kept in a tube at very high pressure and then released through a valve into a low-pressure evaporator. With no force pressing them together, the molecules go on the move, and the liquid boils and then vaporizes. This change of state requires heat. The Freon takes this heat from air in the room.

Then the gaseous Freon passes to the compressor, which pushes its molecules together, and onto condenser tubes, where it liquefies. In the process the heat associated with fast-moving molecules is lost, released to the outdoors. The liquid Freon begins the process again.

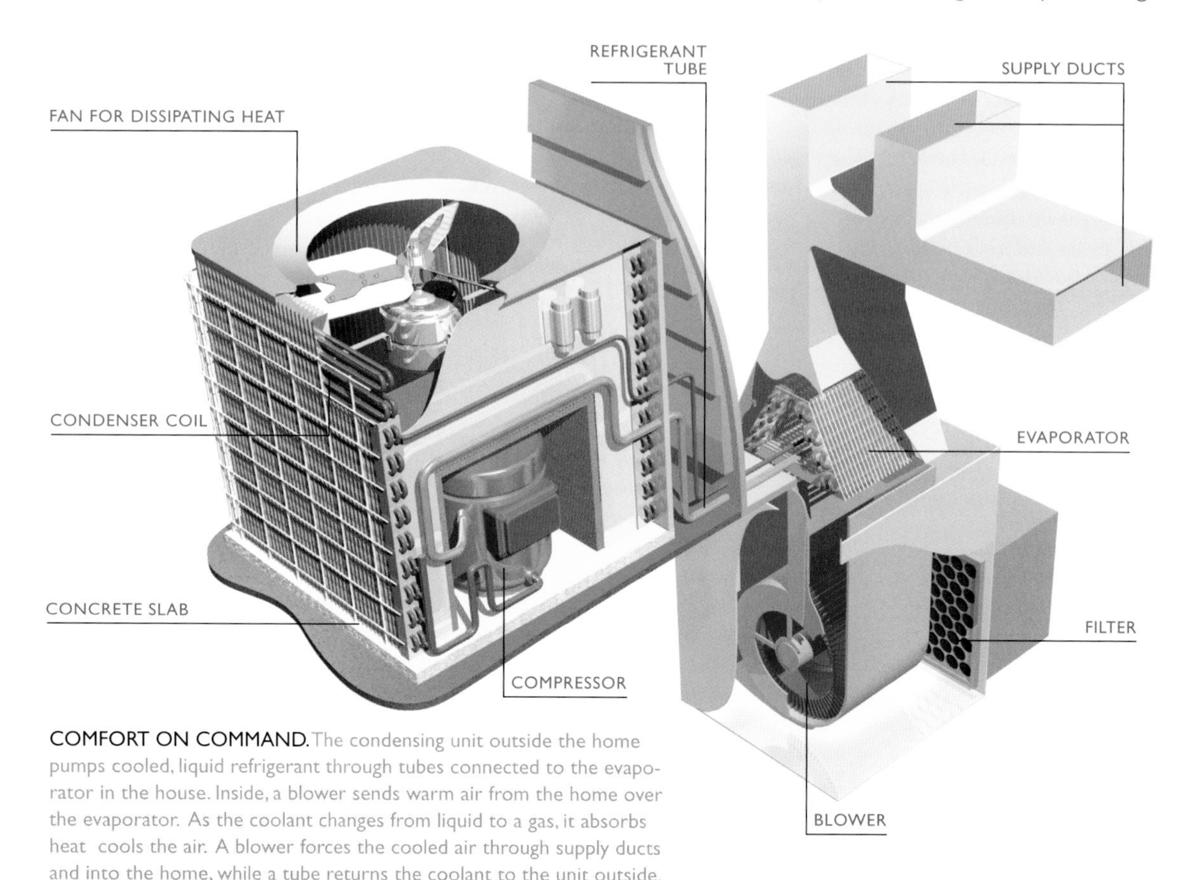

COMFORT ON COMMAND. The condensing unit outside the home pumps cooled, liquid refrigerant through tubes connected to the evaporator in the house. Inside, a blower sends warm air from the home over the evaporator. As the coolant changes from liquid to a gas, it absorbs heat cools the air. A blower forces the cooled air through supply ducts and into the home, while a tube returns the coolant to the unit outside.

See Also

RELATED PRINCIPLES: Electromagnetic Induction, 157 • Law of Conservation of Energy, 30 • Ohm's Law, 164 • Pascal's Law, 80
RELATED APPLICATIONS: Refrigerator, 101

Second Law of Thermodynamics

REFRIGERATOR

Air conditioning was developed in the early twentieth century to combat moldering of industrial supplies and oppressive heat in homes. Around the same time, refrigeration entered the mass market to tackle an arguably more pressing problem: food spoilage.

Of course, for generations people had kept their foods chilled using an insulated box containing a hunk of ice. The refrigerators that occupy nearly all American homes today combine this simple icebox with the same technology used in air conditioning. A refrigerant fluid is made to vaporize via a change in pressure, absorbing heat from inside the fridge, and then condense to a liquid state, releasing heat into the kitchen.

Both the fridge and the air conditioner use pressure differences to promote vaporization and condensation. They do work to overcome a rule governing natural processes: that heat can never flow from a cooler zone (such as inside a fridge) to a hotter one.

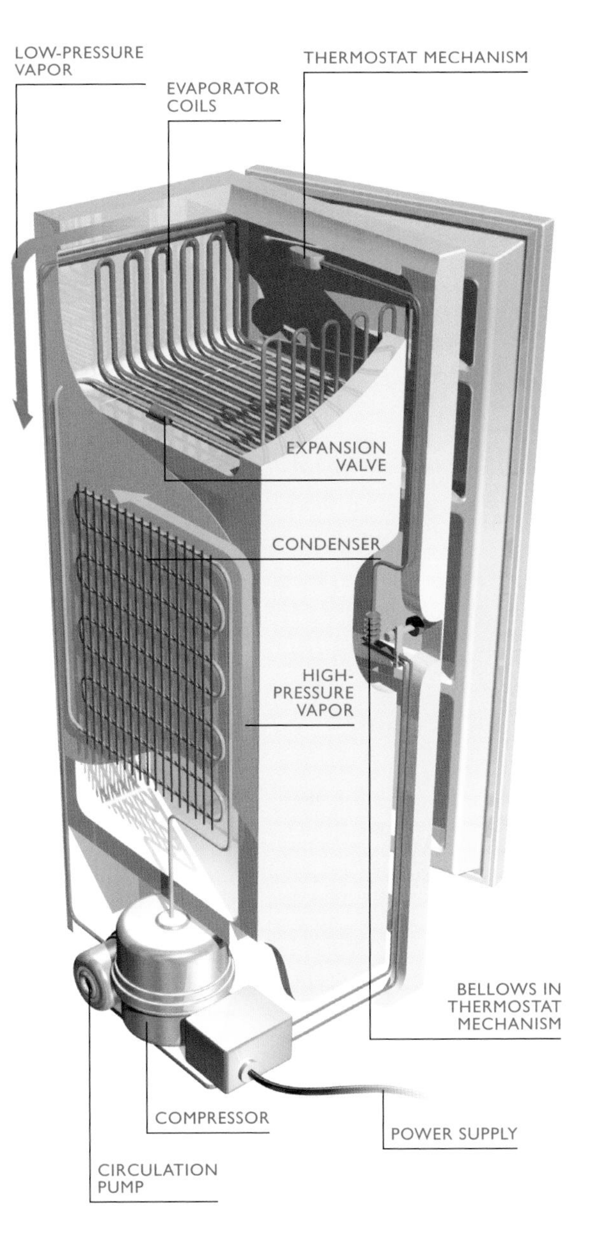

COLD STORAGE. A refrigerator relies on coolant circulated through a sealed system of tubes by a compressor. As the coolant evaporates, it absorbs heat from food inside the refrigerator. The warmed gas returns to the compressor, which sends it to the condenser, where it cools to a liquid and recirculates.

»»See Also

RELATED PRINCIPLES: Electromagnetic Induction, 157 • Law of Conservation of Energy, 30 • Ohm's Law, 164 • Pascal's Law, 80
RELATED APPLICATIONS: Air Conditioner, 100

PRINCIPLE

HEAT GETS AROUND

Convection

Hot air expands and rises. When water gets hot, it, too, becomes less dense and pushes up above colder water, which sinks. These are two ways that a volume of fluid begins to circulate in a current, transferring thermal energy to neighboring particles.

Some of the broadest patterns in our world embody this process, called convection. Oceans continually circulate as deep water warms, rises, cools at the surface, and then sinks. When a coastline heats up in the sun, air over the land expands and rises, creating a low-pressure area; cooler sea air breezes are drawn inland.

Whether you use radiators, air ducts, or baseboard heaters, your home spreads heat by convection: Warm air rises to the ceiling, gradually loses some of its heat to its surroundings, grows more dense, then begins to descend.

Convection

CONVECTION OVEN

In an ordinary oven, whatever is closest to heat coils at the top and bottom tends to cook first. Place a sheet of cookies on the lowest shelf, and you're likely to burn the undersides.

A convection oven surmounts this flaw with the simplest of improvements: a third coil in the back of the oven with a fan that blows the hot air around in the oven. The circulating air penetrates cold spots and bumps against the food itself from different angles. A roasting chicken, for example, browns evenly on all sides. This more efficient transmission of heat means that food can be cooked at a lower temperature or for a shorter time.

»»See Also
RELATED PRINCIPLES: Electromagnetic Induction, 157 • Law of Conservation of Energy, 30 • Ohm's Law, 164
RELATED APPLICATIONS: Electrical Wiring and Power Grid, 166 • Home Heater, 104 • Hot Air Balloon, 105 • Thunderstorm, 106

Convection

HOME HEATER

A furnace uses the energy of gas or oil to heat air. But how does that hot air get where it's supposed to go? How does it warm bathroom tiles and make bedrooms cozy? Most heating systems make at least some use of convection, the circulating movement caused by differences in air temperature and pressure.

Old-fashioned gravity furnaces relied heavily on this phenomenon. Supply ducts connected to the heat source let the air expand and rise through open registers and up stairways. Cooler air naturally dropped down return ducts to the basement furnace. This was a slow-moving and quite inefficient loop.

A radiator spreads heat by convection.

Modern forced-air systems use a fan that blows the hot air through ducts venting into each room. But convection is important here, too. Ducts are often placed at the base of exterior walls, so hot air sweeps up along the wall and into the center of the room; cold air drops and enters a return duct, often placed in a similar location on the opposite wall.

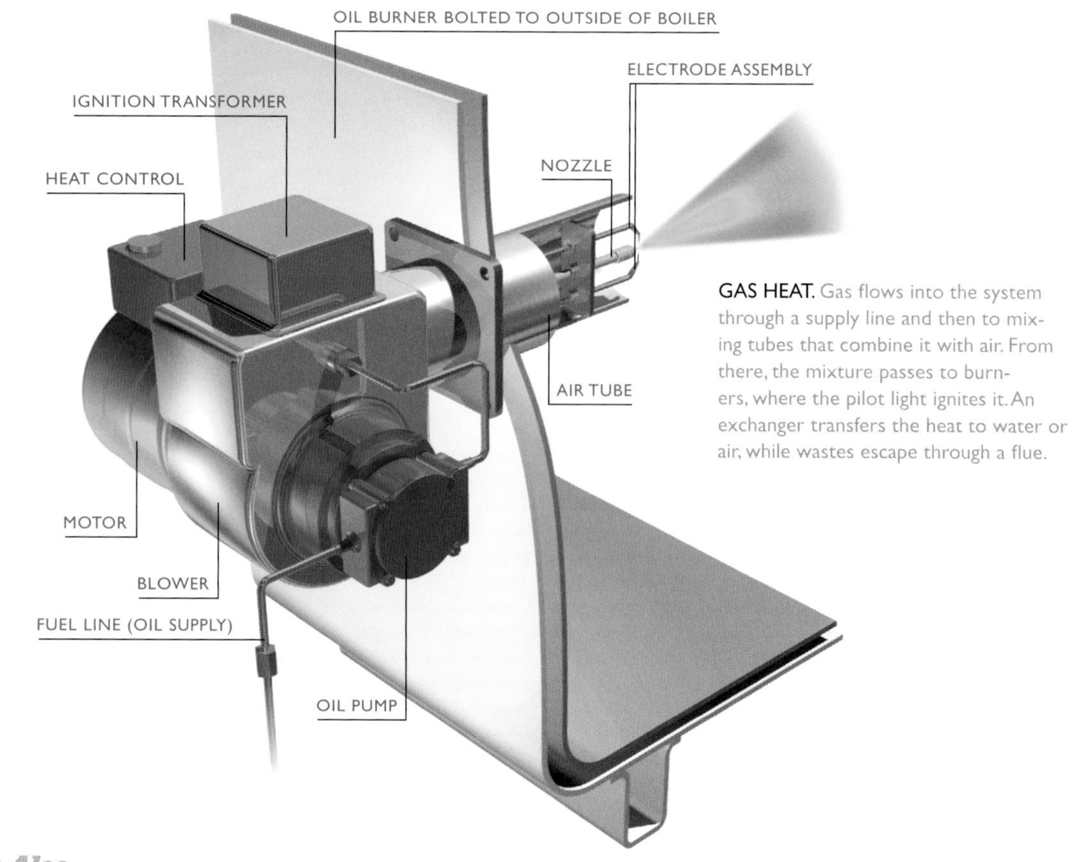

GAS HEAT. Gas flows into the system through a supply line and then to mixing tubes that combine it with air. From there, the mixture passes to burners, where the pilot light ignites it. An exchanger transfers the heat to water or air, while wastes escape through a flue.

»»See Also

RELATED PRINCIPLES: Law of Conservation of Energy, 30 • Ohm's Law, 164 • Pascal's Law, 80 • Rapidly Expanding Gases, 74

RELATED APPLICATIONS: Convection Oven, 103 • Electrical Wiring and Power Grid, 166 • Home Plumbing System, 82 • Hot Air Balloon, 105 • Thunderstorm, 106

Convection

HOT AIR BALLOON

Hot air balloons, those big floating ovals of vividly colored nylon, rise up by convection. A small propane flame at the base of the balloon heats the air near the opening. As the air warms, it expands; its molecules spread out. The hot air in the balloon is less dense—weighs less—than the surrounding cool air. So it rises above the cooler air like a bubble floating up through water.

Indeed, the force lifting the balloon can be determined the same way as the buoyant force pushing an object up in liquid: It's equal to the weight of the volume of cool air the balloon displaces. The biggest balloons have the strongest lift.

Of course, the balloonist must also have a way to return to Earth. The pilot does this by pulling a cord that opens a valve at the top of the balloon, letting hot air escape. The volume in the balloon decreases, and the balloon drops.

See Also

RELATED PRINCIPLES: Bernoulli's Principle, 68 • Geothermal Energy, 175 • Law of Conservation of Energy, 30 • Pascal's Law, 80 • Rapidly Expanding Gases, 74 • Universal Law of Gravitation, 122
RELATED APPLICATIONS: Convection Oven, 103 • Home Heater, 104 • Thunderstorm, 106

Convection

THUNDERSTORM

Many people can sense when the weather is building toward the dark clouds, lashing rain, forks of light, and cracks of sound that make a thunderstorm. It starts, most often on a spring or summer day, with warm, moist air.

This warm, moist air rises above cooler air, sometimes nudged upward by a sea breeze or cold front. As the warm air rises, its vapor condenses into liquid droplets to form clouds. The condensing gas gives up heat, which warms surrounding air. This warm air in turn pushes upward. As the atmosphere becomes more unstable, vertical motion increases.

The cloud gets taller and taller until it forms a thunderhead. Though its mushroom-like shape is not always clear from the ground, a thunderhead can reach several miles in height. Near the top of the cloud mass, droplets get bigger and heavier and eventually fall as rain or hail, accompanied by rushing downdrafts of colder air. Meanwhile, close by, warm updrafts continue to lift small droplets to great heights.

Amid this turbulence, water particles collide, knocking negatively charged electrons from rising particles. In this way an electrical field forms within the storm; the bottom is negatively charged, and the top is positively charged. Zap! Lightning heats the air to more than 50,000 degrees Fahrenheit in less than a second. This causes air in the lightning channel to expand explosively, violently compacting the air around it—a disturbance called a shock wave that our ears hear as thunder.

»»**See Also**

RELATED PRINCIPLES: Electromagnetic Radiation, 248 • Geothermal Energy, 175 • Specific Heat, 111 • Wave Energy, 58

RELATED APPLICATIONS: Convection Oven, 103 • Home Heater, 104 • Hot Air Balloon, 105 • Sea Breeze, 110

PRINCIPLE

THAT SPINNING FEELING

Coriolis Effect

Earth does not stand still. It is a rotating frame of reference. Anyone charting the course of an object moving freely in the air—whether an airplane, wind system, or ballistic missile—must take into account how Earth's rotation moves its intended destination in what's known as the Coriolis effect.

Our planet rotates toward the east. Points around its girth, the Equator, spin eastward at nearly a thousand miles an hour; locations near the poles move much more slowly, tracing only a small circle in each 24-hour period.

So, for example, an airplane heading from the Equator to a distant point due north will land to the right of the intended destination, which hasn't caught up in its eastward rotation. If the plane travels a straight line in the opposite direction—heading due south from the North Pole—it will again land at a point to the right (as seen by a passenger on the airplane). This time it's for a different reason—the Equatorial destination is moving faster than the take-off point and gets ahead in the eastward rotation.

All this has the odd effect that long journeys through the air, especially along a north-south axis, always appear to veer to the right in the Northern Hemisphere and to the left in the Southern Hemisphere.

DRAIN SPIN

It's a favorite gee-whiz science notion that, like cyclonic winds, toilets and sinks drain counterclockwise in the Northern Hemisphere but clockwise in locations below the Equator. Unfortunately, it isn't actually true. The typical hurricane is some 300 miles across, whereas a drain is but a few inches in diameter. For objects traveling such short distances, the effect of Earth's rotation is vanishingly small.

Indeed, it's entirely possible to find that in a single bathroom, the sink drains in one direction, the toilet in another. What matters are the dynamics of the water before it hits the bowl as well as the shape of the bowl and location of the drain.

Coriolis Effect

HURRICANE

Hurricanes begin with wind, and wind starts with high-pressure air pushing toward an area of low pressure. In a major storm, wind sweeps across substantial distances, gaining tremendous momentum along the way. The wind's path toward its destination is affected by Earth's rotation in what's called the Coriolis effect.

Let's say a mass of equatorial air veers north to fill a low-pressure pocket. It will arrive just east—to the right—of the low-pressure system. It misses the mark because the wind originates from a location on the globe that rotates faster than its destination, which hasn't caught up in its eastward rotation. Passing the low-pressure system on the right, the high-pressure winds now actually bend around the low-pressure zone, swirling in a counterclockwise fashion.

In the Northern Hemisphere, cyclonic storms (called hurricanes in some places, typhoons or cyclones in others) always turn counterclockwise around the low-pressure eye. In the Southern Hemisphere, they rotate clockwise.

STORMY WEATHER. This satellite image shows a hurricane swirling around its eye.

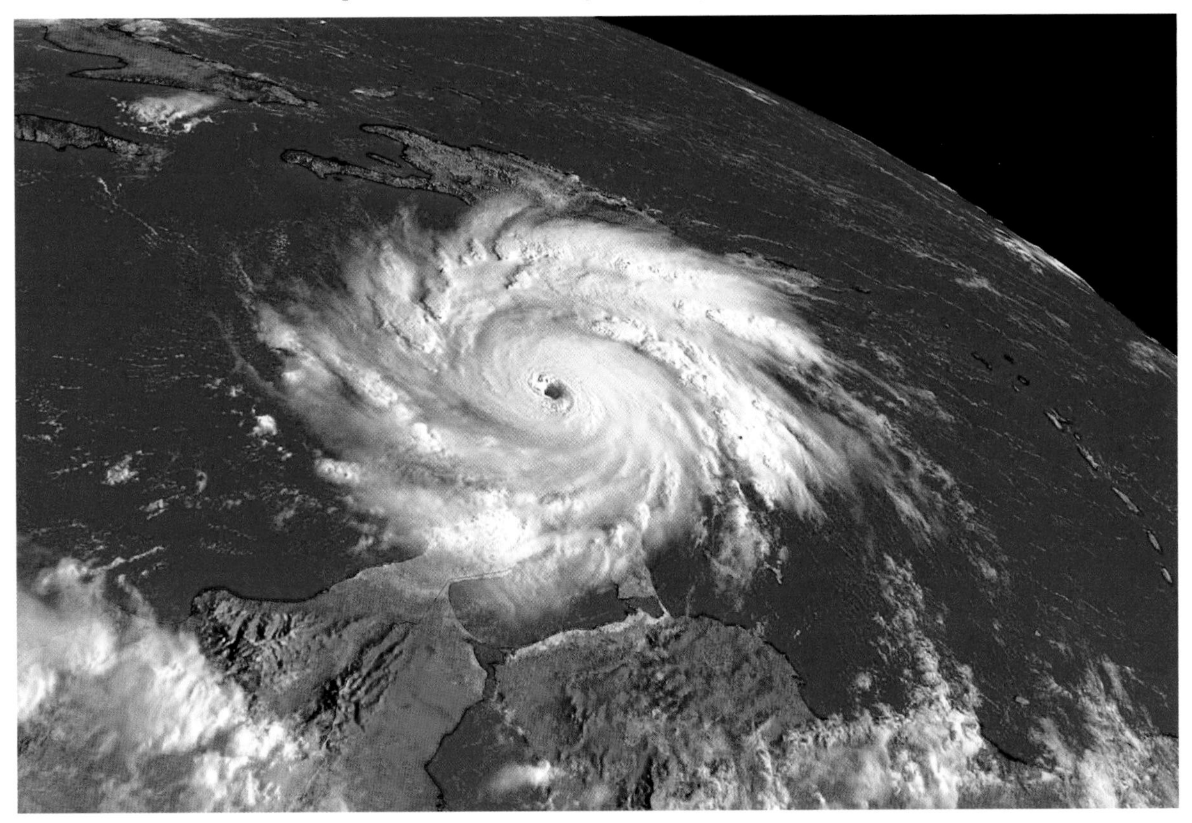

»»See Also
RELATED PRINCIPLES: Convection, 103 • Geothermal Energy, 175
RELATED APPLICATIONS: Thunderstorm, 106

Coriolis Effect

SEA BREEZE

The extraordinarily high specific heat of water—it warms slowly—stands in contrast to the quite low specific heat of soil, especially dry soil, which warms up fast on a clear day. Along a coastline, of course, this temperature difference means a pressure difference: cool, dense, high-pressure air over the water and warm, low-pressure air over land.

A pressure gradient sets the air in motion. A sea breeze pushes inland.

Water's high specific heat also means that, once warm, it is slow to cool. Winds sometimes blow offshore at night, as the land loses daytime heat faster than the water. This is why it really is cooler at the shore.

See Also

RELATED PRINCIPLES: Convection, 103 · Geothermal Effect, 175
RELATED APPLICATIONS: Smog, and Temperature Inversion 111

PRINCIPLE

HOT STUFF

Specific Heat

The specific heat of a substance is the amount of energy required to raise its temperature by a given amount. This depends largely on the substance's molecular structure and the freedom of its particles to vibrate, rotate, and move. It takes nearly five times as much energy to raise the temperature of water by one degree Celsius as it does to heat aluminum the same amount. This variation in specific heat is why a pot on the stove can get too hot to touch well before the water it holds heats up.

In fact, water has the highest specific heat of any common substance. Its presence in human tissues—the delicate brain, for example, is about 90 percent water—provides excellent protection against sudden spikes in ambient temperature.

Specific Heat

SMOG AND TEMPERATURE INVERSION

When a layer of warm air lies over a city like Los Angeles, it can trap thick, pollutant-rich air near the ground, a form of pollution we call smog.

In the usual course of things, the sun warms Earth, and that warm air expands and rises. It cools as it rises, losing about 3.5 degrees Fahrenheit with every thousand feet of elevation.

Downtown Los Angeles engulfed in smog

But certain conditions can disrupt this order. Land warms up in the sun much faster than water and becomes less dense. So in the Los Angeles Basin, for example, cool sea breezes sweeping inland easily wedge themselves beneath this warmer air. At the same time, air warmed on the high deserts around the city is lifted up against mountains ringing the city and slides in on top of the cooler air over Los Angeles.

This warm inversion layer, as it's called, acts like a lid, blocking the vertical circulation of air—and trapping pollutants from a welter of fossil fuel–burning vehicles.

»»» *See Also*

RELATED PRINCIPLES: Convection, 103 • Geothermal Effect, 175

RELATED APPLICATIONS: Sea Breeze, 110

Evaporation

CLOTHES DRYER

Before widespread adoption of the clothes dryer in the twentieth century, people hung out their laundered clothes to dry, exposing as much surface area as possible to sun and wind.

The electricity-powered household appliance merely accelerates these natural aids to evaporation. A tumbler keeps the clothes moving rather than in a heap, providing maximal exposure of surface-water molecules and facilitating their escape from the wet fabric as gas. The machine promotes this escape by applying heat with a wire element and fan; this gets the molecules moving. Another important element is the vent that removes humid air from the tumbler. When the air is full of water vapor, it becomes dense and heavy, in effect pressing against the liquid molecules attempting to escape their liquid bonds. Evaporation declines. The vent keeps the water evaporating until all your clothes are dry.

See Also

RELATED PRINCIPLES: Electromagnetic Induction, 157 • Law of Conservation of Energy, 30 • Ohm's Law, 164
RELATED APPLICATIONS: Electrical Wiring and Power Grid, 166 • Perspiration, 113 • Washing Machine, 24

PRINCIPLE

UP, UP, AND AWAY

Evaporation

Water is a vital part of the human environment in all three of its states: liquid, gas (water vapor), and solid (ice).

When molecules on the surface of a body of water change from liquid to gaseous form, it's called evaporation. Evaporation from oceans, lakes, and rivers provides most of the moisture in our atmosphere.

That transformation requires heat. As they warm, the water molecules are excited into vigorous motion. Some escape the tenacious bonds holding them to other liquid molecules; these escapees are vapor. The evaporated water molecules from lakes, oceans, soil, and other sources rise up into the atmosphere and form what we know as clouds.

Evaporation

PERSPIRATION

It's not really sweat that cools you down on a hot day, it's the evaporation of sweat. For surface liquid molecules to leap into the air as gas, they must overcome the force that binds them together. In the process, they use up energy in the form of heat, drawing this heat from the liquid surface. The water that remains—the perspiration left on your skin, that is—is cooler.

Why is it you can sweat away a sultry afternoon without feeling relief? The thick, saturated air has little room to receive escaping liquid particles, so they stay put. You're wet, but not cool.

In very hot climes, it is common on a sweltering day to consume hot liquids, such as soup or tea. The reason is to facilitate perspiration, which when it evaporates makes you feel cooler.

»»See Also
RELATED PRINCIPLES: Second Law of Thermodynamics, 98
RELATED APPLICATIONS: Clothes Dryer, 112

PRINCIPLE

LOSING STRATEGY

Third Law of Thermodynamics

Though there's some heat in the countertop on which rests a piping hot pan, it's the pan that will heat the countertop and not vice versa; heat always flows from hotter to colder regions.

No part of the universe could end up absolutely without heat energy and molecular motion because it would take heat from elsewhere. But if a pure substance that's absolutely regular in its atomic structure could reach the theoretical temperature called absolute zero, it also would be devoid of entropy.

One definition of entropy is the number of configurations the jostling and twitching atoms and molecules of a substance may take—randomness. Because of this random and unpredictable behavior by molecules within a substance, energy transfer in the real world is not perfectly efficient; some energy invariably remains unavailable for useful work, another definition of entropy.

But in absolute cold, molecules would cease their random jiggling; there'd be no heat to flow from one place to the other, activating molecules along the way. The crux of the third law of thermodynamics is that, as the temperature of a pure, perfectly regular substance falls to absolute zero, its entropy also approaches zero. In a nearly fixed, inert substance randomness subsides.

ABSOLUTE ZERO

The word cold describes something that is not a quality in itself but rather an absence of something: heat.

Heat is energy, and energy drives motion. The whole universe is made up of molecules dancing with the energy of heat. It's a raw, random motion. The faster the molecules move, the warmer the substance they make up.

Even a jug of ice water harbors so much molecular motion that if you gently place a drop of ink on the surface, it will diffuse evenly throughout. In theory, absolute zero is a point where all motion—and hence all heat—ceases to exist. Quantum physics says that absence of motion is impossible. Absolute zero is about −273 degrees Celsius (−460 degrees Fahrenheit).

Third Law of Thermodynamics

SUPERCONDUCTIVITY

Chilled to extraordinarily cold temperatures—colder than −321 degrees Fahrenheit—certain metals conduct electricity with absolutely no resistance. The current can flow undiminished in a loop forever, entropy seemingly vanquished in one hundred percent efficiency.

An electrical current consists of electrons streaming through a medium toward a positive charge. There are snags along the way. The conductive medium is like a lattice of atoms that may have irregularities and that vibrates with the energy of heat. What's more, the atoms in the lattice, having given up loose electrons to the stream, are now positively charged and attract the electrons tumbling past. Akin to the friction that stops a sliding object, these snags result in a loss of electrical energy.

Resistance typically diminishes as a conductive material's temperature drops. But superconductivity is different; when susceptible materials reach a critical cold threshold, resistance abruptly vanishes.

To rush toward their destination with utmost efficiency, the streaming electrons organize themselves into pairs. As one electron passes near the positively charged atoms of the lattice, the atoms bend inward toward the electron, temporarily increasing the positive charge around it. This creates a weak attraction between the electron and the one just behind it, and they are drawn together through the lattice. Locked together, the electrons simply ride roughshod over the lattice, losing no energy. Because heat energy readily breaks them apart, such unusual pairings of electrons form only at a critical cold threshold—the tipping point for superconductivity.

See Also

RELATED PRINCIPLES: Electromagnetic Induction, 157 • Macromolecular Chemistry, 288 • Nuclear Fission, 179 • Ohm's Law, 164 • Quantum Mechanics, 298 •
RELATED APPLICATIONS: Nuclear Reactor, 180 • Particle Accelerator, 150

HIGH-TEMPERATURE SUPERCONDUCTIVITY

In the 1980s scientists discovered that certain ceramics become superconducting at unexpectedly high temperatures—a leap forward that could make this ultra-efficient technology far more practical.

Ordinary superconductors work only in the extreme cold achieved by applying liquid helium, which is costly to produce. They're used in such sophisticated and expensive technologies as medical magnetic resonance imaging (MRI) but will never be a sensible way to transmit power to homes or run electronics.

The newer superconductors do their thing at temperatures as high as −211 degrees Fahrenheit. That's still ultracold, but it's achievable using liquid nitrogen, which is cheaply derived from liquid air.

Of course, the field's holy grail is the material that could eliminate resistance at room temperature—for now, only a dream.

PART 2
NATURAL FORCES

CONTENTS

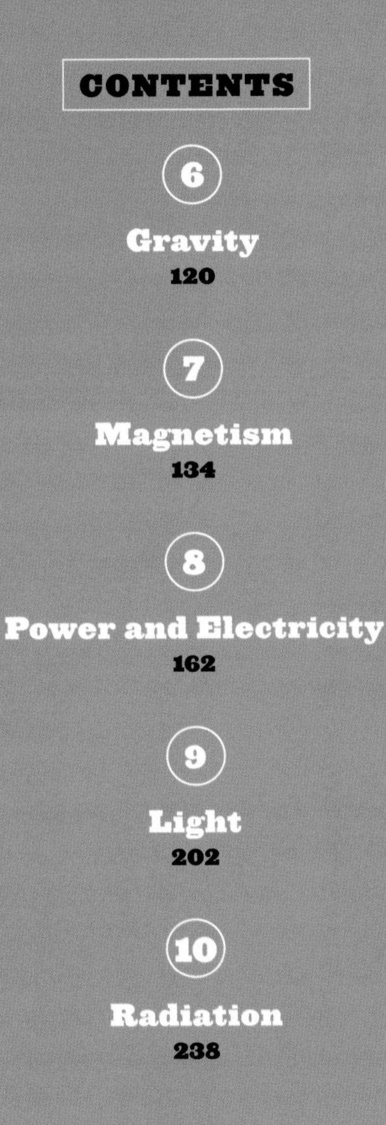

CHAPTER 6
Gravity

When we think of gravity, we tend to think of Sir Isaac Newton sitting under an apple tree: an apple falls on his head, and the idea of gravity is born. The story probably didn't happen quite that way, but Newton did observe falling objects, and he did wonder what mysterious force drew them to the ground and why some items seemed to fall faster than others. Newton's law of universal gravitation, published in 1687, holds that everything in the universe exerts a pull on other objects (called attraction), and the pull increases with mass and proximity.

Others expanded upon Newton's work, including Pierre-Simon LaPlace and Albert Einstein, and today, gravity has helped us to build such things as suspension bridges, elevators, escalators, and earthquake-proof buildings. By understanding the pull gravity has on the tides, we can use water to generate power. Understanding the effect of gravity helps us hit a bull's-eye on a firing range. It explains why planets have an elliptical orbit. We've applied our knowledge of gravity to baseball, using it to calculate the distance of home runs. Because the law of gravity is universal, it applies not just to life on earth but the universe beyond.

PRINCIPLE

STRONG PULL

The Universal Law of Gravitation

From Newton, we understand that gravity is largely determined by mass. Think of gravity as a measure of attraction between two objects. The more mass an object has, the greater its attraction, or gravitational pull. In addition, the smaller the distance between two objects, the stronger the gravitational pull. This pulling force sucks things toward an object's center of mass. That's why our spherical Earth, which is quite massive, draws objects powerfully toward its core. The pull it exerts is gravity, which keeps us from flying off into space. This law is considered universal because it's constant, regardless of where you are in the universe. There's gravity on Earth, as there is on every other planet and celestial body.

Universal Law of Gravitation

PARABOLIC CURVE

If you want to understand what a parabolic curve is, just throw a baseball. The arc the ball travels as it leaves your hand and falls back to the ground takes the shape of a parabola, or high arch, because of gravity. We can calculate this trajectory if we know two things: a fixed point (the focus—your hand or the point where the ball makes contact with a bat) and a fixed line (the directrix—the distance the ball travels). Any point on the trajectory is an equal distance from the focus and the directrix. Why is this important? Well, besides letting us calculate just how far a home run was actually hit, parabolic curves have important implications in everything from suspension bridges to car headlights to satellite dishes. Many constructed arches resemble parabolic curves, but they are not true parabolas because mathematically their points are not at an equal distance from a directrix and focus. The Arc de Triumph in Paris is just one example of an arch that is not a true parabolic curve even though it appears to be one.

»»See Also

RELATED PRINCIPLES: Laws of Motion and Energy, 18
RELATED APPLICATIONS: Curveball, 72

Universal Law of Gravitation

BULLET TRAJECTORY

The path a bullet follows as it leaves a gun is called its trajectory. But a bullet doesn't rise, or arc, as it leaves a gun. Instead, it immediately begins to drop because of the effects of gravity (and the loss of the gun barrel's support); it also slows down due to air resistance. So how does a person shooting a gun manage to hit a target? Even though the bullet doesn't travel in an arc, the shooter still has an arc to take into account. Skilled shooters often aim at a point slightly above the target while holding the barrel perfectly straight (although less-experienced shooters hold the barrel slightly raised and aim a little low). This point can be calculated—and is, in the case of high-tech military artillery—but nobody takes the time to do that on a shooting range. This skill comes with practice and a little trial and error.

»»See Also
RELATED PRINCIPLES: Laws of Motion and Energy, 18
RELATED APPLICATIONS: Surfing, 59

Newton's second law tells us that whenever acceleration occurs, some force is involved. Because a curved trajectory is by definition acceleration (a shift in speed or direction), some force must be at work. That force, Newton explained, is gravity.

Universal Law of Gravitation

ELLIPTICAL ORBITS OF PLANETS

Why do planets have an elliptical, instead of a circular, orbit? In a word, gravity. That the orbits of planets were elliptical (in the shape of an ellipse, or oval) was first proposed by Johannes Kepler in his three laws of planetary motion. It wasn't until Newton developed his law of universal gravitation, though, that we understood why. An elliptical orbit is the result of a complicated tug of war between the gravitational pull of different celestial bodies, like the sun and the planets. The bigger the celestial body, the more pull it exerts on smaller objects around it. One massive body, acting alone, might create a circular orbit. But if there's more than one massive object, as in our solar system, each orbiting object will have an elliptical orbit. In addition, the farther away from the sun a planet is, the less it feels the sun's gravitational pull.

Some planets have a more complicated orbit than others, which was partially explained by Einstein, who later refined Newton's explanation, adding the effect of the curvature of space–time. An object, like the sun, has mass, and this mass essentially deforms space-time, causing it to curve. The work of Kepler, Newton, and Einstein helped us understand the complex mathematics involved in planetary orbits. And the orbit of a planet can change. All celestial bodies—including the sun—are moving; so, as the gravitational pull changes, the orbits change, becoming more or less elliptical in response to the changes in velocity and mass.

»»» ***See Also***

RELATED PRINCIPLES: Laws of Motion and Energy, 18

RELATED APPLICATIONS: Floating Continents, 92

Universal Law of Gravitation

ESCALATOR

An escalator is basically a conveyor belt with steps. It has two wheels at each end that drive a set of chains in a pulley-like fashion. The steps also have two sets of wheels that move along their own inner tracks and are positioned in such a way that each step is always level—which is especially important when someone is getting on or off a moving escalator. Obviously, gravity is exerting a constant force on you; however, the force the escalator exerts as it moves you up or down perfectly counteracts gravity, keeping you in place on the step.

This conveyor-belt technology—the same that has moved coal, sand, and grain for years—is also used in another kind of people mover. The moving sidewalk complete with handrails and other safety devices is a horizontal version of the escalator.

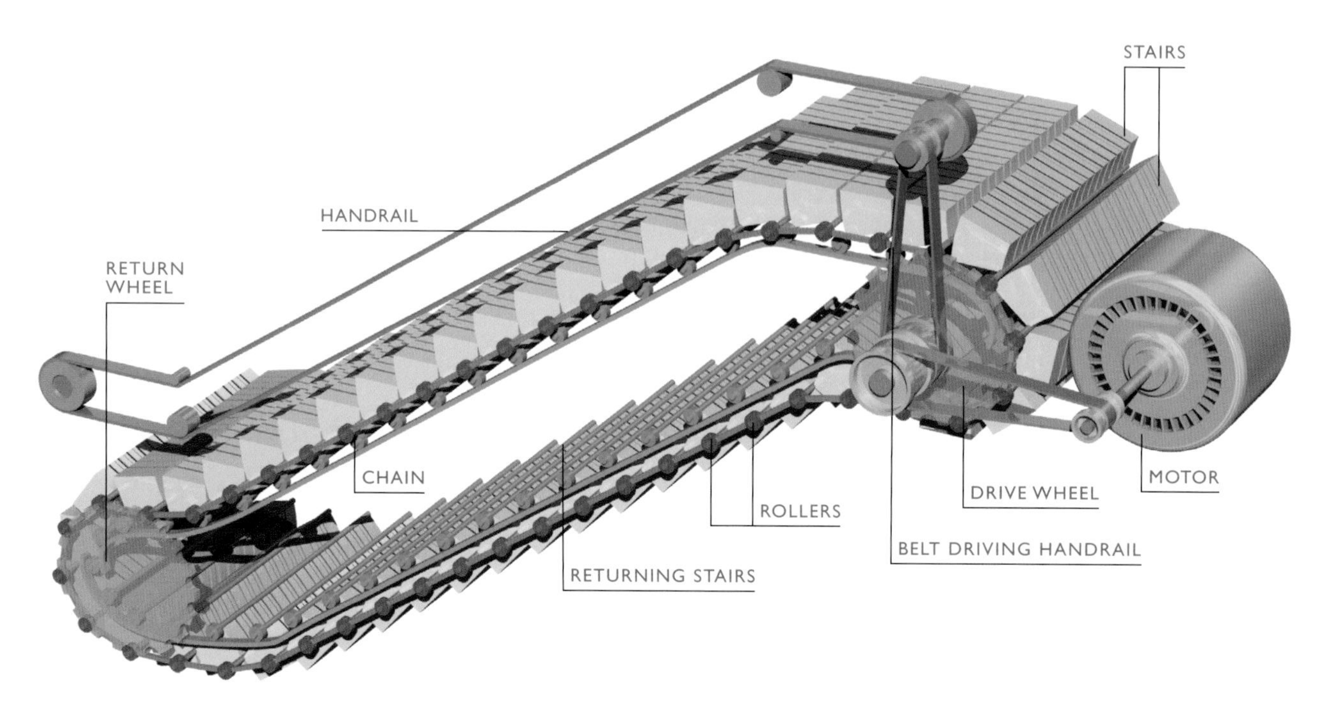

»»See Also
RELATED PRINCIPLES: Laws of Motion and Energy, 18
RELATED APPLICATIONS: Gears, 48 • Lever, 4

Universal Law of Gravitation

ELEVATOR

The most interesting—and important—aspect of an elevator isn't even in it. It's the counterweight that makes it all possible. The counterweight hangs opposite the pulley system and balances the weight of the car in which the passengers ride, plus an additional 40 to 45 percent of the car's weight. An electric motor with a braking system drives an arrangement of cables and pulleys. In some ways, the design of the traditional elevator is actually inefficient; these days, hydraulic systems have replaced the old pulleys. At times, the counterweight is heavier than the car it's lifting—such as when the car has no passengers in it. This arrangement creates excess energy, which dissipates as heat. Recent designs are revamping elevators to harness the power of gravity and divert this energy to a building's electrical grid.

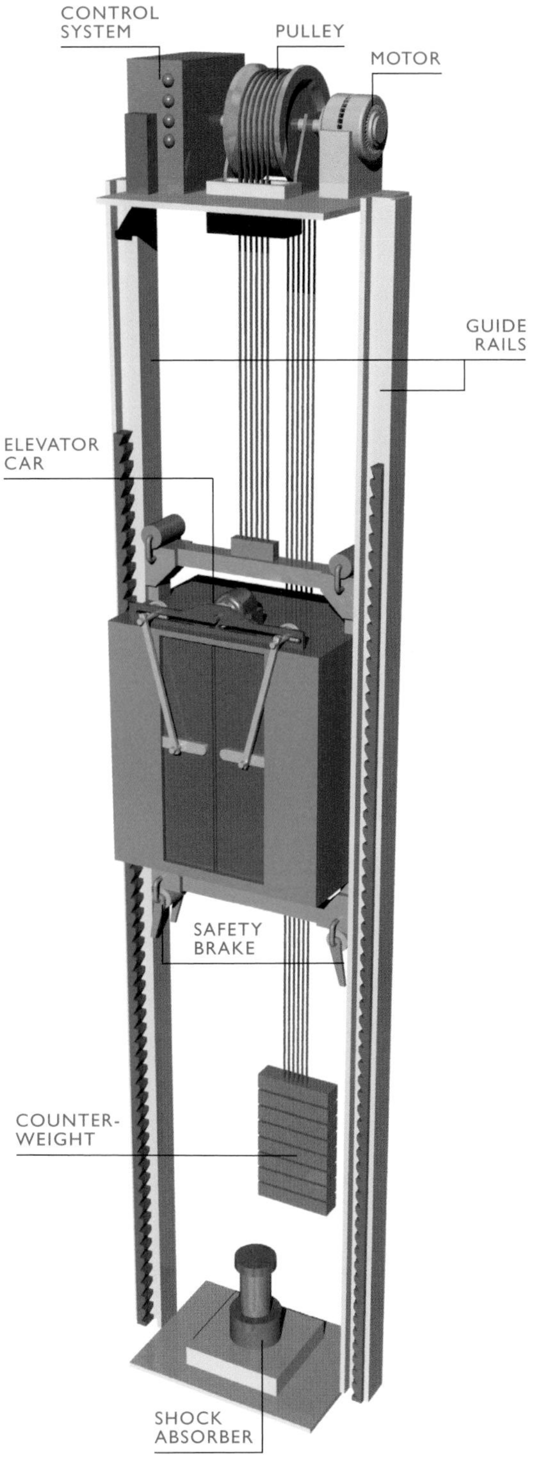

»» See Also
RELATED PRINCIPLES: Laws of Motion and Energy, 18
RELATED APPLICATIONS: Pulley, 47

Universal Law of Gravitation

SUSPENSION BRIDGE

When you think of bridges, you might picture a classic suspension bridge, such as the Brooklyn Bridge in New York or the Golden Gate Bridge in San Francisco. One look at the slender, graceful design of these structures can make you wonder how they stay up, much less support traffic.

Suspension bridges date back to the fifteenth century and comprise a deck, towers (called abutments) at either end, and a network of cables that form a parabolic curve (a parabola is a curve where any point is at an equal distance from a fixed point and a fixed straight line). A main cable runs horizontally from tower to tower, and vertical cables run off the main cable, supporting the weight of the deck and transferring it to the towers. The main cable is actually anchored beyond the abutments, to prevent the bridge structure from giving in to the force of compression. The weight of the bridge itself (called dead weight) pulls inward on the towers; however, the cables counteract this with an equal force. In other words, compression from the bridge's deck and tension from the cables are balanced.

Because of their design, suspension bridges have many advantages over other kinds of bridges. For one thing, they can span great distances. The longest suspension bridge in the world is the Akashi-Kaikyo Bridge in Japan, covering an impressive 12,831 feet. However, suspension bridges have disadvantages, too, which engineers must take into consideration. For example, if the main cable is made of chain links or just one strand of cable, and a link in the chain breaks or the cable snaps, the bridge loses its tension and comes down. Also, if the deck of the bridge is too thin, it can lose its stability in heavy winds and shake itself to pieces, which is what happened to the Tacoma Narrows Bridge in 1940.

See Also

RELATED PRINCIPLES: Laws of Motion and Energy, 18

RELATED APPLICATIONS: Climbing T[illegible] Crane, 49 • Gyrosc[illegible] 21

Universal Law of Gravitation

TIDES

The ebb and flow of the tides may be something we take for granted, but without gravity, the tides wouldn't exist. In a 24-hour period, there are two high tides and two low tides. These occur every day, although the times vary due to something called slack—the period of time in between tides when water is still. How long this slack period lasts depends on the moon's gravitational pull. Even though the moon's gravitational pull is only one ten-millionth that of Earth's, it combines with Earth's centrifugal force to create tides. (The sun also has an effect on tides, but it's not nearly as strong as the moon's effect.) The strength of the moon's effect on tides is based on the moon's phase (whether it's a full moon or a new moon) and distances between Earth, the sun, and the moon in their orbits. Tides are at their maximum when Earth, the sun, and the moon are aligned. The Bay of Fundy in Nova Scotia boasts the highest tides in the world, with a range of 44.6 feet between high tide and low tide.

PULL OF THE MOON. The moon's gravitational field pulls so strongly on Earth that it raises the ocean water directly facing it. We experience the result as tides.

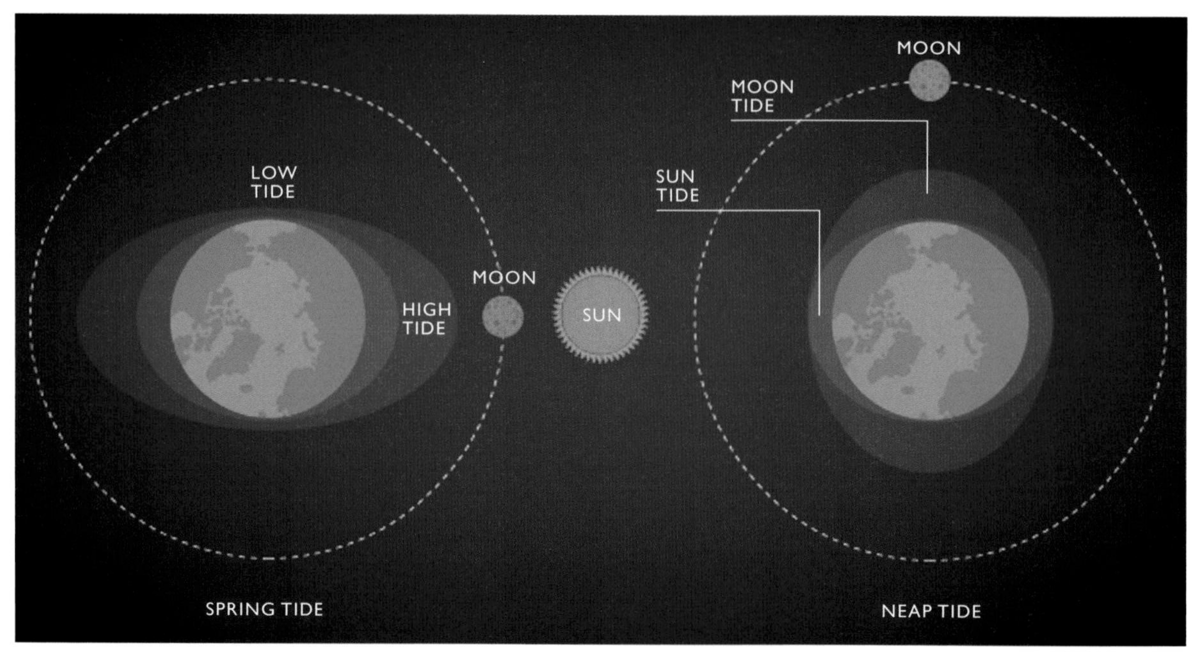

See Also
RELATED PRINCIPLES: Laws of Motion and Energy, 18
RELATED APPLICATIONS: Wave Power Generator, 58

Universal Law of Gravitation

TIDAL POWER GENERATOR

TIDAL POWER. The energy of the rising and falling tides is captured by underwater propellers and converted to electricity.

The power produced by tides as they move in and out can be used to generate power, as the ancient Romans were well aware. Similar to a hydroelectric dam, a barrier is erected across a tidal basin. As the tide rises, water enters the basin, where it spins turbines and generates power. There are some environmental impacts that need to be addressed, such as silt buildup and reduced tidal flow, but the potential these generators provide as a clean source of energy is amazing, with an estimated 80 percent efficiency. The technology is expensive to implement, though. France is the only country currently using it on a large scale (one plant in France can provide power to 240,000 homes), and the hope is that more countries will follow. There's one requirement: there must be an increase of at least 16 feet between low tide and high tide in order for a tidal power generator to work.

»»*See Also*
RELATED PRINCIPLES: Laws of Motion and Energy, 18
RELATED APPLICATIONS: Hydroelectricity, 172 • Wave Power Generator, 58

Universal Law of Gravitation

EARTHQUAKE-RESISTANT BUILDINGS

During an earthquake, the majority of injuries and deaths occur to people who are trapped in collapsed buildings, so designing structures that can withstand earthquakes is vitally important. But is it possible? Yes! (Buildings that date back to the sixth century are still standing despite the fact that they're located in earthquake-prone areas, such as the Hagia Sophia in Istanbul, Turkey.) In order for a structure to resist the massive power of an earthquake, it has to withstand the lateral, or side-to-side, motions earthquakes produce; the vertical motions are accounted for because buildings, by their very design, already counteract the effect of gravity.

The ideal earthquake-resistant building is one that is symmetrical in shape, with little ornamentation, such as cornices or buttresses, that could break off during a tremor. A diaphragm, a middle tier of softer material that allows the building to wobble without breaking, provides side-to-side flexibility. Cross-bracing helps too by providing vertical stability, and shear walls (walls made up of braced panels) provide lateral stability. Some structures even have frames in which the joints provide stability, yet allow the columns and beams to bend in response to an earthquake's force. In addition, some buildings are attached to their foundations; other designs have the building essentially "floating" on a system of cylinders or springs. Lead-rubber bearings are a favorite choice for designers. A lead core makes the bearing stiff and strong in the vertical direction; alternating layers of rubber and steel bands provide strength horizontally. A damping system of gel-like pads connected to a heavy weight on top of the building can also help absorb the force of the earthquake.

SAFER BUILDINGS. The Tokyo Skytree tower in Tokyo, Japan, shown here in the final stages of construction, was designed with numerous earthquake-resistant features.

»»*See Also*
RELATED PRINCIPLES: Laws of Motion and Energy, 18
RELATED APPLICATIONS: Ball Bearing, 52 • Earthquake, 61

The Transamerica Pyramid in San Francisco incorporates a unique truss system that protects it from earthquake damage.

CHAPTER 7

Magnetism

The power to attract or repel is all in a magnet's atoms. Electrons and protons are always in motion. Because any moving charge generates a magnetic field, each electron and proton has one. Usually these fields are haphazard, but in materials such as iron, they tend to line up within specific regions called domains. When ferrous (iron-containing) rock is molten, all the domains align with Earth's own magnetic field; this is how natural magnets are formed.

We have harnessed magnetism in our everyday lives, based on our understanding of not only magnetism, but also electromagnetism and electromagnetic induction. Applications for magnetism include the chips in credit and PIN cards, explosive detection, , and compasses. (Did you know that migrating birds are a type of "living compass"?)

Magnets are used in television screens, including CRT (cathode ray tube), LCD (liquid crystal display), LED (light-emitting diode), and plasma. Magnetic resonance imaging (MRI) has enabled us to make medical diagnoses that were previously unimaginable. We have used electromagnetism to improve our transportation system in the form of maglev (magnetic levitation) trains and ships and electric cars. The electric motor has made daily life easier with the development of such tools as power drills, garbage disposals, and electric razors. Magnetism has changed not just how we look at our Earth, but how we understand the entire universe.

PRINCIPLE

OPPOSITES ATTRACT

Magnetic Fields

Magnetic fields have two polarities—usually called north and south—that attract their opposites and repel their own kind, depending on how the charged particles (electrons and protons) are aligned in the magnet. We've known about this force for a long time—as far back as the ancient Greeks, even if they didn't quite understand it. Columbus used a magnetic compass on his famous sea journey (although he didn't end up where he had expected to), and in 1600, William Gilbert, physician to Queen Elizabeth I of England, took the idea even further by proposing that Earth itself was a giant magnet. We have since learned that he was right.

Magnetic Fields

COMPASS

Compasses expanded our ability to navigate—and therefore explore—our world without losing our sense of direction. Before the invention of these simple devices, voyagers had to use the stars and nearby landmarks to navigate their journeys. The ancient Chinese were the first to employ water-based compasses—magnetic needles floated in bowls of water. The needles were magnetized by stroking them with a lodestone, a naturally magnetized piece of the mineral magnetite. At first, the needles were used in fortune-telling boards before being put to a more practical (and reliable) use. A compass works by sensing the magnetic field that is produced by Earth's core, which contains iron. But consider this: Earth's magnetic field is weak on the surface, which makes sense, given that it's spread out across the entire surface of the planet. In order for the magnetic field to have any effect on the compass at all, the compass must be light and have an almost frictionless bearing.

»»See Also
RELATED PRINCIPLES: Electromagnetism, 147
RELATED APPLICATIONS: Computer Hard Drive, 149

Magnetic Fields

MAGNET EARTH

The magnets that we're most familiar with are made of iron or ceramic. However, magnets made from rare earth alloys combine different elements to produce what is called a permanent magnet—and these are super strong. Two of the most common alloys are samarium cobalt and neodymium iron boron. Because the iron in the magnets can rust, they're often coated with gold plate, nickel, zinc, copper, or some combination of those. How does a rare earth magnet compare to a traditional magnet? Well, a neodymium iron boron magnet is approximately ten times stronger than the little horseshoe-shaped magnet used in basic science classes. Magnets are used in everything from computer hard drives to self-powered flashlights to wind turbine generators.

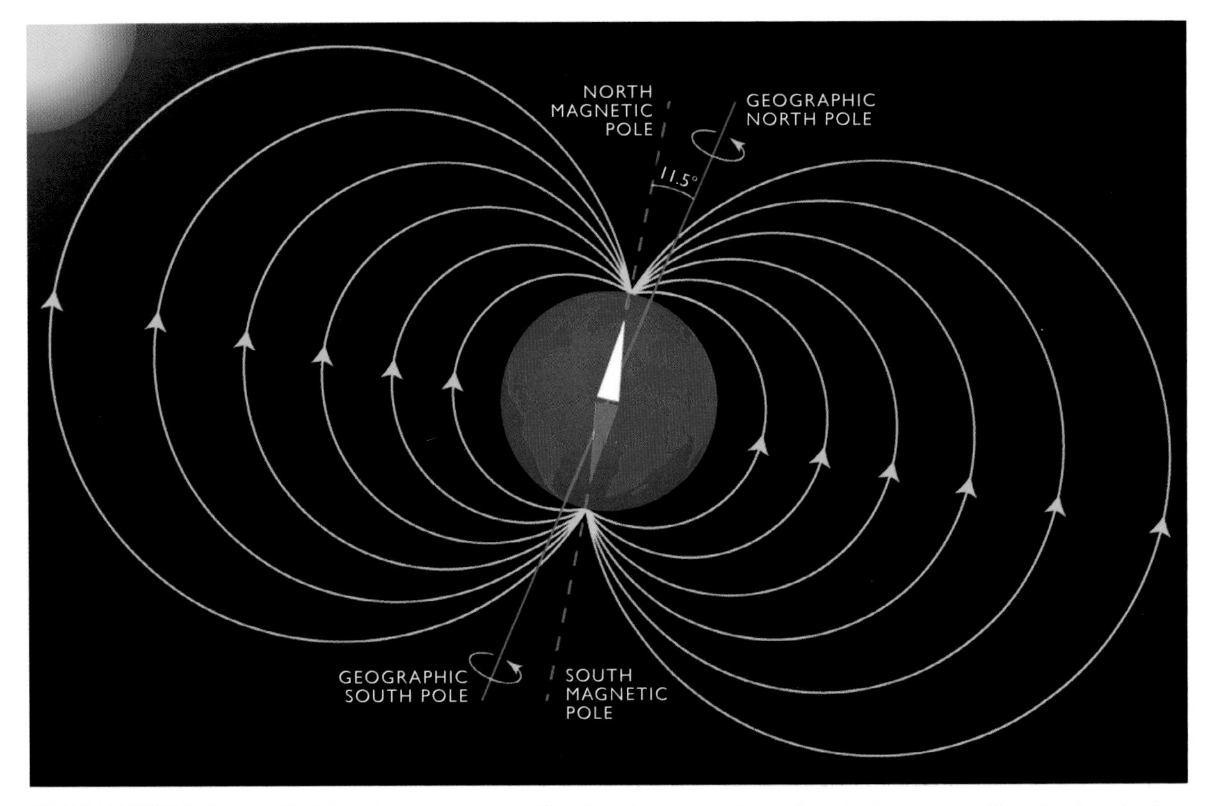

GIANT MAGNET. In this simplified depiction, our planet acts like a huge bar magnet that is about 11 degrees out of line with the axis of spin. Deep in the core, huge blobs of molten iron circulate—driven most likely by convection, or the transfer of heat through a fluid by molecular movement—forming a titanic dynamo, or electric generator, that creates a current, which in turn produces Earth's magnetic field.

»»See Also
RELATED PRINCIPLES: Newton's Third Law of Motion, 25
RELATED APPLICATIONS: Maglev Train, 155

Magnetic Fields

LIVING COMPASSES

What do cows standing in a field, a flock of migrating birds, and bacteria have in common? They all have an unerring sense of where true north (also called magnetic north) lies. How they do this is still a mystery, but scientists believe it may have something to do with a built-in compass the creatures possess. A recent study on a certain kind of bacteria discovered that the organisms have small particles of magnetite (iron oxide) inside them, and this may very well be responsible for their ability to find north. The particles in these bacteria line up with earth's magnetic field. Scientists have also found these magnetic particles near the brains of bees, birds, and trout.

Sea turtles migrate by means of a built-in compass.

»»See Also

RELATED PRINCIPLES: Electromagnetism, 147

RELATED APPLICATIONS: Electromagnetic Spectrum, 147

LANDMINE DETECTION

Metal detectors and dogs can both detect mines, but both methods have their disadvantages: false positives and physical limitations, respectively. The solution is a system that detects a very specific component of a landmine—its chemical signature. One prototype of such a system employs a chemically treated film that's applied to the ground and then viewed through ultraviolet light. If the chemical signature is not present, the film remains fluorescent. If an explosive is present, however, a dark circle forms around the area on the film. Another exploratory method uses microwaves and remote-vibration sensing.

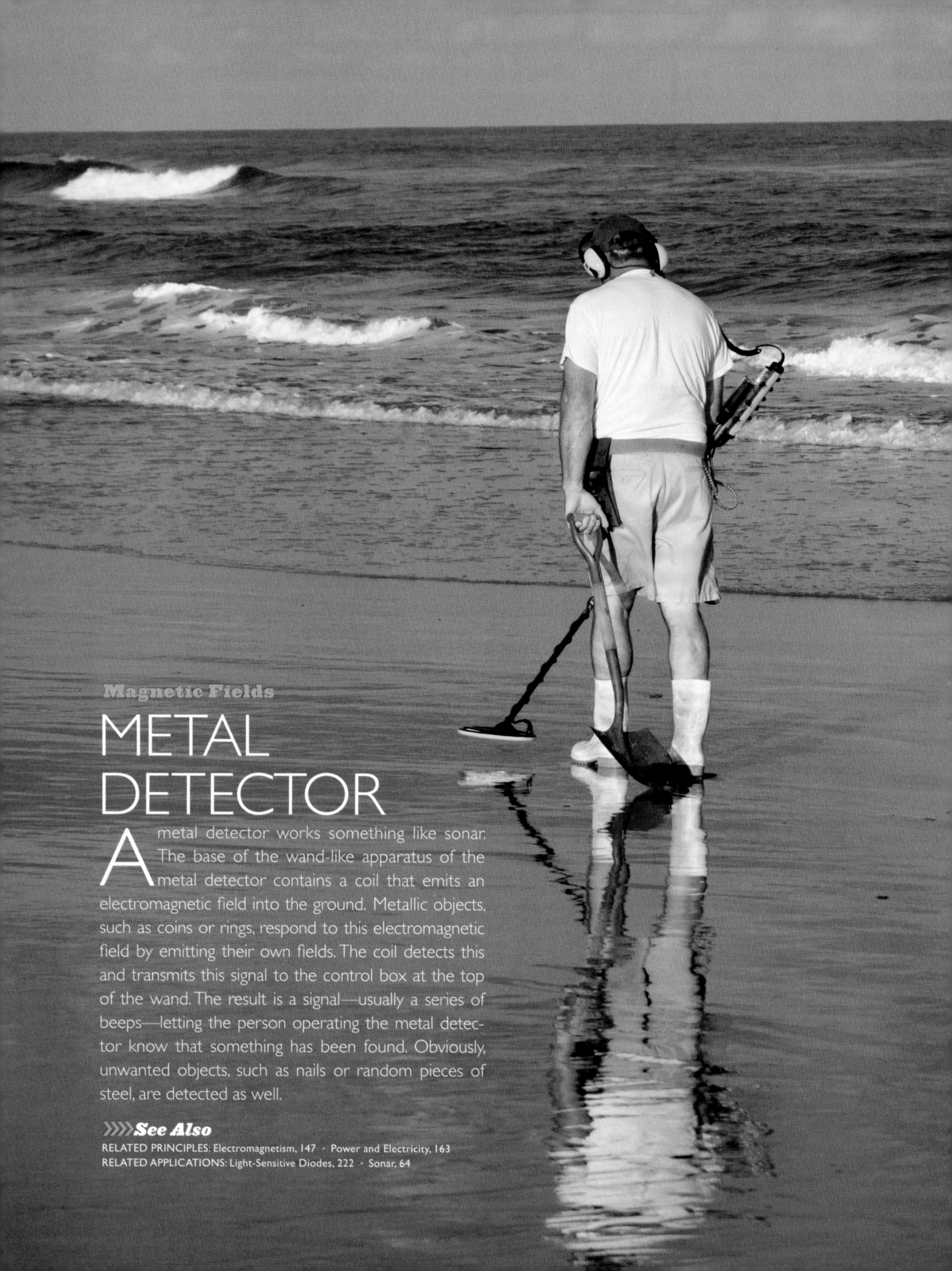

Magnetic Fields

METAL DETECTOR

A metal detector works something like sonar. The base of the wand-like apparatus of the metal detector contains a coil that emits an electromagnetic field into the ground. Metallic objects, such as coins or rings, respond to this electromagnetic field by emitting their own fields. The coil detects this and transmits this signal to the control box at the top of the wand. The result is a signal—usually a series of beeps—letting the person operating the metal detector know that something has been found. Obviously, unwanted objects, such as nails or random pieces of steel, are detected as well.

See Also

RELATED PRINCIPLES: Electromagnetism, 147 · Power and Electricity, 163

RELATED APPLICATIONS: Light-Sensitive Diodes, 222 · Sonar, 64

Magnetic Fields

CREDIT CARD

Did you know that the dark stripe on the back of your credit card—called a magnetic stripe, or magstripe—is actually a type of magnet? It consists of hundreds of tiny magnetic particles contained in a film of plastic. Magstripes have three tracks that contain varying amounts and types of information in specific formats based on established banking standards. The first track contains proprietary information relevant to the card issuer and often you, the cardholder. The second track is primarily used by financial institutions and contains account information, among other things. Most credit cards only use the first two tracks, as there are no standards governing how the third track should be used.

Although credit and debit cards are ubiquitous today and integral to Internet commerce, they have only been around since 1950 when Diners Club was issued as the first general credit card that could be used with multiple merchants.

»»See Also
RELATED PRINCIPLES: Binary Code, 190
RELATED APPLICATIONS: Bar Code Scanner, 230

CHIP AND PIN CARD

Recently, some banks in the United States have moved away from magstripes. Instead, they're issuing cards containing tiny microchips that are activated with a personal identification number (PIN). These microchip cards, already popular in Europe, store the same type of information that a credit card does. When the card is inserted into a reader, two electrical contacts create a current that transfers the information through a point-of-sale system for verification and processing. The readers for these cards are expensive, but the security and universality advantages of these cards make them worthwhile.

Magnetic Fields

MAGNETIC RESONANCE IMAGING (MRI)

We've been able to harness the power of magnetism in so many aspects of our everyday lives, right down to the ability to look inside the human body and analyze it in ways never thought possible. With MRI, a magnetic field is created inside a tube with a metal coil, and radio waves are bounced off the human body. Hydrogen atoms, which are abundant in the body, typically spin in random directions. However, when subjected to the magnetic field produced by an MRI, these atoms align in a uniform pattern according to the direction of the field, north to south. However, not all hydrogen atoms align in this way. When a radio wave is then applied, these unaligned atoms spin in an opposite direction. When the radio wave is removed, these atoms return to their normal position, releasing energy as they do so. It's this energy that is read by the MRI machine, and the result is translated into a three-dimensional (3-D) representation of the body part being examined. MRI has proved to be an invaluable addition to medicine since its invention in 1977, allowing noninvasive detection and diagnosis of a wide range of conditions, from injuries to cancer to multiple sclerosis. Unlike computed tomography (CT) scans or x-rays, MRI scans don't involve any potentially harmful radiation, which is another bonus.

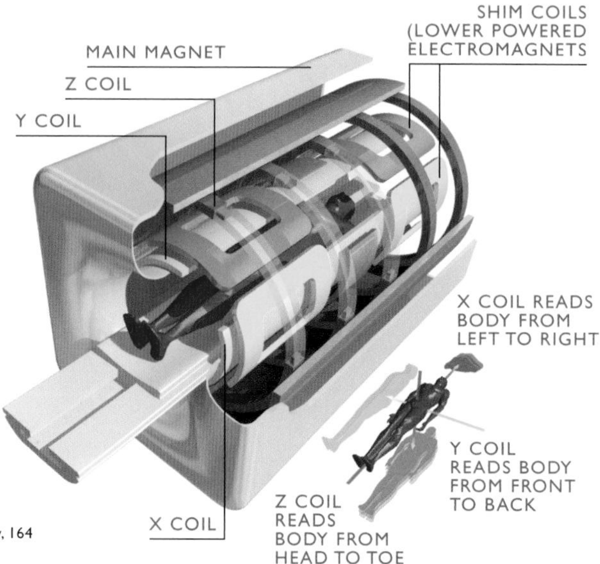

»»» *See Also*
RELATED PRINCIPLES: Electromagnetic Induction and Faraday's Law, 157 • Ohm's Law, 164
RELATED APPLICATIONS: Computed Tomography, 256

FUNCTIONAL MRI

A functional MRI (fMRI) is a special kind of MRI scanner that focuses entirely on the brain. It measures the brain's activity, pinpointing areas of high and low activity, by measuring the level of oxygen in the blood. As hemoglobin levels change in the blood, the response to magnetic fields changes (hemoglobin is the molecule in blood that carries oxygen). When certain areas of the brain are stimulated, more blood flows to those regions, so oxygen levels are higher. The measurement of blood flow, blood volume, and oxygen use is called the blood-oxygen-level-dependent (BOLD) signal. A computer can process this signal and create a 3-D image of the brain.

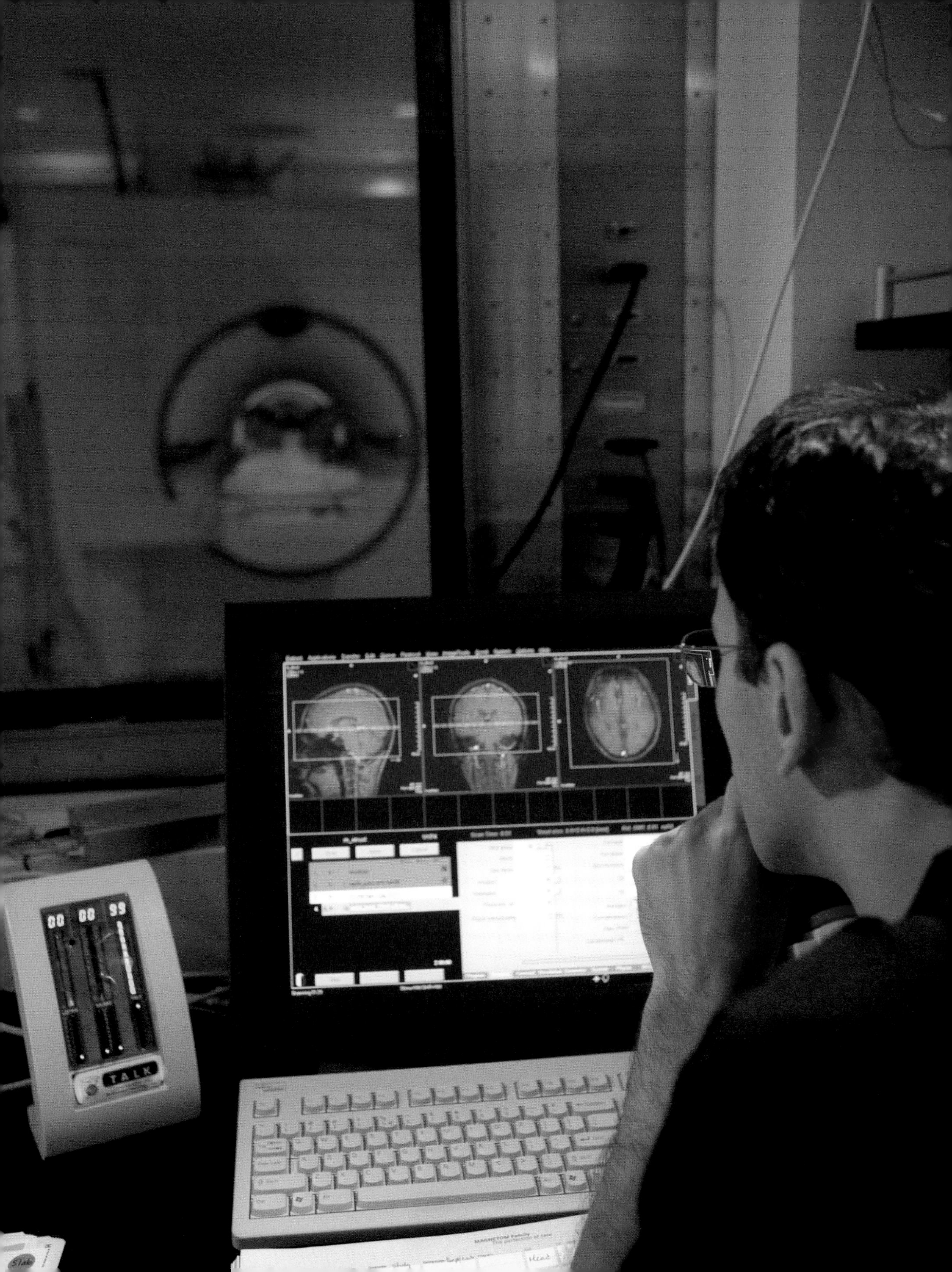
TALK

Magnetic Fields

COLOR TELEVISION

Nobody buys anything but flat-panel TV screens these days, like plasma and LCD screens, but for 50 years the CRT screen ruled the world of color TV. In a CRT tube, the cathode, or negative terminal, is a heated filament in a glass tube that's sealed to create a vacuum. The ray is a stream of electrons that pour off the heated cathode into the vacuum. Because electrons are negative, they flow toward anodes, or positive terminals (remember, opposites attract). The electrons are focused by a focusing anode into a beam, then accelerated by an accelerating anode. This high-speed beam of electrons flies through the vacuum of the tube to a flat screen at one end of the tube. The screen is coated with phosphors, material that when exposed to radiation emits visible light, or light that can be seen by the human eye; in this case, the radiation is the beam of electrons. The phosphors, arranged in dots or stripes, glow red, green, and blue when struck by the beam. Copper coils wrapped around the tube create magnetic fields that steer the electron beam to any point on the screen. The colors that appear on the screen depend on how the beam is fired and how it passes through a secondary screen called a shadow mask.

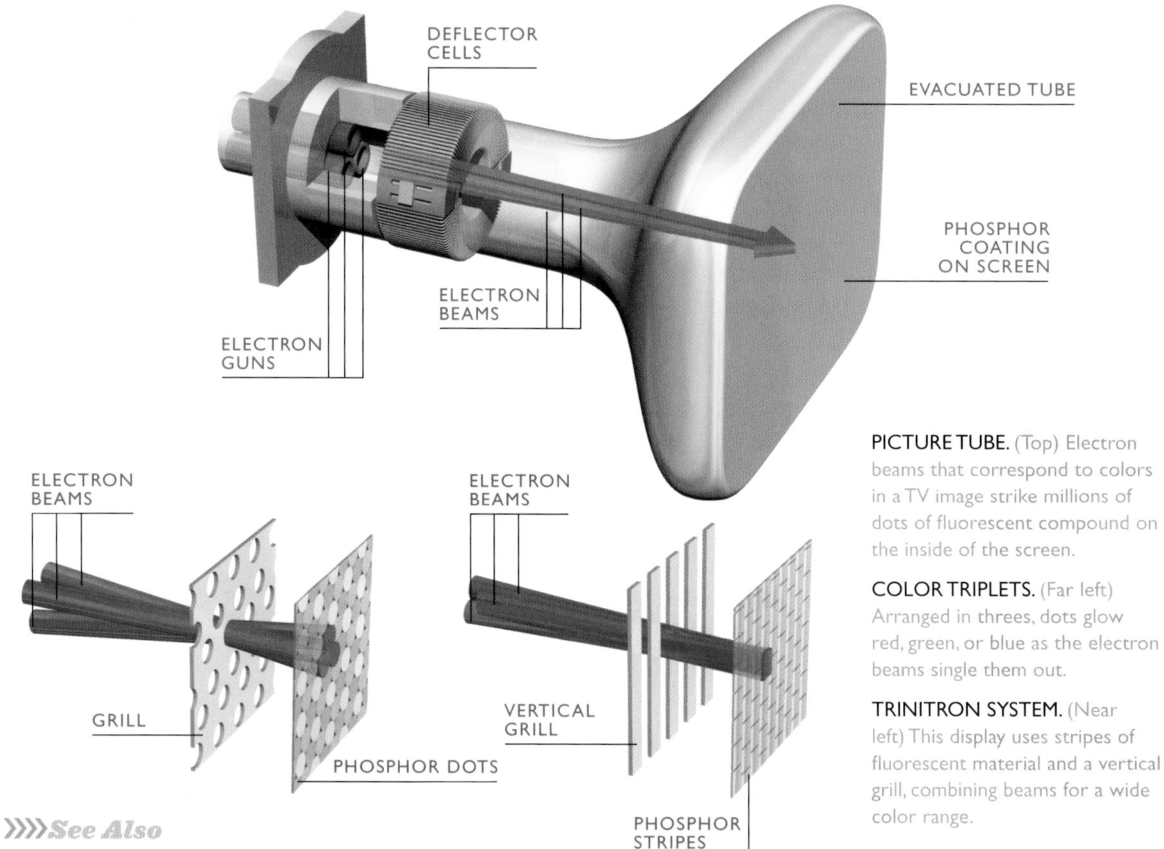

PICTURE TUBE. (Top) Electron beams that correspond to colors in a TV image strike millions of dots of fluorescent compound on the inside of the screen.

COLOR TRIPLETS. (Far left) Arranged in threes, dots glow red, green, or blue as the electron beams single them out.

TRINITRON SYSTEM. (Near left) This display uses stripes of fluorescent material and a vertical grill, combining beams for a wide color range.

»»See Also

RELATED PRINCIPLES: Power and Electricity, 163 • Radiation, 239
RELATED APPLICATIONS: Movies and TV, 208

Magnetic Fields

PLASMA, LCD, AND LED SCREENS

The bulk and size of CRT screens were a big factor in why their popularity faded. The three most common replacements—plasma, LCD, and LED screens—are a fraction as deep and use less energy. All three use technology similar to a CRT to display colors, although with some variations. Electrons are activated in such a way that the energy they release can be harnessed and used to light the screen. The primary difference is in how the electrons are stored. A plasma screen uses plasma, gas filled with xenon and neon atoms. LCD screens light up tiny fluorescent lights instead of dots, and LEDs activate electrons contained between two pieces of polarized film. In addition to TV screens, LCD screens are common in computer monitors, electronic displays, and other devices. LED screens are often found in traffic lights, lighting displays, and indoor lighting. Another promising technology is the organic light-emitting diode (OLED), which, because it allows thinner, lighter displays of high clarity, is already favored for handheld devices.

»»See Also
RELATED PRINCIPLES: Power and Electricity, 163 • Radiation, 239
RELATED APPLICATIONS: Light-Sensitive Diodes, 222

Inside a computer hard disk drive.

PRINCIPLE

A CHARGED INTERACTION

Electromagnetism

Electromagnetism involves two primary forces—electricity and magnetism—that are intimately related. Magnetism is the flip side of electricity. Electrical charges in motion produce magnetic fields, and magnetic field variations produce electrical current. When you run electrical current through a wire, it creates a magnetic field. An even stronger field can be generated by wrapping the wire in a coil around metal. The electromagnetic force is at work in such applications as speakers, computer hard drives, and particle accelerators.

ELECTROMAGNETIC SPECTRUM

Visible light is only a tiny slice of the electromagnetic spectrum, or the various types of electromagnetic radiation, or rays. The types of rays are arranged based on their wavelength, the distance between the troughs of a wave and how far a wave can travel, and their frequency, or strength. Radio, with wavelengths from a few feet to well over a mile, is at one end of the spectrum. Gamma rays and x-rays, with wavelengths less than the size of the nucleus of an atom, are at the other end. The higher the frequency, the greater the energy. While you can't feel a radio wave that allows you to have a conversation on your cellular phone, infrared waves are strong enough to warm your skin, ultraviolet waves can give you a painful sunburn, and enough gamma rays would vaporize your cells.

Electromagnetic waves are amazingly versatile. Microwaves, for example, can heat food and signal satellites. Shorter, near infrared, rays are the ones used by a TV's remote control.

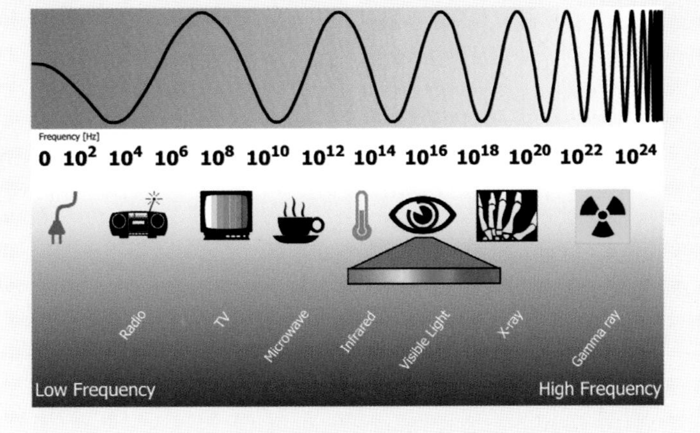

Electromagnetic waves by frequency

Electromagnetism

LOUDSPEAKERS AND HEADPHONES

A vibrating, magnetized coil produces the sound you hear through speakers. The human ear—specifically, the eardrum—vibrates in response to sound waves, which the brain translates into sound. Speakers—and earphones and headphones are just tiny speakers—are like the ear and deliver sound in a similar way.

A cone forms the body of the speaker, and a diaphragm is stretched across the widest part, like the eardrum in the human ear. The smaller end of the cone is connected to an iron coil and a magnet, which in turn are connected to wires—and these connect to a stereo or iPod. When the stereo is turned on, it generates electric signals, which pass through the wires and magnetize the coil. The coil vibrates back and forth in a motion that moves the diaphragm. This motion is then transformed by the vibrations of the diaphragm into sound.

GOOD VIBRATIONS. In a speaker—including headphones and earphones—the incoming current is used to deflect an electromagnet attached to the diaphragm or speaker cone. The greater the current, the louder the sound that is generated. More rapid vibrations of the diaphragm generate higher pitch.

See Also

RELATED PRINCIPLES: Power and Electricity, 163 • Wave Energy, 58
RELATED APPLICATIONS: Noise-Cancelling Device, 62

Electromagnetism

COMPUTER HARD DRIVE

The hard drive in your computer has one job: to store programs and data until you need them. The device was first invented in the 1950s—the adjective hard was later used to distinguish them from "floppy" disks when they emerged in the 1960s. Hard drives employ a magnetic medium coating a disk of either glass or aluminum that has been highly polished as a means of storage. Data is stored in this medium in tiny pits called domains. Electronic controllers operate a read-write mechanism that assembles the domains into bits—ones and zeros that represent one of two states, either on or off—called binary code.

THE HARD DRIVE. A computer's hard drive consists of aluminum disks coated with magnetic material. Read-write heads floating over the disks convert coded electrical signals and write them in a digital code. The heads are electromagnets whose north and south poles align magnetic particles on the disks to form digital ones and zeros.

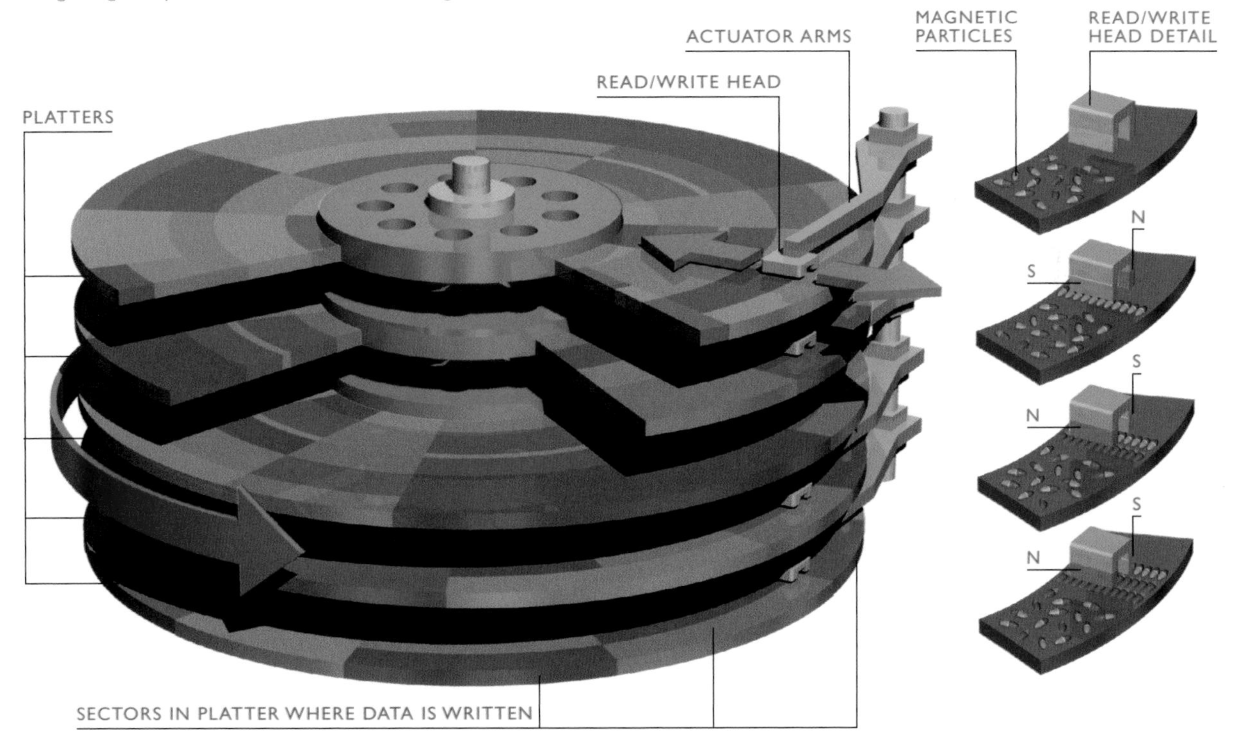

See Also

RELATED PRINCIPLES: Binary Code, 190 • Power and Electricity, 163
RELATED APPLICATIONS: Credit Card, 141 • Integrated Circuit, 191

Electromagnetism

PARTICLE ACCELERATOR

In a particle accelerator, magnetic fields accelerate subatomic particles to high speeds and then trap these particles in a beam. This beam can then be focused for low-level uses, such as x-ray machines.

High-level particle accelerators, used for research in theoretical and experimental physics, focus the beams in opposing directions at velocities approaching the speed of light. In these types of accelerators, when these beams intersect, the particles in them collide and smash apart. Scientists use special tools to detect the smashed pieces and determine what makes up an atom, for example (this is why you may also hear particle accelerators referred to as "atom smashers" or "particle colliders"). The knowledge gained from studying these subatomic particles has led us to better understand such things as the formation of matter, how the universe came into being, the nature of energy, and dark matter.

Antimatter annihilation

There are five different types of particle accelerators: linac, cyclotron, betatron, synchrotron, and storage ring collider. They vary in shape and the path the particles travel within them. There are also differences in the intensity of the strength of the magnetic fields and the purposes for which each device is used.

In less than 100 years since the first collider was built in the 1930s, the accelerator energy available to experimenters has increased by a factor of nearly a million. That capbility has made possible the discovery of a huge number of subatomic particles by physicists.

»»See Also
RELATED PRINCIPLES: Binary Code, 190 • Power and Electricity, 163 • Quantum Mechanics, 298
RELATED APPLICATIONS: Spectrum of Wavelengths, 248 • Superconductivity, 116

LARGE HADRON COLLIDER

When the Large Hadron Collider (LHC) went live in 2008, it became the largest particle accelerator in the world. It stretches 17 miles underground, spanning the border between France and Switzerland. It contains 9,300 magnets and can sample and record the results of up to 600 million proton collisions per second. The LHC has helped prove the existence of the Higgs boson particle—the "missing link" of particle physics.

BLACK HOLES

Before the LHC was turned on, there was speculation on the Internet that the particle collisions would produce a black hole that would destroy Earth (or the entire universe). Scientists pointed out that cosmic ray collisions with energies comparable to those at the LHC have been going on in the atmosphere since the beginning of Earth, and we're still here. If a black hole actually did form, it would either dissipate into what is called Hawking radiation, which is harmless, or the black hole would grow so slowly that it would not be a threat.

Electromagnetism

ELECTRON MICROSCOPE

To try to see the unseeable: that's the purpose of all microscopes. A simple optical microscope uses a lens or lenses and light to magnify an object and make it visible to the human eye. An electron microscope uses beams of rapidly moving electrons focused by magnetic lenses instead of rays of light on a specially prepared specimen in a vacuum-sealed chamber. The result is a level of detail and magnification that is truly astonishing—100 million times better than what we can see with the human eye unaided. There are two types of electron microscopes: transmission and scanning. Scanning electron microscopes form pictures of metallically coated specimens as electrons are bounced off them. Transmission electron microscopes pass electrons through specimens and onto reflective plates. Both microscopes create a detailed picture of the specimen on a monitor. Scanning electron microscopes are used when the surface of a specimen needs to be examined in detail; a transmission electron microscope is used when detail on internal structures is needed. (The state of the science in microscopy is embodied in the scanning tunneling microscope, which uses wavelike properties of electrons to glimpse beneath an object's surface and was developed in 1981.)

With the electron microscope, researchers have been able to explore bacteria, viruses, molecules, and even atoms in ways that were unknown prior to its invention in 1926 (the first working electron microscope was not introduced until 1938 at the University of Toronto).

Electron microscope

Scanning electron microscopy of pollen grains from a daffodil, a paper birch tree, ragweed, impatiens, mallow, marsh marigold, forsythia, pussy willow, and sunflower

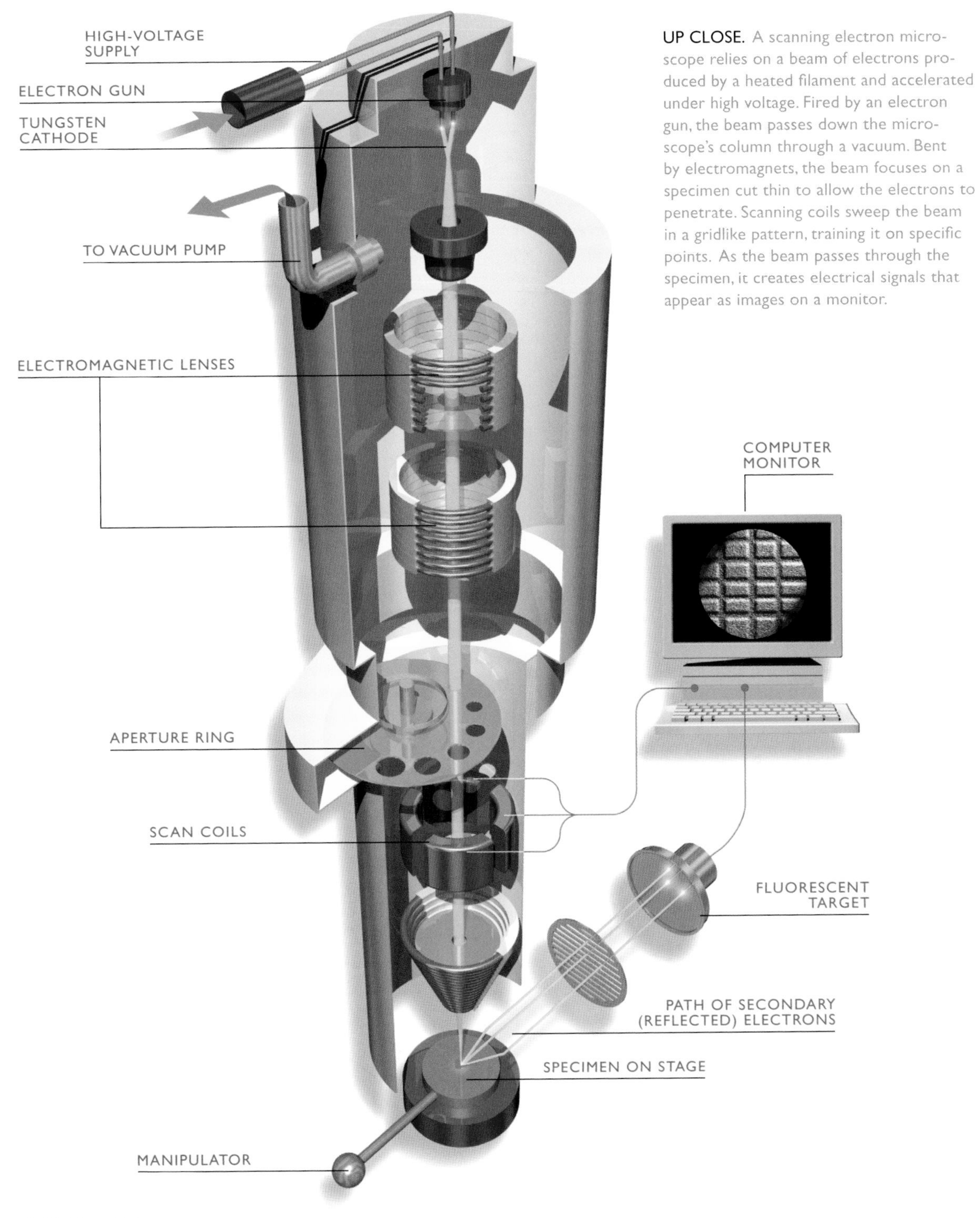

UP CLOSE. A scanning electron microscope relies on a beam of electrons produced by a heated filament and accelerated under high voltage. Fired by an electron gun, the beam passes down the microscope's column through a vacuum. Bent by electromagnets, the beam focuses on a specimen cut thin to allow the electrons to penetrate. Scanning coils sweep the beam in a gridlike pattern, training it on specific points. As the beam passes through the specimen, it creates electrical signals that appear as images on a monitor.

»»See Also

RELATED PRINCIPLES: Binary Code, 190 • Power and Electricity, 163 • Quantum Mechanics, 298
RELATED APPLICATIONS: Hologram, 234

Electromagnetism

MAGLEV SHIP

Maglev, which is short for magnetic levitation, is a concept in which a magnetic field levitates a vehicle—in this case, a ship—and propels it using lift and thrust. While the concept has been applied successfully to trains, right now its application to ships is still theoretical. The principle is simple yet ingenious. A maglev puts a twist on the traditional motor; instead of wires conducting electricity, the ocean acts as the conductor. Electricity is generated between electrodes attached to the bottom of the ship. Powerful magnets then emit a magnetic field into the water. The resulting force pushes against the seawater, driving the water backward and the ship forward. American engineer Stewart Way first came up with the idea in 1960. His 10-foot-long, submarine-like vessel ran at 2 knots (2.3 miles per hour) for about 12 minutes.

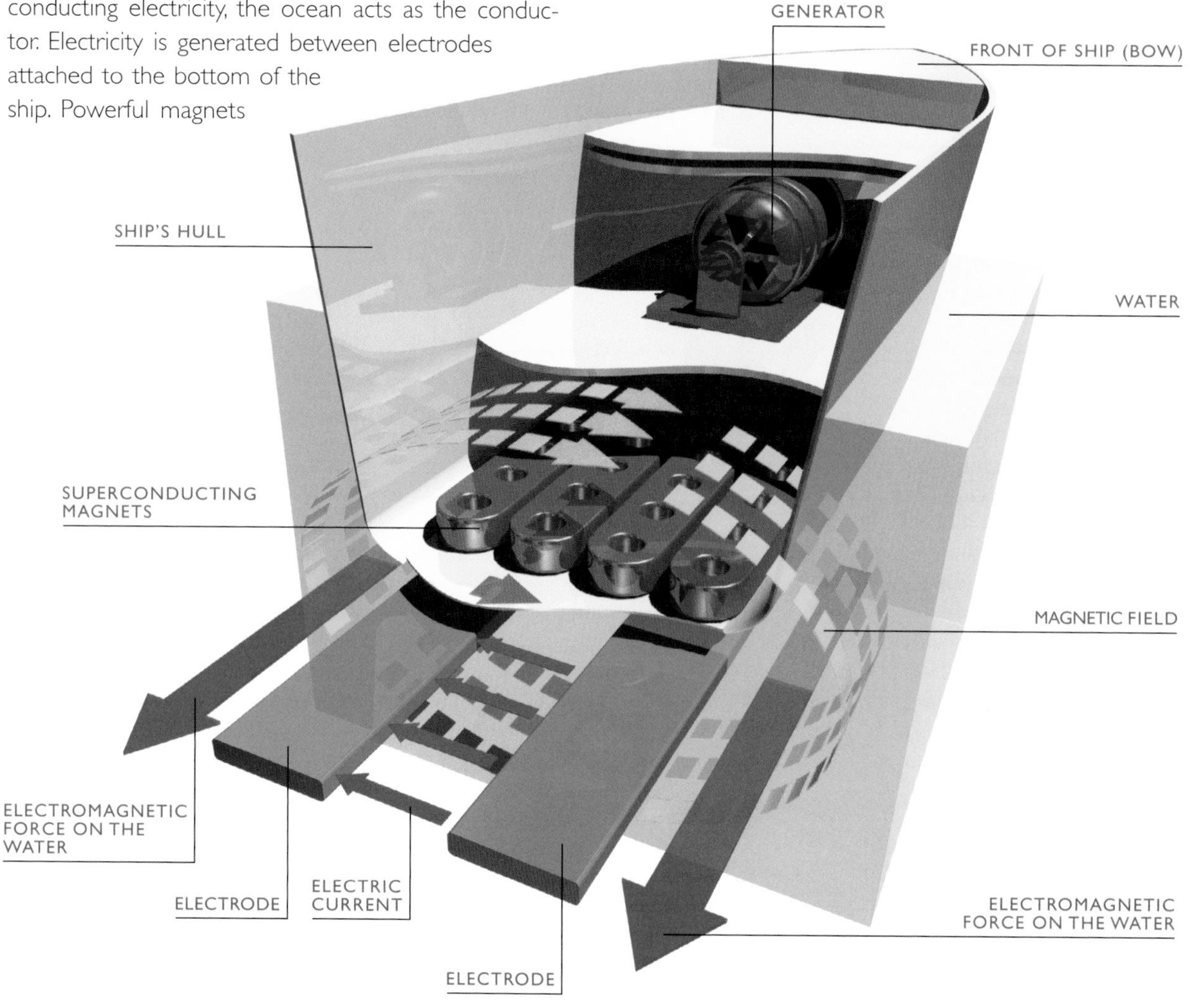

See Also

RELATED PRINCIPLES: Laws of Motion and Energy, 18 • Mechanical Advantage, 38 • Power and Electricity, 163 • Thermodynamics, 96
RELATED APPLICATIONS: Magnet Earth, • 137 Sailboat, 68 •

Electromagnetism

MAGLEV TRAIN

A maglev train might seem mysterious in action, but the concept is easy to understand. Magnets are induced in the track by the motion of the train and disappear when the train is gone. The induced magnets repel magnets attached to the bottom of the train. The train rises just a few inches above the track, and individual magnetic fields of the guideway propel the train forward. Maglev trains travel at nearly twice the speed of conventional trains, use less energy, and cost less to maintain and run—partly because their operating mechanism involves almost no friction. They also emit no pollution and have safety benefits over regular trains.

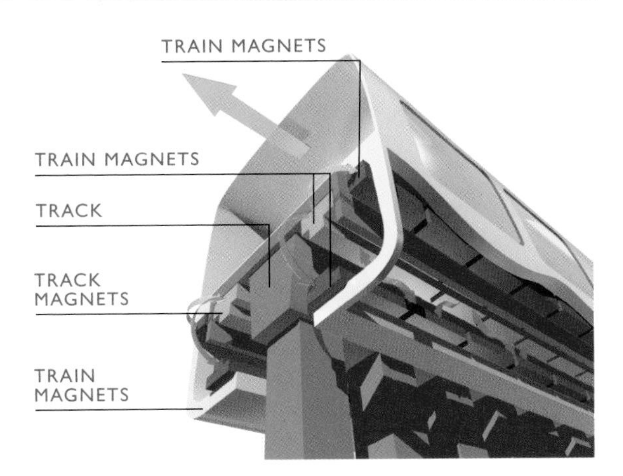

»»See Also
RELATED PRINCIPLES: Faraday's Law of Induction, 157 • Laws of Motion and Energy, 18 • Mechanical Advantage, 38 • Power and Electricity, 163 • Thermodynamics, 96
RELATED APPLICATIONS: Hovercraft, 69 • Squid and Octopus, 25

Detail of a car engine

PRINCIPLE

ELECTRIC YIN YANG

Faraday's Law of Induction

Anything that lets electrons flow in one or more directions is called a conductor, and this is a key component of electromagnetic induction. The term refers to two different ways of creating voltage and generating a current: a conductor can be placed in a magnetic field that fluctuates, or it can move through a magnetic field that is static. Electric motors work—whether they're powering a car, an electric drill, a garbage disposal, or an electric razor—because electric currents, which give off a magnetic field, are surrounded by another current that has an alternating magnetic field, which also generates a changing electrical current in a conductor placed within it, which has its own magnetic field—and the cycle continues. English scientist Michael Faraday is credited with having discovered this principle in 1831 and working out the equation that describes how an induced magnetic field will respond to a change in magnetic flux.

Electromagnetic Induction and Faraday's Law

GENERATOR

An alternating current (AC) generator takes mechanical motion, or energy, and converts it into electricity. It consists of magnets, a spinning coil that is usually wrapped with wire, connecting rings, and carbon conductor brushes. As a motor spins the coil, electrons in the wire respond to the magnets, creating an electromagnetic current. The AC generator gets its name from the fact that as the wire spins, electrons flow first in one direction and then the other. This alternating motion means that the current can be stepped up or stepped down as needed, and it's this characteristic that allows AC to be transmitted over long distances—something that direct current (DC), in which the current flows in only one direction, can't do.

»»See Also

RELATED PRINCIPLES: Laws of Motion and Energy, 18 • Mechanical Advantage, 38 • Power and Electricity, 163
RELATED APPLICATIONS: Regenerative Braking, 31

Electromagnetic Induction and Faraday's Law

ELECTRIC MOTOR

Take the motion of an AC generator and reverse it, and you've got yourself an electric motor. A generator takes mechanical motion and converts it to electricity; a motor takes electricity and converts it into mechanical motion. The electric motor comprises a wire looped around a coil, which creates a circuit, and two magnets. One magnet is typically fixed, and the other rotates. The forces created repel each other, which creates energy, which then spins an axle, for example, and drives the blades of a garbage disposal or the motor of an electric razor. The first commercial electric motor was introduced in 1873; today, they're found in thousands of household devices.

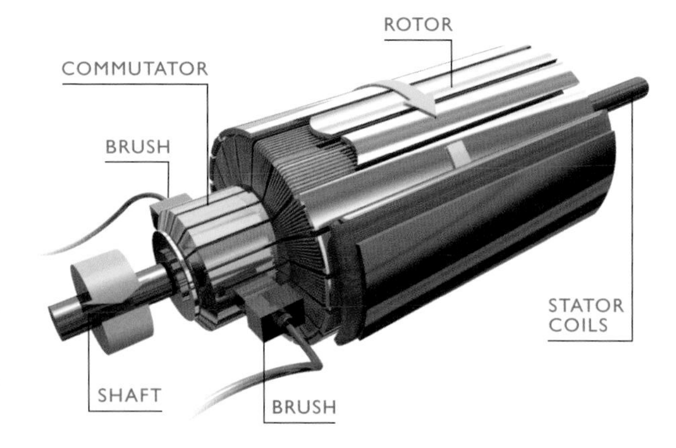

»»See Also

RELATED PRINCIPLES: Laws of Motion and Energy, 18 • Mechanical Advantage, 38 • Power and Electricity, 163
RELATED APPLICATIONS: Electric Car, 160 • Electric Razor, 159 • Garbage Disposal, 159 • Generator, 157 • Power Drill, 158

Electromagnetic Induction and Faraday's Law

POWER DRILL

We use power drills for everything from hanging pictures to building decks. Although some power drills come with dozens of bits, they have only five basic components: the safety latch, the reversing switch, a torque adjustment, a chuck (this keeps the drill bit from moving), and the handle. Squeezing the handle activates the electric motor, and the drill bit turns. The pressure it generates against a surface is what creates the drill hole. An electric drill derives its power from a variable-speed, fan-cooled universal motor that turns its gears and the chuck holding the drill bit. The rotating commutator in the motor receives electricity through rubbing carbon brushes.

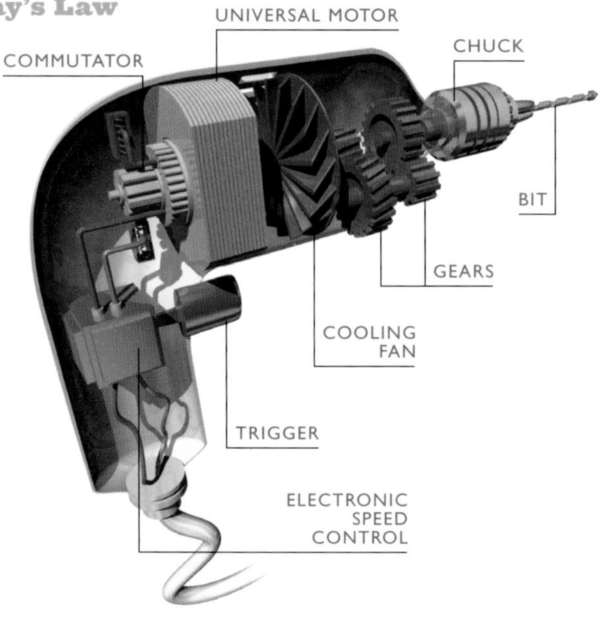

»»See Also

RELATED PRINCIPLES: Laws of Motion and Energy, 18 • Mechanical Advantage, 38 • Power and Electricity, 163
RELATED APPLICATIONS: Electric Car, 160 • Electric Motor, 158 • Electric Razor, 159 • Garbage Disposal, 159 • Generator, 157

Electromagnetic Induction and Faraday's Law

GARBAGE DISPOSAL

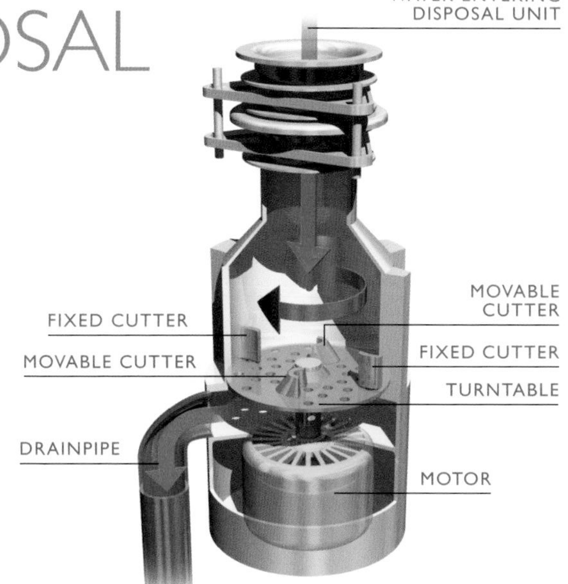

How did we ever live without garbage disposals? This convenient household device is attached to the plumbing beneath your kitchen sink, where it contains a cylinder with a series of blades in it. When we flip the switch to turn on the disposal, it completes a circuit that creates a current that drives an electric motor. The motor, in turn, causes the blades to spin. The blades grind food scraps to bits against a flat surface, which are then flushed down the drain and into the sewer pipes. The garbage disposal was invented in 1927 and came on the market in the United States in the 1930s. Its usage was initially prohibited by many municipalities out of concern for its effect on sewage systems. Intense lobbying on the part of manufacturers ultimately won out.

»»See Also
RELATED PRINCIPLES: Laws of Motion and Energy, 18 • Mechanical Advantage, 38 • Power and Electricity, 163
RELATED APPLICATIONS: Electric Car, 160 • Electric Motor, 158 • Electric Razor, 159 • Generator, 157 • Power Drill, 158 •

Electromagnetic Induction and Faraday's Law

ELECTRIC RAZOR

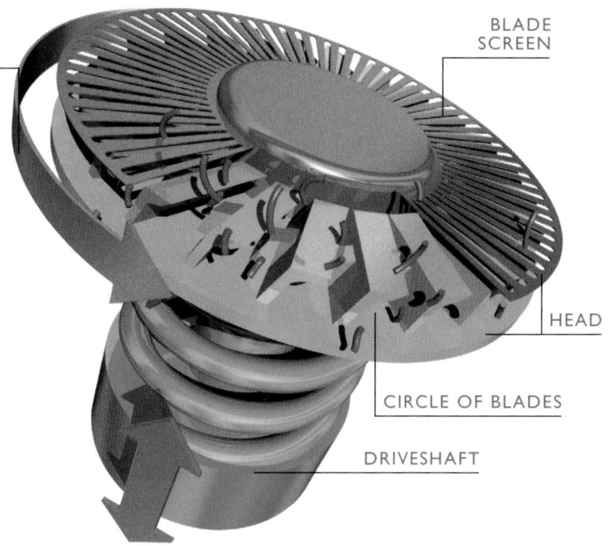

While some people swear by old-fashioned razors, electric razors do have their proponents. There are two different kinds—foil and rotary—but they work the same way. A circle of blades rotates on a springy driveshaft, which is powered by an electric motor. The malleability of the driveshaft is what allows it to follow the contour of the skin being shaved. The foil in the razor glides across the skin, capturing the hair follicles, and then the blades slide against the foil and cut the hair. With a rotary razor, three or four round heads house cutting blades that spin and cut the hair.

»»See Also
RELATED PRINCIPLES: Laws of Motion and Energy, 18 • Mechanical Advantage, 38 • Power and Electricity, 163
RELATED APPLICATIONS: Electric Car, 160 • Electric Motor, 158 • Garbage Disposal, 159 • Generator, 157 • Power Drill, 158

Electromagnetic Induction and Faraday's Law

ELECTRIC CAR

With electric cars, energy in the form of electricity is stored in a battery and converted into mechanical energy when needed. But how does an electric car work? An electric motor is powered by a controller, which in turn is powered by an array of rechargeable batteries. When you push down on the accelerator in an electric car, variable resistors (the technical term is potentiometers) provide a signal to the controller, telling it how much "charge" to take from the batteries. When the car needs "refueling," you plug an onboard recharger into a regular electrical outlet. Electric cars may use either an AC or a DC motor—the AC ones are slightly more complicated.

Electric cars still have relatively limited driving range compared to gasoline-powered cars, and it takes a lot longer to charge up a huge battery than to fill up a tank with gas. But fans of electric cars never pay for gas and never pollute. (The car itself has no emissions. Of course, the electricity plant generates some pollution, but the total is far less than burning gasoline in the car itself.)

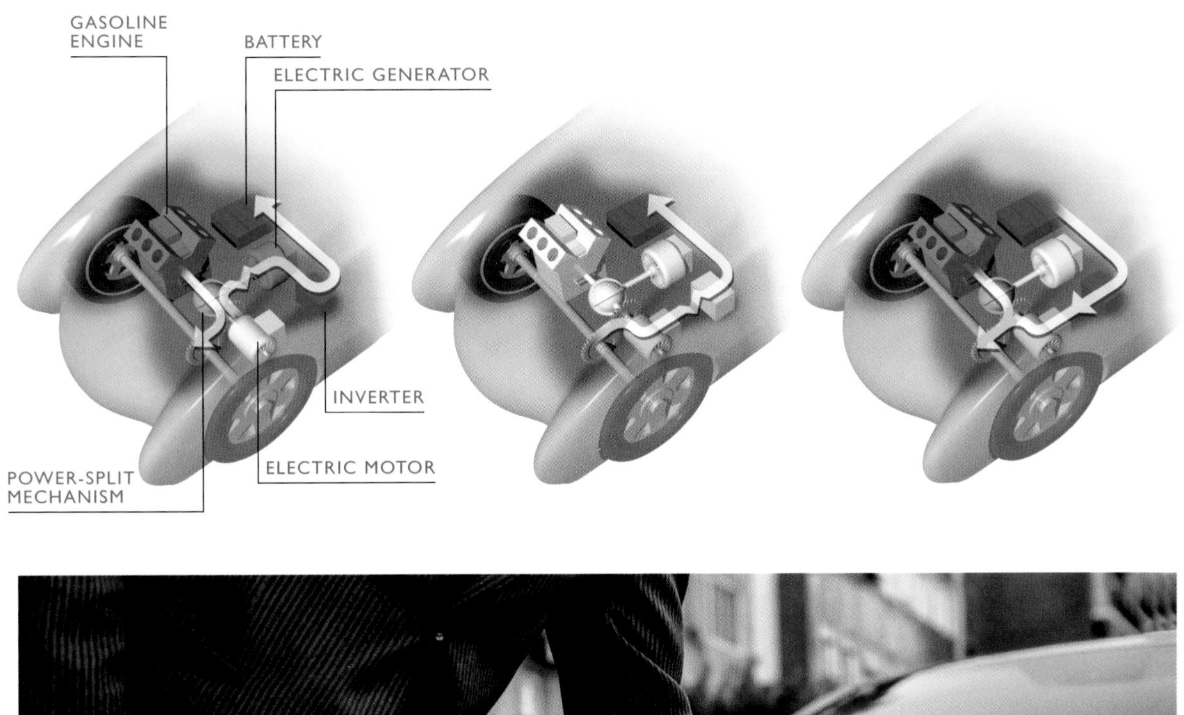

See Also

RELATED PRINCIPLES: Laws of Motion and Energy, 18 • Mechanical Advantage, 38 • Power and Electricity, 163
RELATED APPLICATIONS: Automobile Transmission, 51 • Regenerative Braking, 31 • Wheel, 44

CHAPTER 8

Power and Electricity

Electricity runs our homes, businesses, cars, and mobile devices. And that means our power now needs to be portable. Megastorms like Hurricane Sandy show just how important electricity is when it's tied to the grid, and everyone rushes out to buy generators for their homes. Ben Franklin surely never imagined that we would one day use electricity and magnetism to create images on paper, but that's just what we've done with the photocopier, laser printer, and inkjet printer.

We're also always looking for new sources of energy. Part of the answer might be hydroelectricity, for example, which harnesses the power of water—in rivers, waterfalls, and reservoirs. Generating power using the wind, the Sun, and even nuclear reactions looks promising, too.

PRINCIPLE

THREE OF A KIND

Volts, Amps, and Ohm's Law

Named after Bavarian mathematician and physicist Georg Ohm, Ohm's law demonstrates the relationships between voltage, current, and resistance. Voltage is measured in volts, current is measured in amperes, or amps, and resistance is measured in ohms. All three elements directly interact and affect each other, and if you know any two of the values, the third can be calculated. So why is Ohm's law important? This simple equation explains the physics of electricity by analyzing the behavior of circuits. It has also been applied to the human circulatory system, allowing us to better understand blood flow.

Ohm's Law

TWO- AND THREE-PRONG PLUGS

Grounded plugs, which have three prongs, have a direct connection to the ground; they discharge excess electricity to prevent electric shock. If you look at a three-prong plug, you see two metal tabs—the left one is larger than the right—and a round prong beneath them. The left tab is the "neutral" tab, the right tab is the "hot" tab, and the prong is the "ground." (Two-prong plugs don't have this prong.) Power flows from the hot tab to the neutral tab, creating a circuit when an appliance, for example, is plugged into it. Two-prong plugs and three-prong plugs work identically, so why is one grounded? In the event of a short circuit, the ground wire allows the current to pass safely to the ground without shocking anyone; instead, a fuse blows or a circuit breaker trips. Three-prong plugs are especially important for appliances that have metal cases. If a wire in an appliance with a three-prong plug happens to come loose and touches the case, the third prong prevents the electricity from flowing into the metal and shocking you.

»»See Also
RELATED PRINCIPLES: Radiation, 239 • Thermodynamics, 96
RELATED APPLICATIONS: Joule-Kelvin Effect, 98

Ohm's Law

AC/DC: EDISON VERSUS TESLA

If there's a battle between alternating current and direct current, it began with Nikola Tesla and Thomas Edison. When electricity was first being widely distributed to homes, Edison's DC system was the delivery method of choice. Obviously, he wanted this system to remain in place so he could continue to collect royalties on it; but Tesla, who was an advocate of AC electricity, claimed that DC was inefficient. AC, he pointed out, could carry more current farther, and the current could be modified and made stronger or weaker to meet specific needs. Edison claimed that AC was inherently dangerous; he went so far as to electrocute an elephant to prove it.

The main difference between AC and DC is how current flows. With DC, current always flows in the same direction, from positive to negative, and it can't be changed. With AC, the current can be reversed (or alternated) 50 to 60 times a second, depending on where in the world you live. (Hertz is the unit of frequency for this, named for the German physicist who first produced radio waves artificially and who figured out their periodicity.) Tesla determined that 60-hertz was best for power transmission; however, when the first power station was built in Europe in Germany, the engineers went with 50 hertz. Most of the rest of the world followed suit.

Although AC dominated, there were pockets of cities where DC continued to be used through the 20th century. These included Greenwich Village in New York City, where it was not until 2007 that Consolidated Edison ceased to supply DC altogether.

Nikola Tesla

WIRELESS ELECTRICITY. Among Tesla's many inventions was the magnifying transmitter, also known as the Wardenclyffe Tower. The wireless energy transmitter was abandoned due to expense and supposed lack of profitability.

»»See Also

RELATED PRINCIPLES: Radiation, 239 • Thermodynamics, 96

RELATED APPLICATIONS: Hydrogen Fuel Cells, 188 • Lithium-Ion Battery, 187 • Wind Turbines, 176

Ohm's Law

ELECTRICAL WIRING AND POWER GRID

Flick a switch, and a light comes on. Most people usually don't give the matter another thought—unless the light fails to come on. However, a vast and complicated grid delivers this power to us. It starts with the spinning turbine at the power plant. The power plant produces AC power, which then goes to a transmission substation, where it's converted to a very high voltage for long-distance transmission through a three-wire system. It's "stepped down"—that is, the voltage level is decreased—at distribution grids at a power substation, where circuit breakers and switches split the power off where it's needed and at the appropriate voltage, stepping it up and down as needed. Some grids also use distribution buses, which contain arrays of switches, and regulator banks, which prevent voltage from going too high or too low, to send power. Neighborhood power lines carry 7,200 volts of electricity and are stepped down by transformers on utility poles to feed each home just 240 volts.

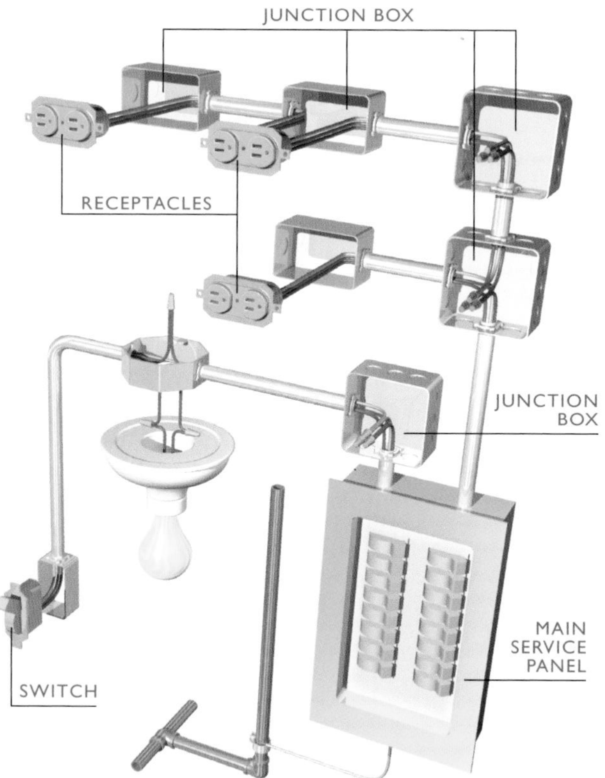

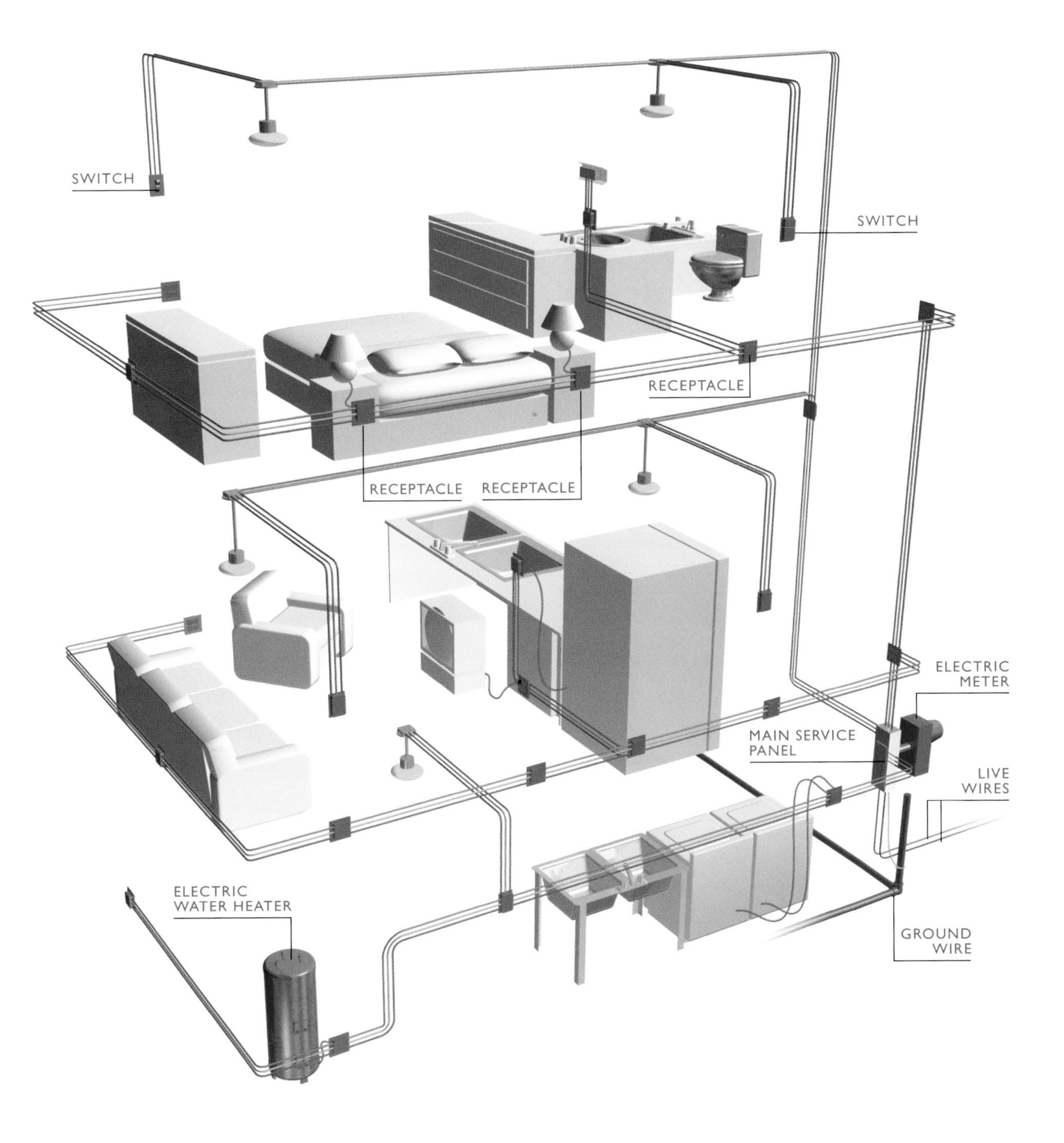

»»See Also
RELATED PRINCIPLES: Radiation, 239 • Thermodynamics, 96 •
RELATED APPLICATIONS: Joule-Kelvin Effect, 98

Ohm's Law

PHOTOCOPIER

The modern photocopier was born in 1959 with Chester Carlson's Haloid Xerox 914. This machine made copies on ordinary, untreated paper. It was a big improvement over earlier devices, such as carbon paper, the mimeograph, and the hectograph, which were messy to use and time consuming. The Haloid Xerox 914 spawned both the Xerox Corporation and a new term—xerography (from the Greek words for dry and writing).

Today's high-speed, clean photocopiers employ two basic principles: that materials with opposing electrical charges attract each other, and some elements conduct electricity better when exposed to light. An element called selenium is the key. Selenium is actually a poor conductor of electricity—until it's exposed to light. Within the photocopier machine, a halogen lamp illuminates the original document. Selenium atoms react to electrical changes between white and dark areas of the document to be copied. White areas lose their electrical charge; dark areas don't. The pattern that emerges is translated into an electrical version of the image. Then, specially charged ink powder, called toner, is brushed onto the drum. It's attracted to the charged (dark) areas and fused to the paper by heat. The result: a nearly perfect copy of the original image. The quality is often affected by the type of paper used.

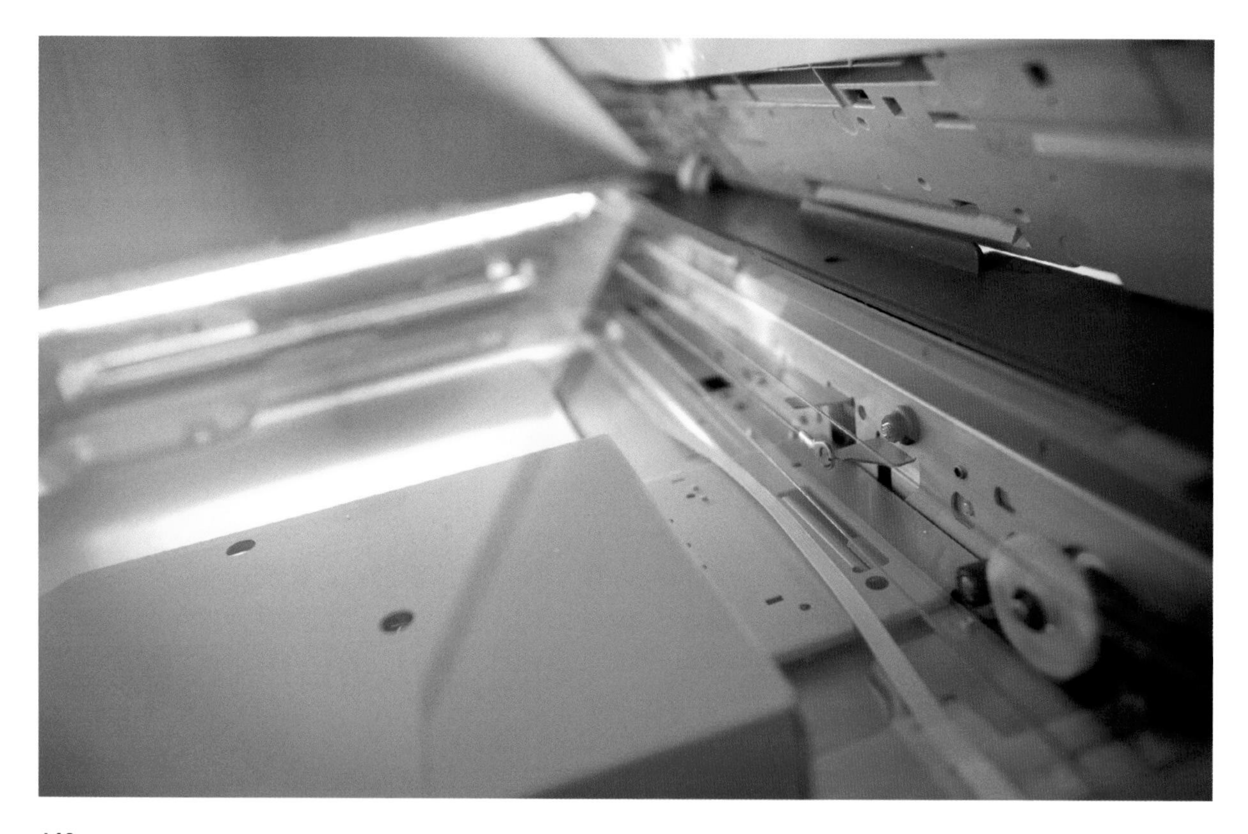

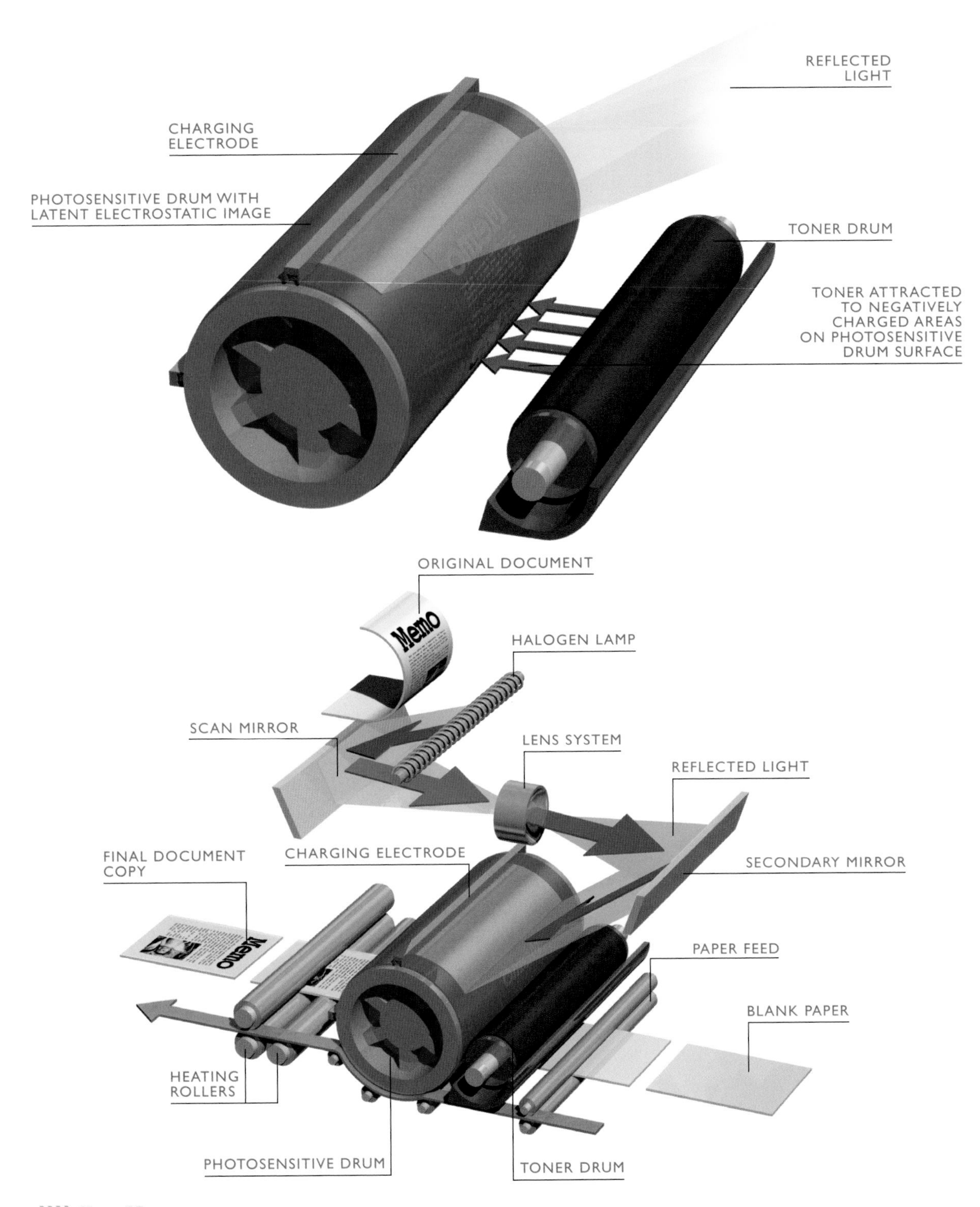

See Also

RELATED PRINCIPLES: Light, 203 • Mechanical Advantage, 38 • Radiation, 239

RELATED APPLICATIONS: Credit Card, 141 • Thermoplastic and 3-D Printing, 295

Ohm's Law

LASER PRINTER

A laser printer works much like a photocopier. First, the computer sends information to the printer in the form of data, telling it what is to be printed. Next, a laser beam in the printer "draws" an image of this data onto a photosensitive drum. The surface of this drum carries a charge, but when the laser beam hits the drum, these areas lose that charge. A roller gathers toner (which carries an opposite charge) and applies it to the drum. The toner particles stick to the charged areas and are then transferred to sheets of paper passed through a fuser. Heat and pressure in the fuser fix the image-carrying toner to the surface. Finally, a special rotating brush cleans any residual toner from the drum. This helps prevent "ghost" images and makes sure that any subsequent copies are clean.

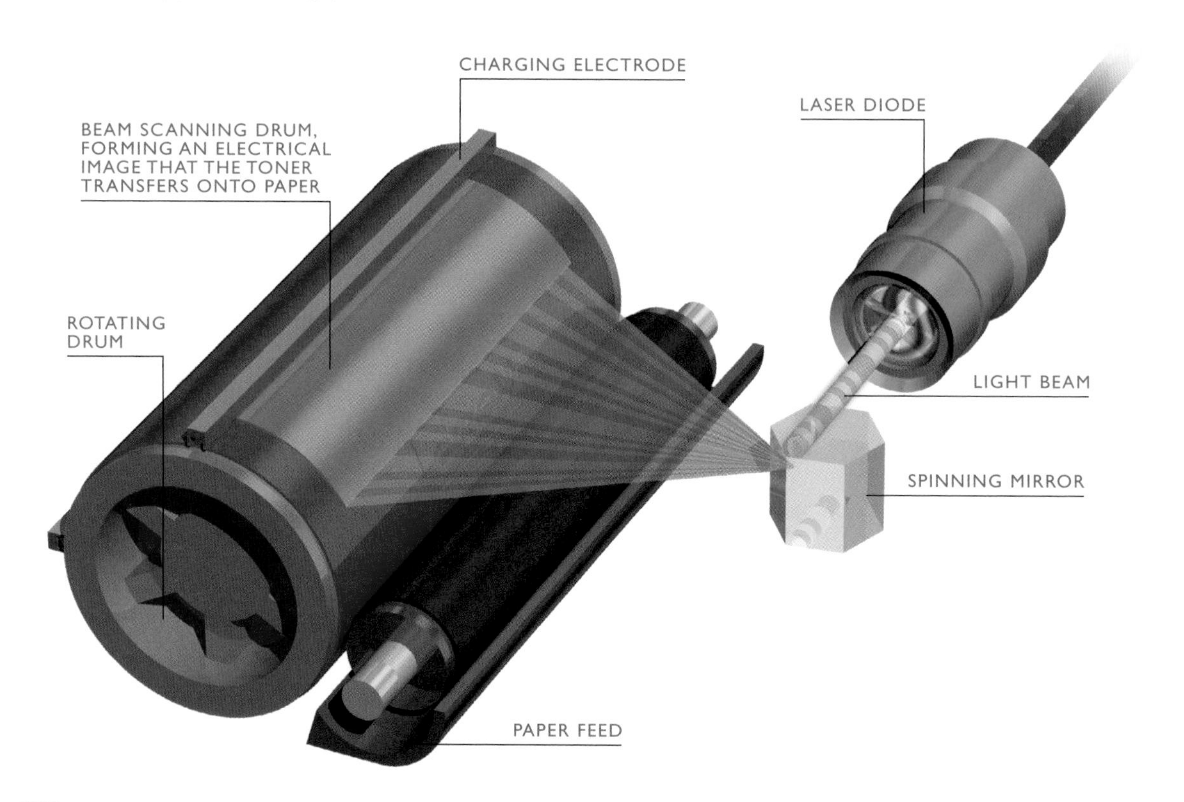

»»See Also

RELATED PRINCIPLES: Light, 203 • Mechanical Advantage, 38 • Radiation, 239

RELATED APPLICATIONS: Laser, 232 • Laser Surgery, 377

Ohm's Law

INKJET PRINTER

An inkjet printer uses the same electrochemical process as a laser printer to process an image and transfer it to paper. The main difference between the two lies in the method by which the data to be copied is transferred to paper. An inkjet printer uses tiny nozzles to spray dots on the paper. Like an Impressionist painting, these dots coalesce into words, numbers, and images. The quality of an inkjet printer is often measured in something called dots per inch (dpi)—the higher the dpi, the better the printer—and it's a result of how many nozzles a printer has. More nozzles means a printer can move over the page faster to produce a copy. In addition, the size and shape of the nozzles affect the size and quality of the ink drops. Early inkjet printers often produced images that were fuzzy and smeared, but better ink cartridges have solved this problem.

See Also

RELATED PRINCIPLES: Light, 203 • Mechanical Advantage, 38 • Radiation, 239
RELATED APPLICATIONS: Thermoplastic and 3-D Printing, 295

Ohm's Law

HYDROELECTRICITY

Using water as a source of energy is not new. People have been doing it for centuries. For example, we know that the ancient Chinese were using waterwheels as far back as A.D. 31. You can think of a hydroelectric plant, in fact, as a giant waterwheel. Because gravity plays a key role in getting the water from its source to the turbine, many hydroelectric plants are located in the mountains, near waterfalls, or behind dams specially constructed to hold back water, unleashing its force when needed. When water is freed from the reservoir, it goes through an intake valve and a special channel called a penstock, where it's driven to the turbine. The water rotates the turbine's large blades, which causes the shaft in an electrical generator to spin, producing power.

Hydroelectrical engineers are not just focused on harnessing the power of dammed rivers, waterfalls, and streams. Some are also interested in applying the power of waves. Waves actually contain a great deal of energy. There's the power of the water, of course, but waves are the result of wind hitting water—and that wind has power, too, which is stored in the wave. Engineers in Japan and Norway have successfully created wave-to-energy devices called oscillating water columns (OWCs) to release this energy.

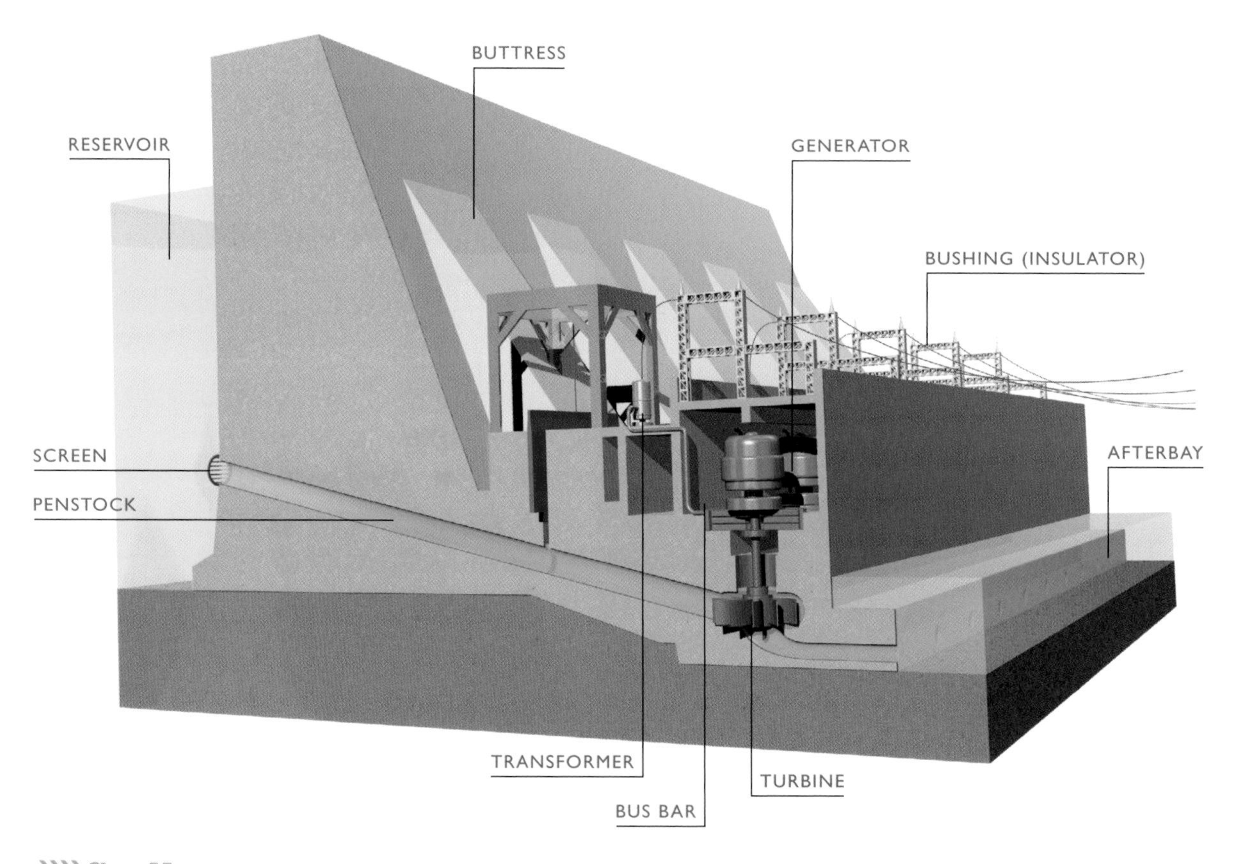

See Also
RELATED PRINCIPLES: Gravity, 121 • Laws of Motion and Energy, 18 • Mechanical Advantage, 38
RELATED APPLICATIONS: Nuclear Reactor, 180 • Photovoltaic Cells, 183

PRINCIPLE

PUTTING EARTH TO WORK

Geothermal Energy

Sources of power are everywhere, even under our feet. Naturally occurring pockets of steam and hot water underneath the ground can be tapped and used to create energy. This hot water and steam are actually the by-products of decaying radioactive elements. Geothermal power is a popular method of heating homes in Iceland, where it's piped in underground through a heat exchanger, which uses the heat captured from the water to heat homes. The used water is injected back down a well into the reservoir to be reheated and used again. Geothermal heat pumps reverse this process in the summer to cool homes, carrying the heat from the house to the ground or water outside, where it's absorbed.

Geothermal Energy

WIND

Air molecules are everywhere and constantly moving around, bumping into each other and into whatever objects happen to be in their path. This bumping creates force—what we know as air pressure. Air pressure is directly related to the number of air molecules present. More molecules equal greater pressure. However, air pressure is not uniform. Local differences in temperature and density produce areas of relatively high or low pressure. When this happens, air molecules quickly move from areas of high pressure to areas of low pressure. Wind is the result. Generally, we feel wind as a gentle breeze. However, when air molecules are forced to move to an area of low pressure very rapidly and in a small space, the wind becomes much stronger.

»»See Also

RELATED PRINCIPLES: Laws of Motion and Energy, 18 • Pressure and Fluids, 56 • Thermodynamics, 96
RELATED APPLICATIONS: Sailboat, 68 • Wind Turbines, 176

Geothermal Energy

WIND TURBINES

We've been using windmills to harness the power of the wind for centuries. A wind turbine, which is just a modern-day windmill, is like a waterwheel. The difference is that wind is the source of energy rather than water. So how does a wind turbine work? It begins with a rotor blade, a propeller-like part. When wind pushes against the rotor blade, the blade starts to spin. A shaft, which is attached to the center of the rotor, also spins as the rotor spins. The wind energy that is captured by the rotor is now transferred to the shaft. The shaft then transfers the energy to a generator. Magnets and conductors inside the generator transform wind energy into electric current, which travels from there along wires and into homes.

Wind farms are not exactly a new concept. In the 1930s, in fact, an estimated 600,000 windmills were in existence in the United States, supplying power to rural areas. Today's more modern versions have computers that can make fine adjustments to the position and angle of the blades, enabling the turbines to function in all wind conditions. There's a significant limitation with this technology, however: wind turbines must be in areas where there's constant wind.

The generator inside of a wind turbine

»»See Also
RELATED PRINCIPLES: Laws of Motion and Energy, 18 • Mechanical Advantage, 38 • Thermodynamics, 96
RELATED APPLICATIONS: Generator, • 157 Sailboat, 68

AIRBORNE WIND TURBINE

The higher up in the atmosphere you go, the stronger the wind. It would be great to capture that wind power, but that would require a turbine with a tower thousands of feet high. Or would it? Scientists and engineers are exploring ways to launch airborne wind turbines into these high atmospheric areas and capture the power there. The designs depict something that looks like an aircraft wing or a giant doughnut tethered to the ground and sent up into the wind. The captured power is transmitted down the tether to a generator. An airborne wind turbine could, in theory, produce three times as much power as a traditional wind turbine.

Largest solar flare on record

PRINCIPLE

CHAIN GANG

Nuclear Fission

In nuclear fission, an atom of uranium-235 is split into two pieces. These pieces each give off two or three neutrons, which in turn strike other uranium-235 atoms, breaking them into pieces that give off neutrons that strike other atoms, and so on, in an endless chain reaction. The mass of the atoms is changing as a result, and that is where the energy comes from. Nuclear fission currently produces one-sixth of the world's electric power, at more than 400 power plants in 30 different countries throughout the world.

PROMISE OF SUSTAINED FUSION

Today's nuclear reactors use fission to create power—atoms of a heavy material, usually uranium, are split apart to create energy. If we could use fusion instead—bring the atoms together to create energy—we would have a nuclear reactor that could yield nearly unlimited power with almost no waste and could run on more common elements. A reactor that uses fusion would give off less radioactive waste. Light elements, like deuterium and tritium, could be used instead of heavy ones ("light" and "heavy" refer to the weight of the atoms). Unfortunately, sustained fusion reactors are still in the experimental stage and are not expected to be in widespread use before 2050.

Nuclear Fission

NUCLEAR REACTOR

Nuclear reactors carefully initiate and control a sustained chain reaction. At the core of a nuclear reactor is the radioactive uranium bundle of rods where the fission takes place. The energy released turns water into steam. The steam causes a turbine to spin, which in turn operates a generator. The magnets and coil inside the generator use the principle of electromagnetism to create current, and lines feed the power to cities, where they power our businesses and homes.

The water in the core of a nuclear reactor actually serves two purposes: it acts as a moderator, controlling the fission process by slowing down the neutrons; it also acts as a coolant, keeping things inside the reactor from getting too hot. Unfortunately, nuclear reactors are not without their drawbacks, as we learned with both the Chernobyl disaster in 1986 and, more recently, with the disaster at the Fukushima-Daiichi power plant in 2011. Although nuclear power plants are a "clean" source of energy, in that they don't emit greenhouse gases, they're not without waste products. They give off excess heat, which dissipates in cooling towers, and they create spent fuel in the form of rods. When this spent fuel is reprocessed to remove and recycle any uranium that remains in it, a liquid residue called high-level radioactive waste (HLW) is left behind, and it must be stored safely.

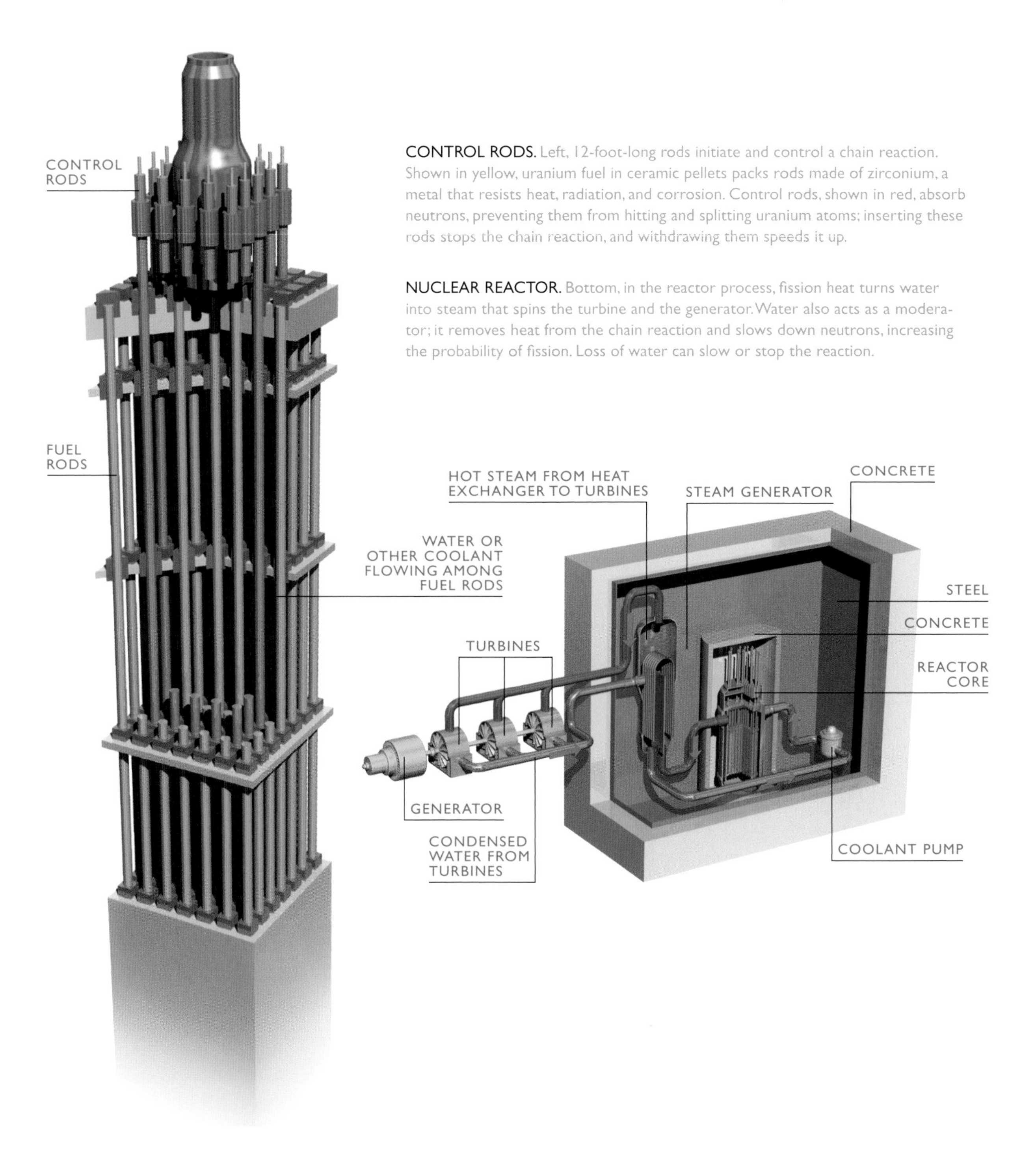

CONTROL RODS. Left, 12-foot-long rods initiate and control a chain reaction. Shown in yellow, uranium fuel in ceramic pellets packs rods made of zirconium, a metal that resists heat, radiation, and corrosion. Control rods, shown in red, absorb neutrons, preventing them from hitting and splitting uranium atoms; inserting these rods stops the chain reaction, and withdrawing them speeds it up.

NUCLEAR REACTOR. Bottom, in the reactor process, fission heat turns water into steam that spins the turbine and the generator. Water also acts as a moderator; it removes heat from the chain reaction and slows down neutrons, increasing the probability of fission. Loss of water can slow or stop the reaction.

»»See Also
RELATED PRINCIPLES: Elements, 261 · Laws of Motion and Energy, 18 · Mechanical Advantage, 38 · Quantum Mechanics, 298 · Radiation, 239
RELATED APPLICATIONS: Generator, 157 · Wind Turbines, 176

PRINCIPLE

CATCHING RAYS

Photovoltaic Effect

How could we not look up at the Sun and think how good it would be to capture its power? That's what's behind the principles of solar energy. One of those principles is the photovoltaic effect, which was first described by French physicist Edmond Becquerel in 1839.

The photovoltaic effect refers to a process in which sunlight is converted to electricity—what we call solar power. The conversion takes place at the atomic level. When a certain type of material is exposed to light, it absorbs photons, or tiny particles of light, and gives off electrons. These electrons are then captured and used to produce mechanical energy.

By the 1960s, the technology was widely used by the NASA space program. Today an estimated quarter million homes in the United States get their power in this way.

Photovoltaic Effect

PHOTOVOLTAIC CELLS

A photovoltaic cell, also called a solar cell, is the device used to harness the power of the Sun and use it as a form of electricity. Crystalline silicon photovoltaic cells are the most common. What makes a photovoltaic cell unique is something called the built-in electric field. This field supplies the voltage that drives the current down the wire and to a lightbulb, for example. Most photovoltaic cells convert visible light, but some can transform infrared light or ultraviolet light. The advantages of photovoltaic cells are enormous: once set up, they can run practically indefinitely, there's no pollution, and they require very little maintenance.

Photovoltaic cell

»»See Also

RELATED PRINCIPLES: Laws of Motion and Energy, 18 • Light, 203 • Quantum Mechanics, 298 • Radiation, 239

RELATED APPLICATIONS: Lithium-Ion Battery, 187 • Spectrum of Wavelengths, 248

Photovoltaic Effect

SOLAR ARRAYS

Photovoltaic cells can be grouped together in what is called a solar array or a photovoltaic array. These arrays can be any shape or size and can be configured to produce any level of output desired by adjusting the wiring. Some are small and power just one family home; others span acres and can power hundreds of homes. In fact, although the technology is quite expensive, solar arrays are especially useful in rural areas where stringing power lines would be difficult. The components of a typical solar array include the photovoltaic cells themselves, the module (the frame that holds the cells), the mounting hardware, a charge controller that regulates voltage, a means of storing the energy, and an inverter that changes direct current to alternating current when necessary.

Most solar arrays remain in a fixed position—generally facing south—but some have tiny motors and can be turned to track the path of the Sun. Before solar cells were widely adopted as a source of energy for homes and businesses, NASA was using them to power its spacecraft. They're still used today to power orbiting satellites and the International Space Station.

See Also

RELATED PRINCIPLES: Laws of Motion and Energy, 18 · Light, 203 · Quantum Mechanics, 298 · Radiation, 239
RELATED APPLICATIONS: Electrical Wiring and Power Grid, · 166 Wind Turbines, 176

SOLAR THERMAL CONVERSION. Another way to generate electricity from sunlight is to focus the light by means of an array of mirrors on a liquid-filled reservoir. The heated liquid runs a generator that produces electricity.

Hydrogen
ULTRA POWER
fuel
cell

Hydrogen
fuel
cell

Hydrogen
fuel
cell

PRINCIPLE

POWERFUL CONVERSION

Electrochemistry

Electrochemistry refers to the process of converting stored chemical energy, through reactions between chemicals, into electricity. At its heart, it involves the exchange of ions, or particles that are electrically charged, either positive or negative. Certain materials attract or repel ions better than others, so the choice of materials in electrochemistry is key. For the process to work, you need more electrons at one end of a circuit than the other; otherwise, electricity will not conduct. The most common, everyday example of electrochemistry in action is the battery. However, hydrogen and even biofuels are being explored as alternatives.

Electrochemistry

LITHIUM-ION BATTERY

Lithium-ion batteries are extremely popular these days—they can be found in a wide range of devices, from toys to cell phones to laptops. They have a lot of advantages over other rechargeable batteries: they're light, have a high energy density because lithium is so highly reactive (ions move at a very high voltage in a lithium battery), and can handle hundreds of charge-discharge cycles. However, they also have their disadvantages: their overall life is short (about two to three years), they're sensitive to high temperatures, they're expensive, and although rare, the pressurized pack can fail and the battery can burst into flame, as the engineers working on the Boeing 787 Dreamliner found out when a lithium-ion battery caught fire on a Japan Airlines plane parked at the gate in Boston. This and other battery flare-ups led the Federal Aviation Administration and other international regulators to ground the fleet. After modifications to insert protection between each of the lithium-ion cells, Dreamliners were airborne again in a matter of months.

»»See Also

RELATED PRINCIPLES: Elements, 261 • Laws of Motion and Energy, 18 • Macromolecular Chemistry, 288 • Quantum Mechanics, 298
RELATED APPLICATIONS: Cell Phones, 243 • Lithium, 265

Electrochemistry

HYDROGEN FUEL CELL

Hydrogen is the simplest of the elements, and the most plentiful, so it was only a matter of time before we began exploring it as a potential fuel source. A hydrogen fuel cell is similar in principle to a battery, except it doesn't need recharging; it recharges itself while operating. As long as the necessary chemicals for the conversion process are flowing into it, the battery will continue to function and produce electricity. In a hydrogen fuel cell, hydrogen is combined with oxygen from the air to produce water and in turn electricity. The only by-products of a hydrogen fuel cell are heat and water. Because hydrogen doesn't need to be extracted from fossil fuels (it can be extracted from water using sunlight), the most obvious application for these fuel cells is in cars, and engineers are exploring designs in which hybrid cars use both a hydrogen fuel cell and a gas engine.

See Also
RELATED PRINCIPLES: Elements, 261 • Laws of Motion and Energy, 18 • Quantum Mechanics, 298
RELATED APPLICATIONS: Electric Car, 160

Electrochemistry

BIOFUEL

Biomass is a catchall term referring to organic matter. When we take this organic matter and convert it into a source of energy, it's called a biofuel. The two most common types of biofuel in use today are ethanol and biodiesel. Both are already present in gasoline in small amounts as an additive that helps reduce vehicle emissions. Ethanol, which is also called grain or pure alcohol, is extracted from sugars and starches that have been fermented. Biodiesel is created by combining alcohol with an animal fat, vegetable oil, or recycled cooking grease. The advantage to biofuels is that they're renewable, unlike traditional fossil fuels, and the plants used in their production essentially offset the carbon dioxide vehicles emit. However, some argue that the process of growing the plants used to produce biofuels is labor- and resource-intensive and outweighs the benefits.

DIVERSE SOURCES. Scientists are creating biofuels from a variety of plants, including corn, switch grass, and sugarcane (left). (Right) A scientist works on corn biofuel in a state-of-the-art fermentation lab.

»»See Also

RELATED PRINCIPLES: Elements, 261 • Laws of Motion and Energy, 18 • Quantum Mechanics, 298
RELATED APPLICATIONS: AC/DC: Edison Versus Tesla, 165

PRINCIPLE

A WORLD OF ZEROS AND ONES

Binary Code

Binary code has revolutionized the way we process and store digital information, and it's at the heart of computing. Unlike our traditional base-10 number system, which relies on decimals, binary code uses a base-2 system. With this system, all values can be expressed as one of two states, represented by ones and zeros. These correspond to matching voltage states, where one equals yes or on and zero equals no or off. The origins of binary code actually go all the way back to the 1670s, when German philosopher and mathematician Gottfried Wilhelm von Leibniz explored a novel way of improving upon Blaise Pascal's "arithmetic machine" so that it could multiply as well as add.

ANALOG TO DIGITAL

Practically all the information our senses process—light, sound, and so on—is analog in nature, which means it's expressed as a continuous signal. To manipulate it digitally—to transmit your voice over a telephone, for example, or to scan an old photograph so you can store it on your computer—it must be converted to a format of ones and zeros. A special circuit is used for this purpose that employs binary code. Analog information is noisy as a result of fluctuations in the signal, and that noise often gets transmitted along with the relevant data. During the conversion process, however, the relevant information is converted into binary code; and once the conversion is made, noise is suppressed.

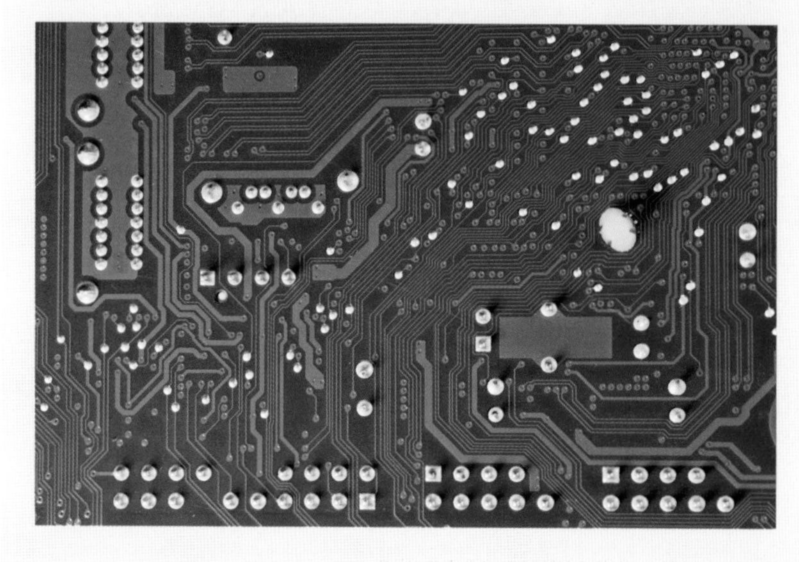

A computer circuit board

Binary Code

INTEGRATED CIRCUIT

Nearly every electronic device used today, from cell phones to dishwashers to television sets, has integrated circuits. Integrated circuits are composed of an electronic circuit, containing a diode (a one-way valve for electricity), one or more transistors (devices that amplify and switch electrical signals), and a resistor (a device that prohibits the flow of electricity). Some (but not most) also have a capacitor (a device that stores an electric charge). These components work together to regulate the flow of electricity through a device, and when they're all made out of the same material, they work even better. Integrated circuits do have disadvantages, though. For example, all connections must remain intact or the device will not work, and speed is definitely a factor. If the components that make up the circuit of the device are too large, or if the wires connecting the components are too long, the device will be slow and ineffective. This problem was solved by constructing the integrated circuit and the chip it sat on out of the same semiconducting material. This meant that wires and components no longer had to be assembled manually, the circuits could be made smaller, and the manufacturing process could be automated. According to Moore's law, the number of transistors on a circuit will double roughly every 18 months, which means that the size of the circuits will continue to shrink. This prediction has held true for decades, and engineers today are managing to construct integrated circuits that are the size of only a few nanometers (1 nanometer = one-billionth of a meter).

BINARY					DECIMAL
8	4	2	1		
0	0	0	0	=	0
0	0	0	1	=	1
0	0	1	0	=	2
0	0	1	1	=	3
0	1	0	0	=	4
0	1	0	1	=	5
0	1	1	0	=	6
0	1	1	1	=	7
1	0	0	0	=	8
1	0	0	1	=	9
1	0	1	0	=	10

1x8 + 0x4 + 1x2 + 0x1

»»See Also
RELATED PRINCIPLES: Binary Code, 190 • Elements, 261 • Health and Medicine, 354 • Macromolecular Chemistry, 288 • Quantum Mechanics, 298 • Thermodynamics, 96
RELATED APPLICATIONS: Cell Phones, 243 • Silicon, 271

Binary Code

DIGITAL STILL CAMERA

Instead of film, a digital camera has a sensor made out of silicon that converts light into electricity. The sensor is made up of a grid of tiny pixels (short for picture elements) that are sensitive to light. There are millions of these pixels in the sensor of a DSLR (digital single lens reflex) camera.

The sensors used by digital cameras are charge-coupled devices (CCDs) or complementary metal oxide semiconductors (CMOSs). Both types convert light into electrons, similar to the way a solar cell works. A CMOS device uses transistors at each pixel and wires to move the electrical charges.

The surface of these sensors contains millions of tiny diodes that capture a single pixel of the subject captured by the lens eye when the shutter opens and closes. More diodes typically mean a higher quality of picture, measured in megapixels.

In a CCD, an analog-to-digital converter turns each pixel's value into a digital value by measuring the amount of charge at each pixel and converting that into binary form. A digital image, then, is basically a string of ones and zeros that represent all the tiny dots or pixels that make up the image.

Special electronic filters make adjustments to color, sharpness, and other aspects, and the size of the picture is reduced. Finally, the image is sent to a temporary storage area and then to a memory card. Once it is digitally stored, it can be manipulated and shared.

»»*See Also*
RELATED PRINCIPLES: Binary Code, • 190 Light, 203 • Quantum Mechanics, 298
RELATED APPLICATIONS: Bar Code Scanner, • 230 Digital Scanner, 223

Binary Code

DIGITAL VIDEO CAMERA

If you think of light as composed of billions of tiny particles, or photons, you can better understand how a digital video camera works. Inside the video camera is a CCD that contains a silicon "film." This film is sensitive to light, and when we focus the camera lens on the scene we are filming, millions of tiny sensors in the CCD read the pattern of light and dark in the scene. Photons striking each tiny pixel on the silicon "film" dislodge electrons, and the pattern of light and dark is stored electronically on a magnetic tape or memory chip.

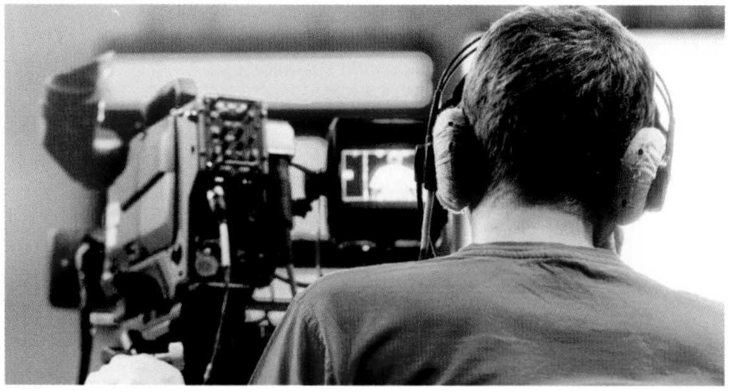

Professional cameraman and HD camcorder in live television

»»*See Also*
RELATED PRINCIPLES: Binary Code, • 190 Light, 203 • Quantum Mechanics, 298
RELATED APPLICATIONS: Bar Code Scanner, 230 • Digital Scanner, 223 • Flexible Oled Displays, 304

Close-up of a digital camera matrix

Binary Code

DIGITAL MUSIC

The ability to convert analog signals into digital output has transformed how we play, store, and listen to music. Actually, the analog sound is converted to a digital binary format and then back into analog so that we can hear it through speakers. A common way to package this music is in the form of compact discs (CDs) and CD-ROMs, the computer equivalent. These discs may seem completely smooth, but they have sequences of microscopic pits and flat regions that contain binary code. In a CD player, a laser beam is focused on the surface of the CD, sort of like how a needle reads the grooves in an old-fashioned phonograph. The beam reads the binary code and reconverts it into sound.

Although the CD was a major breakthrough in terms of music, things really got exciting when we realized the possibilities open to us when we applied computers, the Internet, and MP3 technology to music. MP3 is an audio compression format that shrinks digital files by reducing the number of bytes—and it does this without causing a loss in sound quality. MP3 audio files can be downloaded from the Internet and played using special software, converted into CD files, stored on data CDs, or loaded into portable players.

Not only can we listen to and store music digitally, but we can create it digitally as well. MIDI, short for Musical Instrument Digital Interface, is a technology that has allowed us to create orchestra-quality sound using an electronic keyboard, special hardware, software, controllers, sequencers, power amps, and speakers. A musician can even design and layer sounds, dub and overdub, fill, edit, and play songs, just like an artist works with paint on a canvas.

See Also

RELATED PRINCIPLES: Binary Code, 190 • Quantum Mechanics, 298
RELATED APPLICATIONS: Compact Disc Player, 236

PUBLIC KEY CRYPTOGRAPHY

With all of the private information being transmitted electronically these days, how do we keep it safe? Public key cryptography, or encryption, is the answer. Invented in 1976 by Stanford University professor Martin Hellman and graduate student Whitfield Diffie, public key cryptography lets someone using an unsecured network securely transmit private data, like identity information.

The encryption is based on extremely long strings of numbers, called keys. There are two keys: a private key, which only you can access, and a public key, which can be accessed by anyone. The two keys work together: a message scrambled with the private key can only be unscrambled with the public key, and vice versa. The more numbers in the keys, the more secure the transmission.

TURING MACHINE

If a computer is given enough time and storage space, it can perform any calculation: that is the essence of the Turing machine, named for Alan Turing. This device grew out of the algorithm that is known as the Church-Turing thesis, which is a statement on the nature of computation. It has an infinitely long strip of tape and a head that moves along the tape, reading the symbols on it and changing the values based on a table of rules. The beauty of the Turing machine is that it helps us to understand how the CPU functions inside a computer and what its limitations are. However, there are some problems that not even a Turing machine can solve—such as ones where creativity must be employed to arrive at a solution—and these devices are not necessarily the fastest problem-solvers out there.

Binary Code

PERSONAL COMPUTER

When you think of it, it's amazing how much the introduction of the personal computer (PC) changed our lives. Technology that was once available only to major corporations and government agencies was made available to the average person. In addition, developments in the operating system meant that you didn't have to be a computer scientist with an advanced degree to use it.

The average PC contains a motherboard, processor, central processing unit (CPU), memory, drives, a fan, and cables. Most also have a variety of peripherals attached: the mouse, keyboard, monitor, speakers, printer, scanner, and so on. These hardware components together run software—operating system and any additional programs a user might have, such as a word-processing program.

Today, computers affect nearly every aspect of our lives, and the key to this versatility is the enormous capacity now built into a PC's memory bank, such as in the random-access memory (RAM) that contains programs and the systems that run the computer. RAM provides quick access to data, regardless of where it's stored in the memory. The more RAM, the more efficient and faster the computer. Improvements in the CPU—the central processing unit, or "brain" of the computer—are also key. The most important components of a CPU are the arithmetic logic unit, which calculates operations, and the control unit, which reads and carries out instructions. Nowadays, processors manage around a thousand millions of instructions per second (MIPS).

»»» *See Also*

RELATED PRINCIPLES: Binary Code, 190 • Elements, 261 • Health and Medicine, 354 • Macromolecular Chemistry, 288 • Quantum Mechanics, 298 • Radiation, 239 • Thermodynamics, 96

RELATED APPLICATIONS: Computer Hard Drive, 149

Binary Code

ATM

The digital world has revolutionized how we manage our money, make purchases, apply for a loan, and more—all without having to stand in a teller's line. Consider the automated teller machine (ATM). The magstripe on the back of your ATM card holds information in the form of letters and numbers, stored in binary code. When the card is inserted into an ATM, the data and transaction request are sent electronically to a computer at the financial institution that operates the ATM. There, your account number, balance, and personal identification number (PIN) are verified. If everything checks out, your requested transaction is carried out. ATM machines are connected in large interbank networks that rely heavily on a cipher called Triple Data Encryption Standard (DES) that makes use of keys to transmit information in a highly encrypted state that is resistant to cracking techniques. Triple DES is slow in terms of processing speed, but the security it offers makes this a fair trade-off.

Some financial institutions are starting to replace ATM cards with so-called smart cards—mini-processors that contain electrically charged silicon chips that store thousands of bytes of information, much more than an ATM card currently carries.

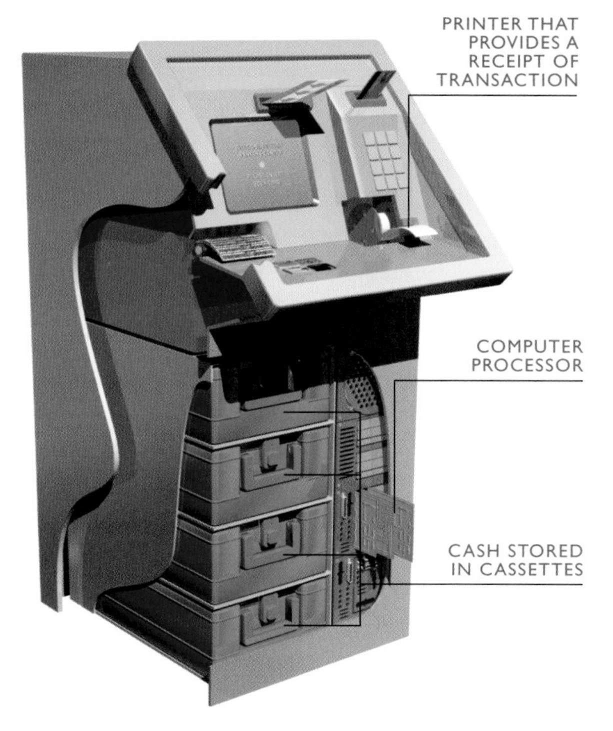

AUTOMATED TELLER. The automated teller machine (ATM)—perhaps the most visible symbol of electronic banking—provides fast cash. The machines not only dispense bills but also take deposits, allow transfers of funds, and provide statements. To activate an ATM, the user inserts a plastic card with a data-rich magnetic strip into a slot and then punches a personal identification number (PIN) on a keypad. A computer verifies the identification number and cash balance, making note of the transaction requested. When customers withdraw cash, electronically counted money comes out of built-in strongboxes, or cassettes.

See Also

RELATED PRINCIPLES: Binary Code, 190 • Friction, 52 • Magnetic Fields, 135 • Mechanical Advantage, 38
RELATED APPLICATIONS: Credit Card, 141

Binary Code

INTERNET

It's no exaggeration to say that the Internet has changed our lives in ways we could never have imagined. What began as a collaboration between universities, government, and industry in the late 1960s has evolved into a vast information infrastructure. Today, on the Internet, we can work from home, meet potential mates, buy groceries, conduct business meetings, and more. The Internet is based on a few key concepts. The first is packet switching, which involves routing data in specially formatted units from source to destination using network switches and routers. Open architecture networking is what enables different providers to use any individual network technology they want, and the networks still work together. Transmission Control Protocol/Internet Protocol (TCP/IP) is what makes open architecture networking possible. TCP assembles a message or file into smaller packets that are transmitted over the Internet, and IP reads the address part of each packet so that it gets to the right destination.

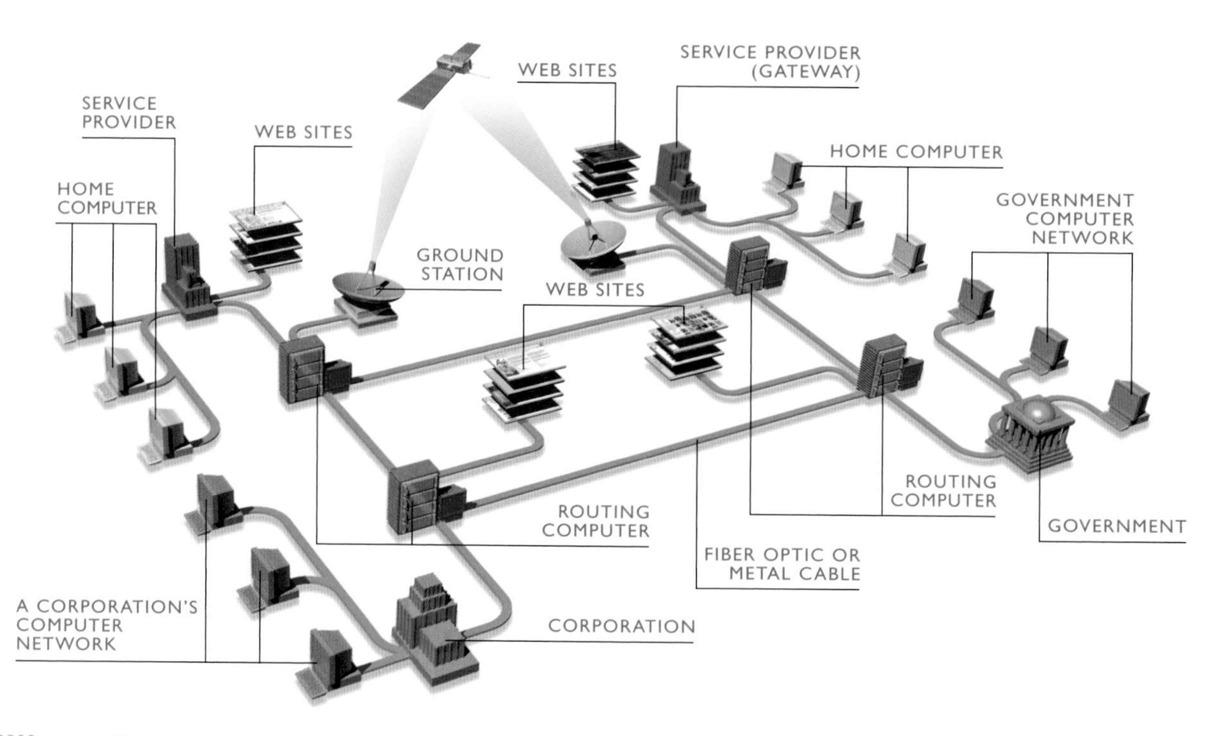

»»*See Also*

RELATED PRINCIPLES: Binary Code, 190 • Elements, 261 • Quantum Mechanics, 298 • Radiation, 239
RELATED APPLICATIONS: Wi-Fi, 246

Binary Code

WORLD WIDE WEB

The World Wide Web and the Internet aren't the same thing. You can have the Internet without the Web, but you can't have the Web without the Internet. In 1990 Tim Berners-Lee invented the Web, creating the first web browser and web pages, which people could access via the Internet. Web pages incorporate hypertext—text displayed on a monitor that contains hyperlinks to other documents—and they must have a specific "domain" name and the "dot parts" that follow it, such as .com, .org, .edu, and .net.

The Internet is a massive network that itself contains thousands upon thousands of other networks. The Web is one service that runs on the Internet. (E-mail, phone apps, and chat programs are some others.) Data on the Internet is broken down electronically into "packets," which are sent out simultaneously, over a variety of routes, from server to server, moving through exchange or access points. But to accommodate a request for information, servers must send the electronic information to the right "computer client"; once the packets get there, all the data contained within has to be collected and reassembled. All of this is the job of the communications "protocols," which are sets of rules and regulations established for the servers. The most important of these are the TCP, which collects and reassembles the fragments of data, and the IP, which handles the proper routing.

»»See Also

RELATED PRINCIPLES: Binary Code, 190 • Elements, 261 • Quantum Mechanics, 298 • Radiation, 239

RELATED APPLICATIONS: Fiber Optic Cable, 229 • Wi-Fi, 246

Binary Code

E-MAIL

E-mail was first invented in 1971 by Ray Tomlinson, a computer engineer at a Boston firm who was charged with trying to come up with uses for ARPANET, an early computer network. At first, only universities and research organizations used e-mail, but its adoption quickly grew. At the heart of e-mail is Standard Mail Transfer Protocol (SMTP). This routing program is what ensures a message gets from sender to receiver. SMTP employs an elaborate letter-and-number code that tells each computer station what it must do with a message so that it can reach its destination. On the other end is something called a Post Office Protocol (POP3) or Internet Mail Access Protocol (IMAP) server that handles the incoming mail. When you send an e-mail, your mail client contacts the SMTP server using a designated port and communicates the sending and receiving addresses as well as the body of the message through binary code. The SMTP server parses out the recipient name and the domain name. Then a Domain Name Server (DNS) gives the SMTP server the IP address for the destination domain. The SMTP server connects with the destination domain, and the message is delivered.

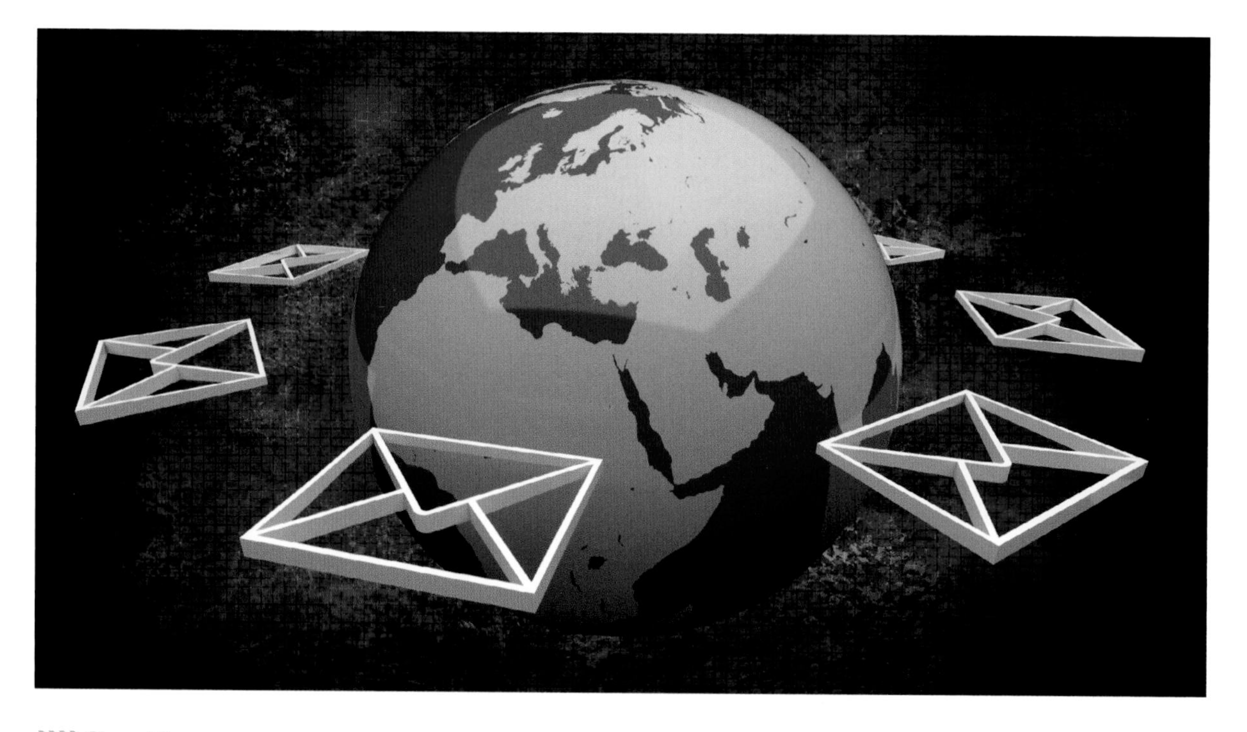

»»**See Also**
RELATED PRINCIPLES: Binary Code, 190 · Elements, 261 · Quantum Mechanics, 298 · Radiation, 239
RELATED APPLICATIONS: Computer Hard Drive, 149 · Credit Card, 141

AUGMENTED REALITY

Virtual reality involves a computer-generated environment. Augmented reality, on the other hand, involves enhancing the real world by enhancing our senses. One of the earliest examples of this technology at work is the yellow arrows announcers use on televised football games when analyzing a play. It has been applied recently to all sorts of consumer electronics, from cell phones to personal digital assistants. For example, augmented-reality apps running on your smartphone can superimpose useful information about whatever its camera is seeing. One app translates foreign-language signs into English, right there on your screen's video image; another shows you what subway lines are beneath your feet.

AUGMENTED REALITY HOTEL. The Holiday Inn London Kensington Forum became the world's first augmented reality hotel in 2012 for the London Olympic Games. The augmented reality experience allowed guests to interact with star athletes via their tablet or smartphone.

Binary Code

CLOUD COMPUTING

Recent advances in computing technology are allowing us to use the Internet to store data, resources, services, and applications. The buzzword for this idea is cloud computing, which basically means accessing computing resources via a remote data center.

To understand the "cloud," think of it in terms of layers. The front-end layers are what you see and interact with—your Gmail account, for example, and the computer software that connects you to the cloud. The back-end layers are the hardware and the software that run the front-end layers, with servers and data storage systems. A central server oversees the entire system, allocating resources and shifting traffic and client needs as necessary. It does this through a set of rules called protocols and special software called middleware. Middleware enables networked computers to communicate, and something called server virtualization helps maximize the output of servers by allowing servers to share hardware, which means fewer machines can do more work.

»»*See Also*
RELATED PRINCIPLES: Binary Code, 190 • Elements, 261 • Quantum Mechanics, 298 • Radiation, 239
RELATED APPLICATIONS: Computer Hard Drive, 149 • Wi-Fi, 246

CHAPTER 9
Light

Albert Einstein won the Nobel Prize in physics in 1921 for introducing the idea that light can be thought of as composed of tiny particles of energy, or photons, rather than as one continuous wave. This idea revolutionized the way scientists thought of light, energy, radiation, and how they interact.

There's the light we can see—the visible spectrum—and the light we can't—ultraviolet, infrared, and so on. Also, light behaves in certain ways, and we can manipulate and use this behavior to our advantage.

For example, we can use it to make our clothes seem brighter and whiter by fooling our eyes and making use of how we perceive color. By reflecting light, we can see objects in space; reflect it even more precisely, and we have the laser, which has been applied to everything from surgery to art to computer technology.

But the uses for light don't stop there. It has been employed in scanners, from digital and bar code scanners to retinal and iris scanners. Light-sensitive diodes are used in security and detection systems in our homes. Thermography helps us predict—and prepare for—the intensity of storms. Night vision goggles help us see clearly in the dark.

Light and vision go hand in hand. How we see is a trick the brain plays on the eyes—or maybe vice versa—and as we age vision naturally declines. However, light can help improve function in the form of eyeglasses, contacts, and even laser surgery.

PRINCIPLE

A COLORFUL EXPLANATION

Absorption of Light

So why, exactly, are we able to see blue skies, pink sunsets, green grass, and all the other colors? When light hits an object, some electromagnetic radiation is absorbed, and some is reflected. Atoms can absorb only those waves whose energy content corresponds with possible energy states those atoms can have, reflecting the rest. Unless an object emits its own light, its color is a function of those rays it discards: A blue jacket, for example, is that color because the atoms of chemical dye in the cloth absorb almost every wavelength except blue. Strawberries appear red because they absorb blue and green wavelengths. Leaves are green because the pigments involved in photosynthesis happen to be most receptive to red and blue light.

Every wondered why the sky is blue? All wavelengths of sunlight get scattered by molecules and dust in the air—but not equally. Blue light with its shorter wavelength scatters more than red or orange, which have longer wavelengths. When you look at the sky and not the sun, you see indirect, scattered light—making the sky appear blue.

HOW WE PERCEIVE COLOR

The human eye has about three million color-perceiving cones, which come in three different types, each favoring one of three ranges of wavelengths: red, blue, or green. We perceive color when the different wavelengths are absorbed and reflected, and each element has a slightly different set of wavelengths it can absorb. The millions of colors the human eye is capable of perceiving are made up of varying combinations of red, green, and blue. Things that appear largely transparent to our eyes, such as air and water, don't have the right energy levels to absorb visible light of any color.

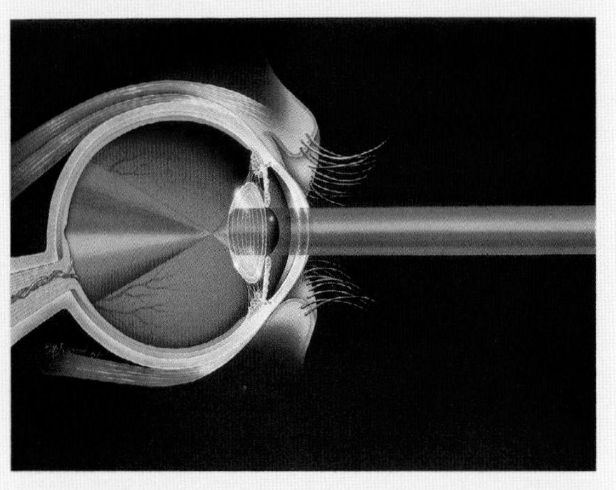

Absorption of Light

VISIBLE SPECTRUM

The light we see is actually the result of interactions between electric and magnetic fields that together form electromagnetic waves. Visible light, however, is only a tiny piece of the electromagnetic spectrum. The spectrum is infinite, ranging from low-frequency radio waves to high-frequency x-rays and gamma rays. The higher the frequency (and the shorter the wavelength), the more energy each wave carries. One of the most amazing scientific breakthroughs is our ability to now "see" in other wavelengths than visible light, which ranges from 400 (very blue) to 750 (deep red) nanometers, or billionths of meters. Astronomy, medical research, military technology, and sciences of nearly every sort now routinely use numerous parts of the electromagnetic spectrum to gain knowledge.

See Also

RELATED PRINCIPLES: Electromagnetic Radiation, 248 • Photovoltaic Effect, 183 • Quantum Mechanics, 298

RELATED APPLICATIONS: Digital Still Camera, 192 • Incandescent Light Bulb, 249 • Spectrum of Wavelengths, 248 • Sunburn and Sunblock, 217 • UV Light and Eyes, 215

Absorption of Light

COLOR-BOOSTING DETERGENT

Can a detergent actually make the colors of your clothes brighter, or is it just a marketing ploy? Actually, there's some truth to the claim. These detergents use optical brighteners, also called optical enhancers. When you pour this product into your washing machine, the detergent molecules do the actual "dirty work," removing stains and lifting dirt from your clothes. They have no effect on a garment's color. (Although some clothes do fade after repeated washings, that is related more to the dye and the age of the garment than to the detergent.) A color-boosting detergent contains additional molecules that create a special chemical reaction on the clothes. They leave behind a microscopic film of fluorescent particles on the garment. When light hits these particles, our eyes perceive the colors as brighter and more vivid than they truly are. You can tell the difference between clothes washed in a color-boosting detergent and clothes that are not by feel: the film left behind on clothes make them feel stiffer compared to those not washed with a color-boosting detergent. There's no difference in cleanliness.

Some people believe that whiter and brighter clothes come at a cost to the environment. Color-boosting detergents can be harmful to animal and plant life, and they're not easily biodegradable, so the hazards they present to the environment can remain in place for quite some time.

DETERGENT MOLECULE

A detergent molecule has a nonpolar and a polar end. The nonpolar end attaches to oily dirt, while the polar end interacts with the surrounding water, ionizing to form positive charges. These charges repel the ones on other soap-encrusted dirt particles, keeping the dirt suspended and allowing it to be rinsed away.

See Also

RELATED PRINCIPLES: Macromolecular Chemistry, 288 · Quantum Mechanics, 298

RELATED APPLICATION: How We See, 227

Absorption of Light

3-D MOVIES AND TV

The 3-D format of many a Hollywood movie these days might seem like a new technology, but it was actually introduced in 1922. It didn't really take off, however, until the 1950s. So how do 3-D film and television work? When we focus on an object, the brain processes how far the light reflected off it had to travel to get to our eyes. With faraway objects, the light travels on a parallel path to both eyes. The closer an object gets, however, the more the light converges instead of being parallel. The closer an object is, the more your eyes have to converge to keep the object in focus. In a sense, each eye is seeing a slightly different "picture" of the object.

Three-dimensional technology fools your brain into thinking the object on screen is closer than it really is by causing your eyes to converge, thanks to the funny glasses you have to wear. They come in two varieties—the ones with red and blue lenses and the more recent polarized lenses—but both work in essentially the same way. If you look at a 3-D movie without the glasses, you see two different images, one slightly off-center from the other. Put the glasses back on, and light is filtered and colors canceled out in such a way that the brain perceives just the one image and the illusion of depth.

SEEING DOUBLE. A cameraman operates a 3-D television camera at a soccer match. The two lenses on the camera record the images slightly off-center from one another. When a viewer watches the program through special glasses, they have the illusion of depth.

»»See Also
RELATED PRINCIPLES: Power and Electricity, 163
RELATED APPLICATIONS: Color Television, 144 • How We See, 227

Absorption of Light

REFLECTING TELESCOPES

Telescopes, regardless of size, work by selecting a portion of the light rays from a distant object and then reflecting and magnifying them with a curved mirror so that they make a disproportionately large image (refracting telescopes, by way of contrast, use convex lenses). The more the rays are bent, the larger the magnified image. For example, when your unaided eye looks at a small tree 100 yards away, it might make an image only a couple of degrees wide because the rays of light coming from opposite ends of the tree are so close together. But hold up a 40-power telescope, and suddenly the tree fills your field of vision, projected as an arc 70 or 80 degrees wide on the back of your eye. Magnifying power is determined by the ratio of those angles of vision, aided to unaided.

In theory, the larger the lens, the larger the image. But there are limits. A glass lens big enough to capture the sparse rays from distant stars would collapse under its own weight. One solution: mirrors. Like lenses, mirrors shaped to specific curvatures can dramatically magnify images. They also possess a structural advantage over lenses: because they only need to reflect light rays, not transmit them, they can make use of steel and other supports. The world's largest optical telescopes are all the reflecting type. Mexico's Large Millimeter Telescope, for example, boasts an active surface area 50 meters (approximately 164 feet) in diameter.

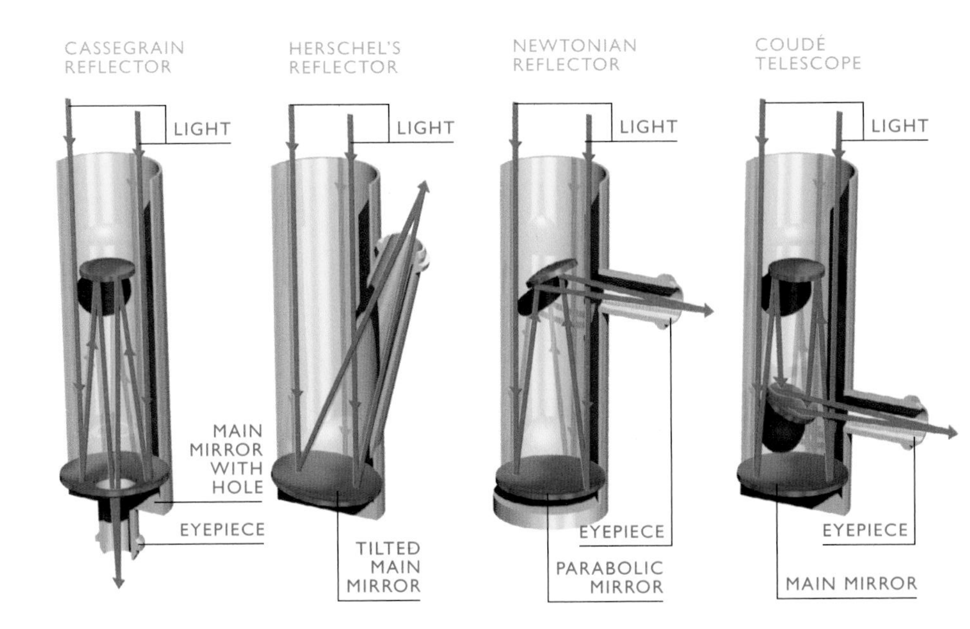

REFLECTING TELESCOPES. These telescopes use curved mirrors to magnify distant objects. In the Cassegrain reflector, a 17th-century creation, light passes through a hole in the main mirror and focuses the image on a secondary mirror. To view the image, one peers through the eyepiece. The 1773 Herschel reflector used a tilted main mirror with an eyepiece looking into it. Isaac Newton's model, only six inches long, had two mirrors; Coudé's versions use three mirrors to focus images to the side.

»»*See Also*

RELATED PRINCIPLES: Electromagnetic Radiation, 248 • Elements, 261
RELATED APPLICATIONS: How We See, 227 • Silver, 269

Absorption of Light

SPACE TELESCOPES

Earth-bound telescopes aren't so useful when it comes to exploring the night sky because Earth's atmosphere absorbs and distorts incoming light, or electromagnetic radiation, and because there are so many sources of human-made light around us that make even a pitch-dark night not so dark. Orbiting observatories have helped us overcome these limitations. By launching telescopes into space, we can see into radiation wavelengths difficult to image with Earth-based instruments, and we can examine high-energy processes in faraway galaxies. The Hubble is probably the most famous of these telescopes. Its replacement, the new James Webb Space Telescope (named after NASA's second administrator), is scheduled for launch aboard an Ariane 5 rocket sometime this decade for a three-month journey into orbit. It will have a larger mirror to give it more light-gathering power, will operate much farther from Earth—a million miles away—and will see deeper into space. The Webb mission is expected to last between five-and-a-half and ten years, and the scientists at NASA hope the telescope will expand upon our current knowledge of the universe and provide us with previously unheard of data.

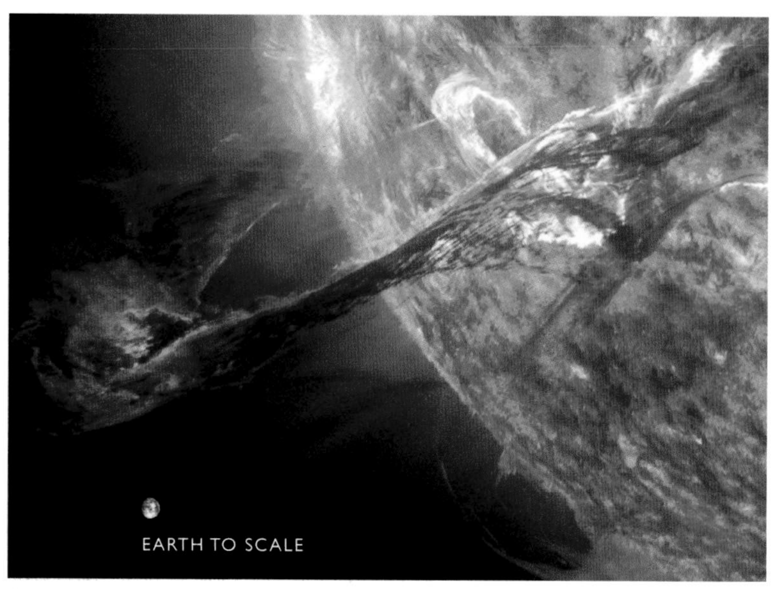

Solar flare

»»See Also

RELATED PRINCIPLES: Electromagnetic Radiation, 248 • Elements, 261 • Laws of Motion and Energy, 18 • Power and Electricity, 163 • Quantum Mechanics, 298 • Radiation, 239

RELATED APPLICATIONS: Silver, 269 • Spectrum of Wavelengths, 248

HUBBLE TELESCOPE

The Hubble space telescope gathers light at its open end from a large, concave primary mirror and reflects it off another mirror into an array of sensors. (The mirror was actually ground to the wrong specifications and had to be fitted with a "contact lens"—an array of several small mirrors that corrected the distortion and repaired any image defects. Thank goodness for the Space Shuttle, which delivered the corrective lens into space.) Sensitive instruments detect x-rays, infrared light, and ultraviolet light, revealing the makeup of far-off celestial structures and systems. Hubble even discovered what may be a "construction site for galaxies," a vast birthplace of stars, 12 billion years in the past, that had been invisible to astronomers.

Hubble telescope

Finger prints on a glass revealed under ultraviolet light

PRINCIPLE

UNSEEN LIGHT

Ultraviolet Light

The frequencies of light occupy a very wide spectrum, and we can see only part of it—the part called visible light. However, there are electromagnetic waves that we can't see, and ultraviolet (UV) is one example. Discovered in 1801, it sits past the violet end of the spectrum and has shorter wavelengths than visible light. Although our human eyes can't detect it, the eyes of some insects can, such as bumblebees. Ultraviolet light comes from the Sun and is classified according to its intensity, with A being the least harmful and C being the most harmful.

Ultraviolet Light

UV LIGHT AND EYES

Just because the human eye can't see ultraviolet light doesn't mean that UV isn't dangerous. In particular, UV-A and UV-B rays pose the greatest threat because these are the ones that tend to get through Earth's protective ozone layer. UV light damages DNA in cells by causing a reaction among molecules of one of the bases in DNA (thymine). Prolonged exposure to the eyes can lead to all sorts of problems, ranging from cataracts to macular degeneration. So does this mean you should spend your time indoors and only come out at night? Hardly. Easy ways to protect your eyes from UV light include wearing sunglasses and broad-brimmed hats when outdoors and seeing an eye doctor on a regular basis to detect problems early.

»»» *See Also*
RELATED PRINCIPLES: Health and Medicine, 354 • Radiation, 239
RELATED APPLICATIONS: Spectrum of Wavelengths, 248 • UV-Blocking Fabric, 308

Ultraviolet Light

UV LITHOGRAPHY

In an effort to make microchips in processors ever smaller and ever more powerful, researchers and engineers are studying a technology called UV lithography or extreme UV technology. A joint venture between the U.S. Department of Energy, Motorola, Advanced Micro Devices (AMD), and Intel is exploring this concept, which, if successful, could result in microprocessors that are 100 times more powerful and capable of storing 1,000 times more information than they're currently capable of now. The current technology used to etch the circuits on microprocessors involves what is called optical lithography, in which an intricate array of lenses refracts beams of light in a precise fashion onto a silicon wafer. Ultraviolet light, however, has different properties than visible light, such as wavelength. This means that to get the level of precision desired, mirrors rather than lenses need to be used to reflect an extremely intense beam of UV light onto the silicon wafer.

Right now, only prototypes of UV lithography machines exist, and there are problems that still need to be overcome. For example, developers are still trying to figure out how to get the exact level of precision required with mirrors, as well as how to compensate for—if not entirely remove—the amount of chemical contamination that is currently inherent in the UV lithography process when UV photons react with gases, leaving behind carbon-like deposits.

Computer circuitry etched on a wafer by means of prototype equipment using extreme-ultraviolet lithography

»»*See Also*

RELATED PRINCIPLES: Macromolecular Chemistry, 288 • Quantum Mechanics, 298 • Radiation, 239
RELATED APPLICATIONS: Integrated Circuit, 191 • Silicon, 271

Ultraviolet Light

SUNBURN AND SUNBLOCK

We use sunscreen because it gives us a thin layer of protective chemicals that soak up ultraviolet radiation before it hits our skin. Earth has its own sunscreen in the form of the ozone layer. Created in the stratosphere about 10 to 20 miles up, ozone absorbs much of the UV radiation that would otherwise reach the planet's surface. UV light is bad news for living things, since its high-energy waves can destroy or severely mutate DNA, the genetic material essential to all cellular reproduction. In fact, a sunburn is painful proof of the damage that UV light can cause. Because the genetic damage caused by UV radiation is additive, sun damage to babies and children is particularly concerning because it can lead to skin cancers in later years.

>>>> **See Also**

RELATED PRINCIPLES: Health and Medicine, 354 · Macromolecular Chemistry, 288 · Quantum Mechanics, 298 · Radiation, 239

RELATED APPLICATIONS: Spectrum of Wavelengths, 248

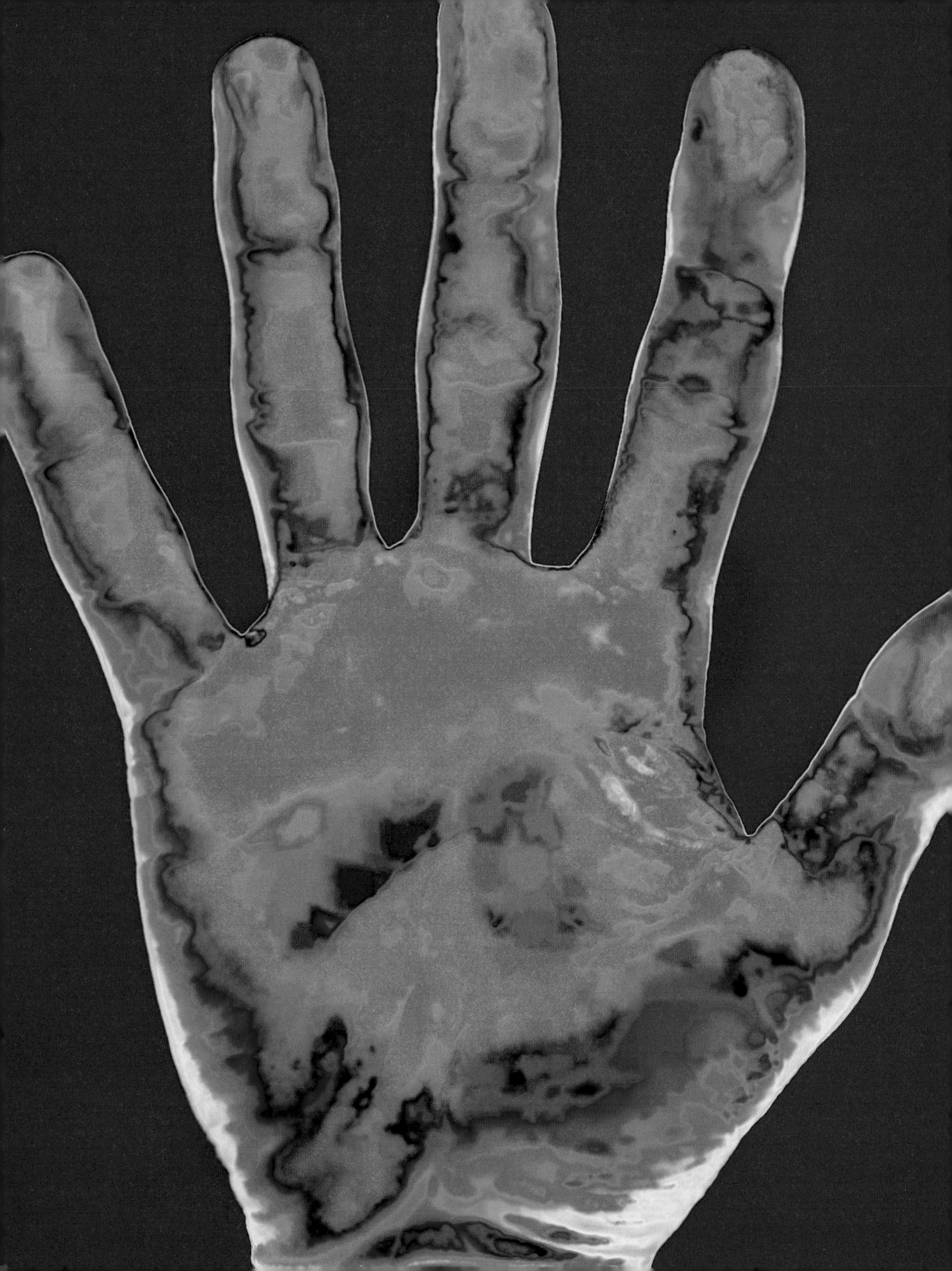

PRINCIPLE

USING HEAT TO SEE

Thermal Imaging

Thermal imaging allows us to see electromagnetic radiation—in particular, infrared light. A special thermal camera picks up electromagnetic radiation within a certain wavelength range (0.0 to 14 micrometers). To make the contrast more obvious, the viewscreen of the camera employs colors to delineate areas of intensity, with white typically being equated to the hottest areas and black the coolest. Thermal imaging has been used for a wide range of applications: the military employs it in night vision goggles, astronomers use it to gather information about stars, and biologists use it to study how certain animals give off heat.

Thermal Imaging

WEATHER MAPS AND THERMOGRAPHY

Objects emit infrared energy proportional to their temperature. That is, they give off invisible light of specific wavelengths that depend on their temperature. Thermal imaging devices use an array of detectors to scan the specially focused light from a lens to obtain temperature information about a visual field: the hotter an object, the more infrared energy it gives off. The thermal information feeds to a signal processor. The longest thermal infrared wavelength is less than one-eight-hundredth of an inch. Traditional weather maps make use of electromagnetic radiation at much longer lengths. Weather radar stations send out electromagnetic waves of a few inches length, about the same as microwaves. The waves either bounce back or bend—that is, reflect or refract—in predictable ways. The reflected signal, captured by a receiver, shows the shape of the storm, the pattern of clouds, or whatever is out there.

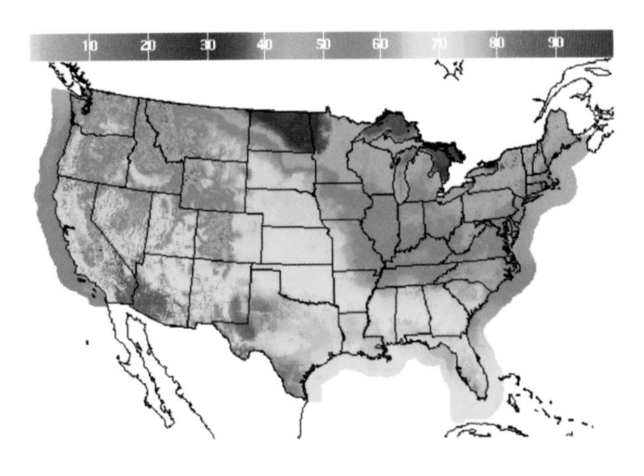

»»See Also
RELATED PRINCIPLES: Laws of Motion and Energy, 18 • Power and Electricity, 163 • Radiation, 239
RELATED APPLICATIONS: Hurricane, 109 • Radar, 251

Thermal Imaging

NIGHT VISION GOGGLES

Humans and other mammals have rods in the retina of their eye—millions of tiny cylindrical elements that contain rhodopsin, a purple pigment that can detect dim light. We also have millions of retinal cones—light-sensitive cells that enable us to read fine print. Nocturnal creatures, however, such as cats, lemurs, and owls, only have rods, which makes them capable of seeing far better than humans in low light.

Since we don't have the eyesight of creatures of the night, technology must come to our aid. Night vision binoculars and scopes are electro-optical instruments that are incredibly sensitive to a broad range of light, from visible through infrared. Light that enters a lens in a night vision scope (this light could also be light from stars or ambient light, which is general light in the environment) reflects off an image intensifier to a photocathode and is converted to an electronic image. Amplified on a viewing screen, the image reveals much more than a night scene observed through a conventional scope. Like cameras, these devices have various magnifications. They're especially valuable, for example, during night operations involving close air support of ground troops.

ELECTRO-OPTICAL NIGHT VISION. This process works by intensifying and increasing available light. As faint light enters a photocathode, its photons accelerate through a vacuum. Next, a microchannel plate multiplies the photons thousands of times. The device converts them to an electronic image and focuses it on a phosphor screen for viewing. Amplified green light then enhances the image's visibility.

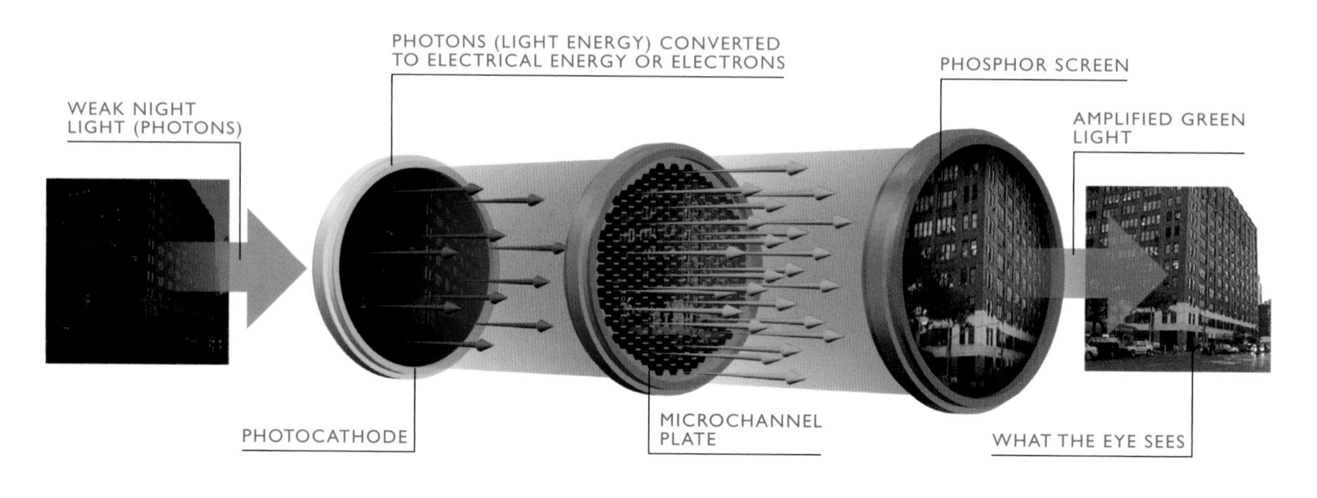

See Also
RELATED PRINCIPLES: Power and Electricity, 163 • Radiation, 239
RELATED APPLICATIONS: UV Light and Eyes, 215

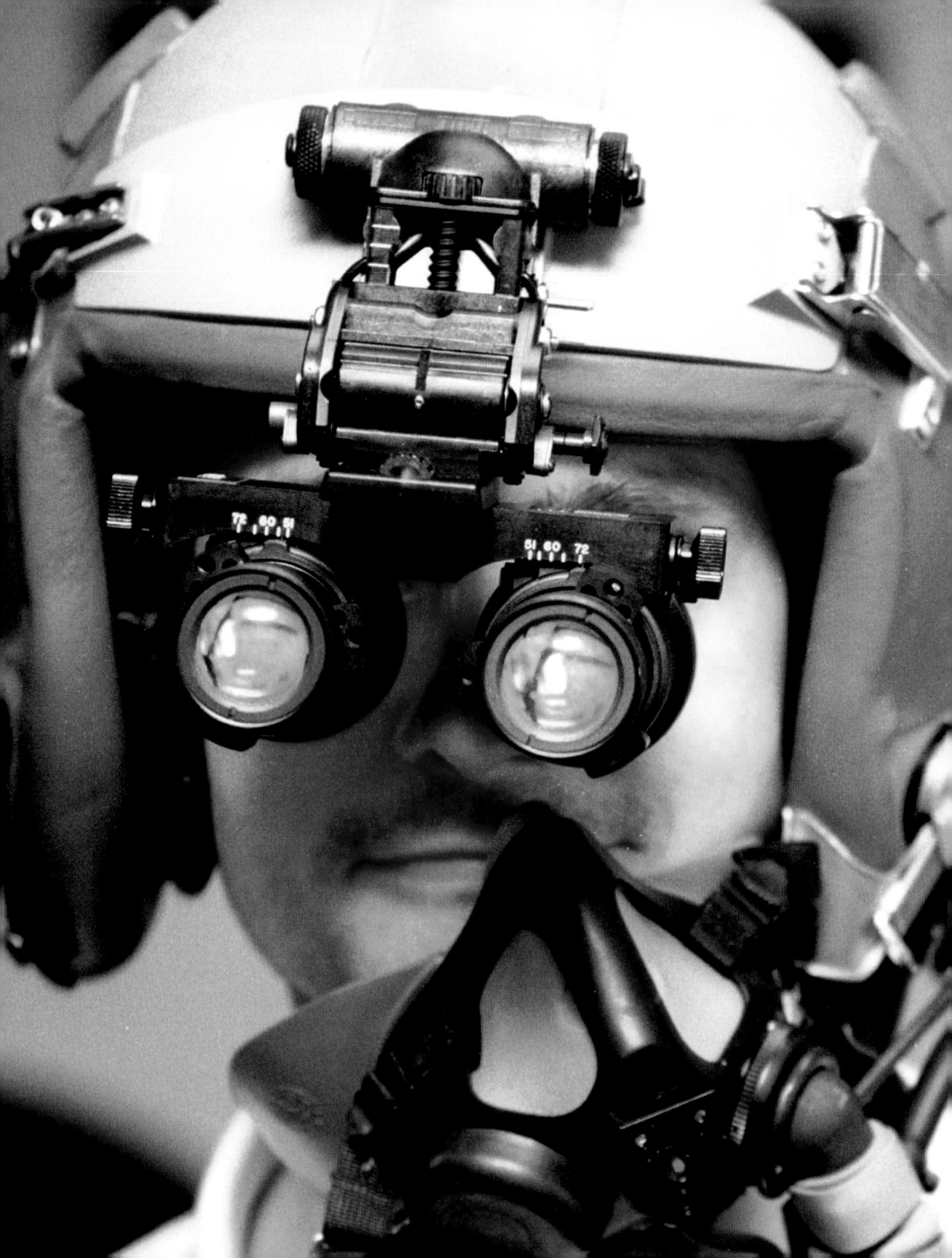
72 60 51
51 60 72

PRINCIPLE

BEAM THERE, DONE THAT

Photoelectric Effect

When certain metals are exposed to electromagnetic radiation above a particular frequency, such as visible or near-ultraviolet light, they will absorb the light and emit enough electrons to create a measurable current. This is known as the photoelectric effect. When coupled to switches, the current can be used in photoelectric devices to open doors automatically, for example, or trigger alarms. A transmitter beam is aimed at a receiver some distance away; as long as light hits the receiver and throws off enough electrons to provide a current, nothing happens. But if something opaque interrupts the beam, the current disappears, which trips a switch and starts the action.

Photoelectric Effect

LIGHT-SENSITIVE DIODES

A smoke detector uses photoelectrics in its light-sensitive diodes. Popularly referred to as an electric eye, this special cell has electrical properties that vary depending on how light falls on it. In a smoke detector, a steady light shines on a dark surface inside; if smoke enters the chamber, it scatters particles of light to a photocell that triggers the alarm. Other security systems use similar photocells to measure light levels so that they can turn lights on and off.

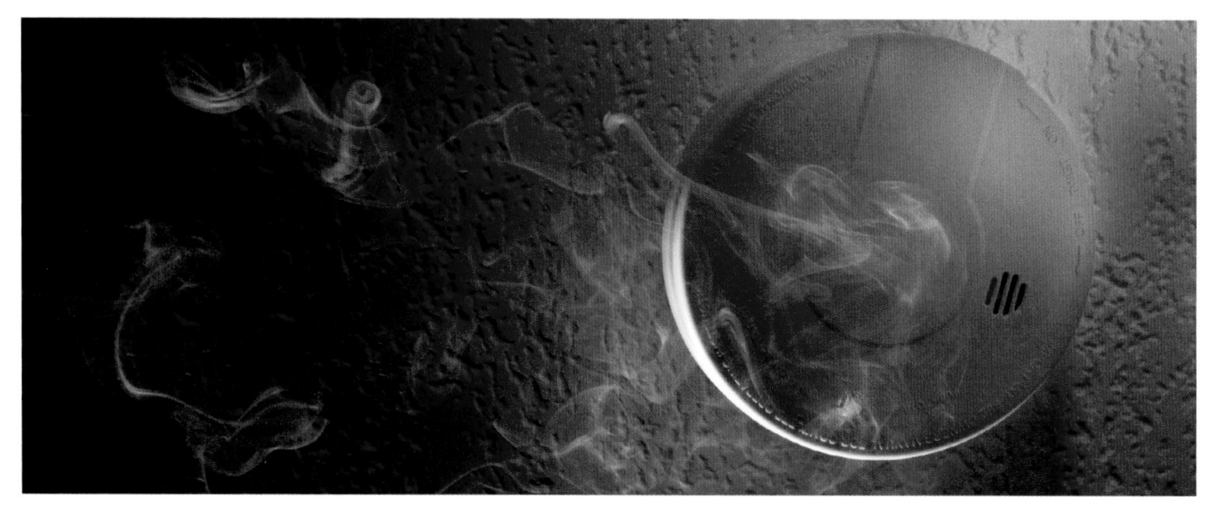

›››› *See Also*

RELATED PRINCIPLES: Power and Electricity, 163 • Radiation, 239

RELATED APPLICATIONS: Integrated Circuit, 191 • Photocopier, 168 • Spectrum of Wavelengths, 248

Photoelectric Effect

DIGITAL SCANNER

A desktop scanner, such as a flatbed model, captures images and text and saves them in a computer. Once stored there, they can be altered, printed, or sent out over the Internet. When documents or pictures are placed under a cover on the scanner's glass plate—much like documents are placed on a copy machine—light from a lamp illuminates them. A motorized system of mirrors, sensors, and an array of extremely light-sensitive diodes that convert light into electrical charge moves over the objects. When the light hitting the scanned object is reflected back, the scanner analyzes it and reconstructs it electronically. Computer software then transfers the scanned material to the computer in a format the computer understands. Digital scanning technology when combined with Optical Character Recognition (OCR) enables text on a page to be converted to an editable document.

»»See Also
RELATED PRINCIPLES: Mechanical Advantage, 38 • Power and Electricity, 163 • Radiation, 239
RELATED APPLICATIONS: Credit Card, 141

Photoelectric Effect

IRIS AND RETINAL SCANS

As any James Bond fan knows, retinal and iris scans are part of high-end security systems that protect access to highly classified or sensitive areas or information. When this system scans your eyeball, a low-level light source shines on it, and a sensor creates a high-resolution image. Since the pattern of blood vessels in the eye and the texture of the iris are unique to each individual, these characteristics can be analyzed and then converted into a mathematical algorithm. Then, when you approach the scanning machine and stare at the designated spot for 10 to 15 seconds, your patterns are compared to the database; if there's a match, access is granted. The ever-increasing processing power of computers has enabled faster recognition and more widespread usage.

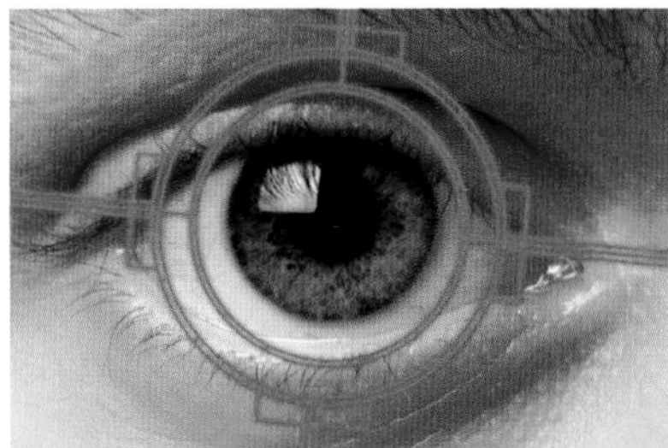

»»See Also
RELATED PRINCIPLES: Power and Electricity, 163 • Radiation, 239
RELATED APPLICATIONS: Flexible OLED Displays, 304 • X-Ray Imaging, 254

BROADWAY

PRINCIPLE

THE STUDY OF LIGHT

Optics

Optics is the science of light and vision. This branch of physics encompasses all kinds of electromagnetic radiation: visible, invisible, infrared, and ultraviolet. Optics is devoted to understanding light—What is it made of? How does it react with various substances and types of matter? And how can we use this understanding of the behavior of light to meet our needs? The uses for optics are widespread, including lasers, scanners, computers, cameras, DVDs, and remote control devices.

Optics

AGE AND VISION

Like it or not, as we age, our vision changes. The muscles in the eye lose their flexibility, making it harder to see things close up or to accurately judge distances. At the back of the retina, almost exactly opposite the pupil, lies a small spot called the macula. At its center is a tiny depression known as the fovea. Composed of densely packed cones but no rods, the fovea is the seat of greatest visual acuity, or sharpness. But with age, the macula often breaks down—what's known as macular degeneration. Blood vessels behind the retina can leak, causing loss of central vision. Sometimes, laser procedures can seal off the leakage; more often, though, nothing can be done, and the condition must be monitored and controlled with medication.

As we age, the eye's lens naturally loses elasticity. Eventually, it can't flex sufficiently to focus on nearby objects. This is why people starting in their 40s hold objects farther and farther away in order to be able to read them. When they run out of arm, it's time to get eyeglasses.

»»» *See Also*
RELATED PRINCIPLES: Electromagnetic Radiation, 248
RELATED APPLICATIONS: Electromagnetic Spectrum, 147 • Eyeglasses and Contact Lenses, 226 • How We See, 227

Optics

EYEGLASSES AND CONTACT LENSES

Optical lenses—whether in the form of eyeglasses or contact lenses—aid vision by focusing, bending, and spreading light rays emitted or reflected by an object. Unlike the lens in a human eye, which can shift its focal length through muscular contraction and relaxation, optical lenses have focal length and power built in, depending on their shape. They generally consist of either two curved surfaces or one flat and one curved. Various surfaces and thicknesses determine a lens's focal power and function, while combinations of different lenses prevent the blurring, distortion, and other problems that can occur with single, thin lenses. For example, in bifocals, the upper and lower parts of a lens are ground differently to correct for both close and distant vision. Trifocals are ground with a center lens for intermediate distance. Intraocular cataract lenses that replace clouded natural lenses are made of a variety of synthetic materials and may be monofocal, which provide vision at a fixed distance, or multifocal, which broaden viewing distance near to far.

Eye specialists have many techniques for measuring the eye's refractive power and for determining your need for corrective lenses. To compensate for nearsightedness, in which near objects are seen clearly, but far objects are blurred, they grind lenses in concave (inward curving) shapes; for farsightedness, in which far objects are seen clearly, but near objects are blurred, the lenses are convex (outward curving). If you have astigmatism, a condition in which light doesn't focus properly on the retina because of a defect in the curvature of the natural lens, you get cylindrical lenses. Prisms, which bend, spread, and reflect light, are used for other defects.

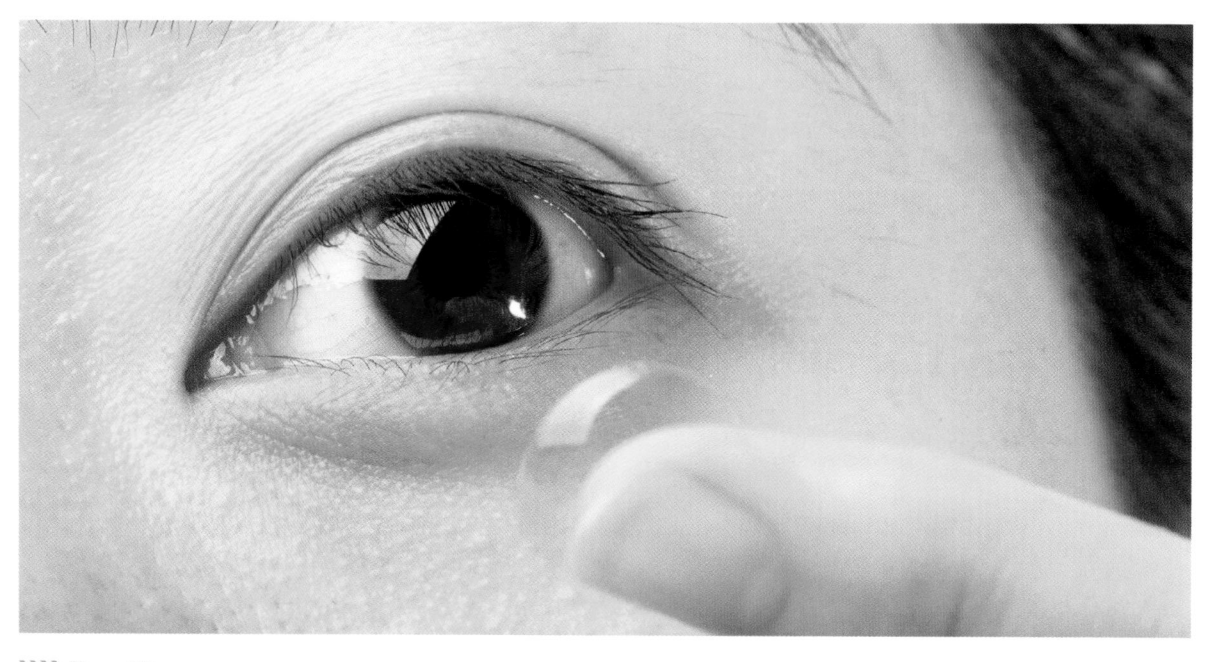

See Also

RELATED PRINCIPLES: Health and Medicine, 354 • Macromolecular Chemistry, 288
RELATED APPLICATIONS: Age and Vision, 225 • How We See, 227

Optics

HOW WE SEE

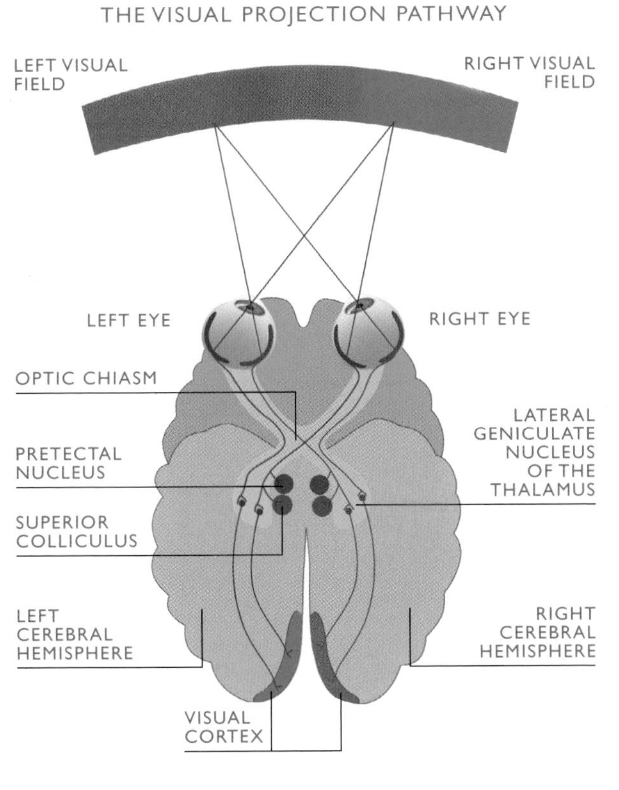

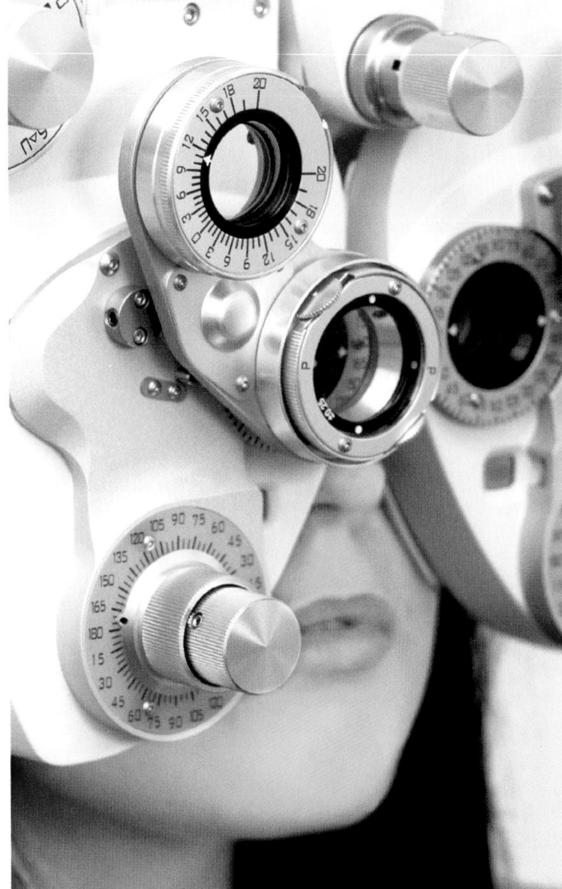

The human eyeball contains about a hundred million rods: complex, light-sensitive nerve cells that see only in shades of gray but perform well in low light. It also has about three million color-perceiving cones, which are most sensitive to bright light. When light hits the retina, photons prompt pigments within the rods and cones to split, sparking a chemical reaction that ultimately results in a signal being sent to the optic nerves.

The optic nerves from each eye extend separately a couple of inches and then come together at a spot called the optic chiasm. There, information from the right half of each eye's visual field is joined and routed to the left side of the brain; left-side images from both eyes go to the right. The visual cortex analyzes vertical versus horizontal, light versus dark, background versus foreground, and so on before combining them all into what we perceive as reality.

»»*See Also*
RELATED PRINCIPLES: Electromagnetic Radiation, 248 • Health and Medicine, 354
RELATED APPLICATIONS: Age and Vision, 225 • Eyeglasses and Contact Lenses, 226

PRINCIPLE

GLASS ALL FULL

Fiber Optics

One of the most important technological breakthroughs of the twentieth century was fiber optics: using efficient glass fibers to carry light signals. One kind of glass is wrapped around another, like the insulation on a wire, producing nearly total internal reflection of the light waves. Because light waves are so small, and because it's relatively easy to get several wavelengths to share the same space, they're a natural choice to replace electrical signals in a new generation of optical, or photonic, computers. These types of computers, which are still in development, use light (in the form of holograms for storage and laser beams for connection) instead of electrical current.

Fiber Optics

FIBER OPTIC CABLE

Each individual fiber strand in a bundle of fiber optic cable can carry billions of signals per second. Electromagnetic waves play a big role in this technology. They consist of electric and magnetic fields that vibrate at right angles to each other and move at the speed of light (186,000 miles a second). The signals and impulses of electromagnetic waves can carry sounds, words, images, and data. The higher their frequency, the more energy they have; gamma and x-rays are at the high end, and microwaves and radio waves are at the low end.

Radio waves exist naturally as radiant energy from the Sun, but when they're generated by electricity in an antenna tower, they transport sound and images. The visible light segment of the electromagnetic field has also been used to communicate—in laser form it beams through fiber optic cable, its digitally coded pulses translated into thousands of telephone conversations that can be handled simultaneously.

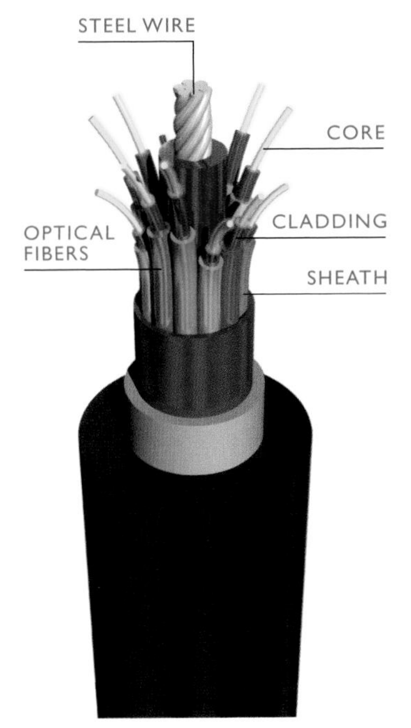

Sound also comes in waves, but they're pressure waves, not electromagnetic ones. Best defined as a measurable, mechanical disturbance, sound still can be translated into electricity, which is a must before it can be sent through a network of fiber optic cable. Although fiber optics is a fast, powerful technology, the signals weaken over long distances, so they have to be boosted at intervals.

»»See Also
RELATED PRINCIPLES: Electromagnetic Radiation, 248 • Health and Medicine, 354 • Macromolecular Chemistry, 288 • Materials and Chemistry, 260 • Quantum Mechanics, 298
RELATED APPLICATIONS: Air Traffic Control, 253 • Silicon, 271

PRINCIPLE

SHORT-TERM PARTNERS

Stimulated Emission

Stimulated emission, which explains how lasers work, was introduced by Albert Einstein. Certain crystals, some gases, and a few liquids give off photons that are perfectly the same if they're stimulated in just the right way, whether by electric current or another light source. In these situations, the atoms of these materials enter an odd, semistable, excited state that can last for several thousandths of a second. When an electron in one of those atoms gets energized by an incoming photon from a neighboring atom, it's stimulated to emit a photon identical in every way to the one that triggered it. The two go flying off to hit other atoms, quickly producing four more identical photons, then eight, and so on, and the population of coherent photons is amplified. The result is light that is in phase with the energy wave, creating a single wavelength.

Stimulated Emission

BAR CODE SCANNER

Bar codes are so common that we might not even really notice them anymore on products we buy. They represent the decimal price and other details (referred to as the Universal Product Code, or UPC) about an item in a series of parallel vertical lines and white spaces. Bar codes can also help us track documents and packages, gene sequences in databases, and books in libraries and bookstores.

When a cashier passes an item over a scanner or under a handheld one, a laser beam set to a specific frequency reads the binary code in the bars. The code is transmitted to a decoder and a computer, and the computer checks files and locates the item and its price.

A matrix bar code, or QR (for Quick Response) code, uses square black dots arranged in a grid on a white background. It's easier to read and has greater storage capacity than standard UPC bar codes.

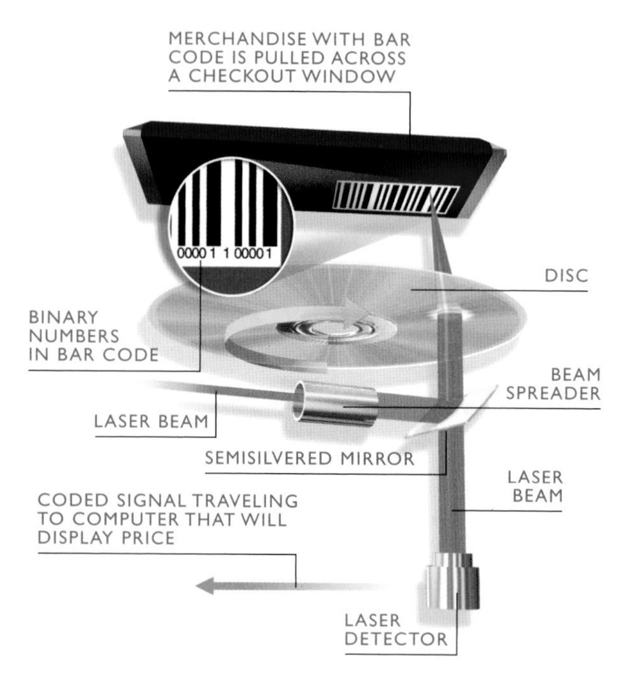

»»See Also

RELATED PRINCIPLES: Binary Code, 190 • Electromagnetic Radiation, 248
RELATED APPLICATIONS: Credit Card, 141 • Photocopier, 168

STIMULATED EMISSION. Laser beams pointed toward the stars illuminate English Bay and the night sky in downtown Vancouver.

Stimulated Emission

LASER

Lasers transmit data over fiber optic cables. They play CDs, slice steel beams, guide missiles, measure distances, fashion suits and semiconductor chips, and bore holes in diamonds in a fraction of a second. They also help heal. Medical lasers can cut or destroy abnormal tissue without harming healthy tissue, shrink or destroy tumors, and cauterize, or seal, blood vessels to prevent excessive bleeding. In addition, "cosmetic" lasers remove unwanted hair, obliterate tattoos and birthmarks, smooth facial wrinkles, and activate tooth-bleaching solutions.

Laser is an acronym for light amplification by stimulated emission of radiation, but its light differs from that of the Sun or a bulb, which radiates in every direction. Laser light beams are concentrated and narrow, they're monochromatic, or one color, and they move in the same direction. They also can focus intense heat on tiny spots. Because of these qualities, they're described as coherent: all of the waves of light are exactly the same frequency, are precisely in phase, and are headed in the same direction.

The types of lasers vary according to their wavelength and the amount of power produced. For example, gases such as helium-neon, argon, carbon dioxide, and nitrogen can be used. Argon is used as a laser medium in eye surgery because the beam's energy can reach the retina without being absorbed by eye fluid. The carbon dioxide laser has a beam that is absorbed by substances containing water. Chemical lasers, such as hydrogen fluoride, are used in laser weaponry. Metal-vapor lasers include helium-cadmium and gold and copper vapor; these are used in high-speed photography and printing and also in dermatology. Solid-state lasers include ruby (used in tattoo removal), Nd:YAG (short for neodymium-doped yttrium aluminum garnet; used in surgery), and holmium YAG (used in dentistry and removal of kidney stones), among others.

»»**See Also**

RELATED PRINCIPLES: Health and Medicine, 354 • Power and Electricity, 163 • Radiation, 239 • Wave Energy, 58

RELATED APPLICATIONS: Electron Microscope, 152 • Laser Printer, 170

Stimulated Emission

HOLOGRAM

Holograms, which are three-dimensional (sometimes two-dimensional) images created by laser light, are not just on Star Trek. You can find them in drivers' licenses, credit cards, computer games, and more. They might not seem so impressive when used this way. However, when applied on a much larger scale, they allow us to examine landscapes and objects in stunning detail from any angle—pretty impressive. One interesting feature of a hologram is that if you were to cut it in half—or even take a tiny piece of it—the entire hologram would still be contained within the fragment, thanks to the science behind how a hologram works.

To create a hologram, four basic parts are needed: a helium-neon laser (often called a red laser); a lens that spreads the laser beam; mirrors that direct the beam precisely; and holographic film, which is similar to photographic film, except it's very sensitive to changes in light (on a microscopic scale, in fact) and records this light at a very high resolution. The laser beam is divided into two parts, and the mirrors are arranged in such a way that the beams follow a specific path. Then, each of the two beams passes through another lens, which transforms the beam into a wide beam of light rather than a narrow beam. One beam, called the object beam, reflects off of the object and onto the holographic film. The other beam, the reference beam, hits the holographic film without reflecting off of anything other than a mirror. The result: a perfect representation of the object in holographic form.

Today, holograms are used by engineers to explore prototypes for new models of automobiles, and in some cases, they have even been used in place of diagnostic medical x-ray imaging. Non–3-D holograms are used to store data (holographic data storage used in cloud computing; in this case, the data is written to a hologram that stores the information like a hard drive or CD drive but at greater capacity).

3-D VIDEO CONFERENCING. (Top) A prototype 3-D video-conferencing pod allows people in different locations to videoconference as if they are standing in front of each other. Since the 3-D video image is visible 360 degrees around the Pod, the person can walk around it to see the other person's side or back.

HARD TO COPY. (Bottom) A 50-Euro banknote includes a hologram in the corner used to thwart counterfeiters.

See Also

RELATED PRINCIPLES: Health and Medicine, 354 • Power and Electricity, 163 • Radiation, 239 • Wave Energy, 58 •
RELATED APPLICATIONS: Cloud Computing, 201 • Reflecting Telescopes, 211

Holographic image of Britain's Queen Elizabeth

Stimulated Emission

COMPACT DISC PLAYER

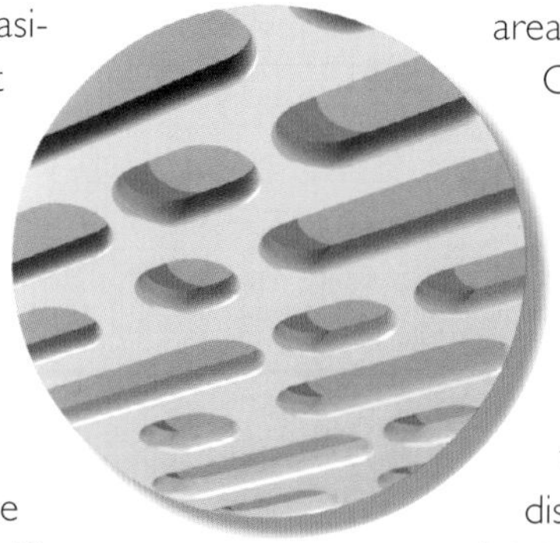

A digital compact disc (CD) is basically just a plastic plate that has been etched with a few billion microscopic pits. The plate is coated with very shiny aluminum and then covered with a transparent protective layer. When you put a CD in a CD player, it spins quickly on a turntable, and a laser in the player follows the grooves in the CD, looking for the microscopic pits. Flat areas reflect the scanning laser light in the CD player back to a detector; the pits scatter the beam. The detector records these scatterings—tens of thousands a second—and sends this information to a computer, which decodes it, turning it into sound waves. Originally, marketed as an indestructible alternative to vinyl albums and tapes, the compact disc player has in turn been supplanted by digital music files.

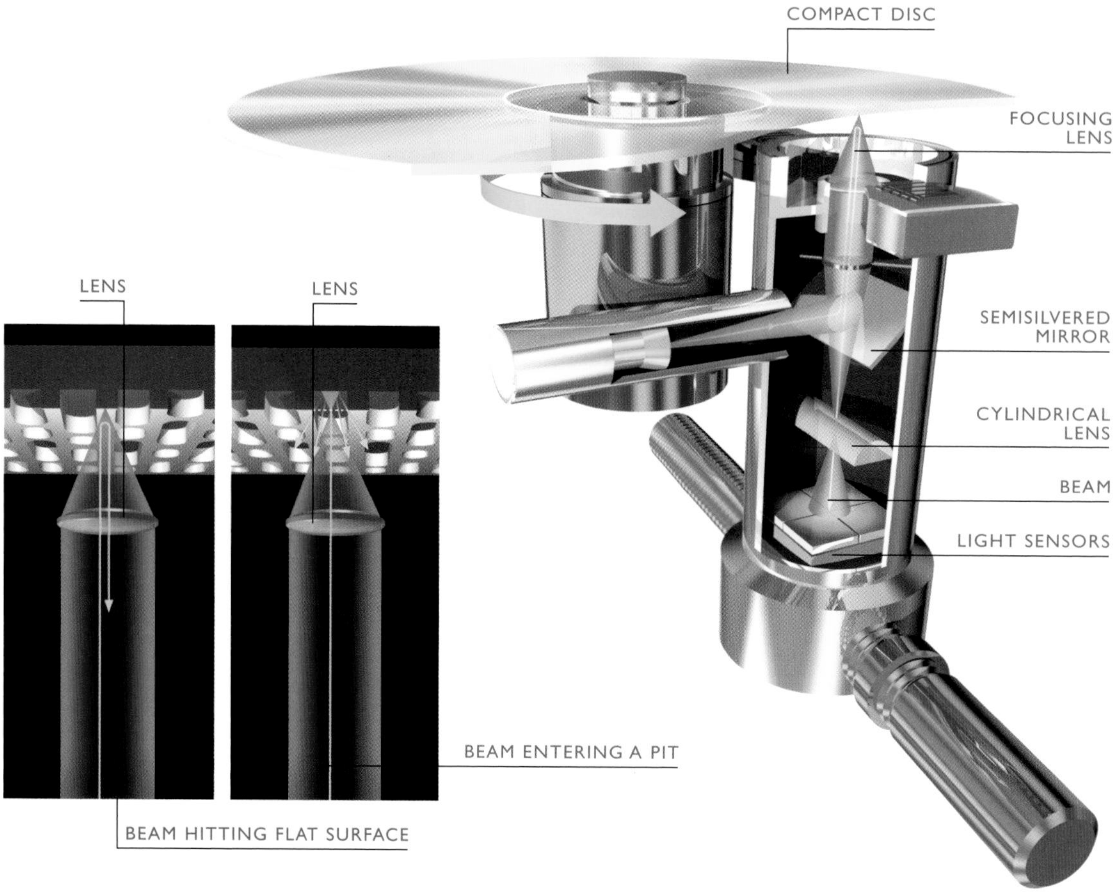

See Also

RELATED PRINCIPLES: Binary Code, 190 • Power and Electricity, 163 • Radiation, 239 • Wave Energy, 58
RELATED APPLICATIONS: Computer Hard Drive, 149 • Loudspeakers and Headphone, 148

INTERLEAVING AND ERROR CORRECTION

Interleaving refers to the process of writing data out of sequence in order to employ multiple data streams, such as audio and video, simultaneously. Special software called a codec is used to tell the player how to properly read and play the file. Interleaving also helps speed up processing in computer memory, although latency (delay in time) is sometimes a problem.

COSMIC RAYS TIME DILATION: Subatomic particles disintegrate through the atmosphere above the London skyline, as revealed by specialized imaging equipment of the European Organization for Nuclear Research (CERN).

CHAPTER 10

Radiation

Light—both visible and invisible—and sound are waves, so they give off energy, especially when reflected off objects. We have harnessed this energy in the natural world and even learned to create this energy ourselves. Radio waves, for example, revolutionized the way we communicate and stay in touch with our world. Cell phones, once a far-fetched concept, have evolved to the point where they're essentially mini-computers. We can determine a person's location with pinpoint accuracy using the Global Positioning System. Our world is becoming more and more wireless, and our wireless devices can communicate with each other seamlessly using Bluetooth technology.

Radar has also led to advances in meteorology and the air traffic control system. We can map the shape and size of storms. We are learning to make air transportation more efficient, which leads to quicker travel times and the ability to put more planes in the air and still ensure passenger safety. We can even cook our food with radar. In addition, this technology has led to scientific breakthroughs like x-rays and CT scans, which lets us see the world on a more detailed, smaller scale. Particle accelerators that use ionizing radiation have led to scientific breakthroughs on an atomic level, as seen with the Large Hadron Collider.

PRINCIPLE

TUNING IN
Radio Waves

Guglielmo Marconi may have invented the wireless radio, but he has Heinrich Rudolph Hertz to thank for it. A German physicist, Hertz was the first to prove the existence of radio and other electromagnetic waves. The hertz, the unit of frequency, is named for him.

Radio waves begin at a microphone in a broadcasting station, where vibrating sound waves are converted into relatively weak electrical pulses. Sound waves vibrate the diaphragm of a microphone, which turns the acoustical energy into a weak electrical signal. This signal is amplified and then added to a carrier wave so it can be broadcast. Each radio station is assigned a carrier wave with a different frequency. An antenna at the top beams the audio-carrying radio waves at the station's assigned frequency.

After the waves leave the transmitter, they're picked up by the antenna in your radio. A tuner then selects the program by matching the receiver to the station's transmitting frequency. The radio wave signals are turned into electrical signals that are amplified and then turned into audio signals. These are sent to the radio's speaker, where the electrical waves are converted back into sound and amplified again.

Radio Waves

RADIO WAVE FREQUENCIES

The distance radio waves travel is determined by their frequency and by electrical atmospheric conditions. Radio waves demonstrate characteristics associated with their frequencies. Impressing sound waves onto radio, or carrier, waves involves amplitude modulation (AM) or frequency modulation (FM). In AM, the amplitude, or the overall strength of the radio wave in terms of voltage or power level, is varied to incorporate the sound information. In FM, the frequency, or the number of times each second that the electric current changes direction, varies.

See Also
RELATED PRINCIPLES: Power and Electricity, 163 • Wave Energy, 58
RELATED APPLICATIONS: Spectrum of Wavelengths, 248

Radio Waves

RADIO

We make radio waves carry signals by changing, or modulating, the waves. Some radio stations send their signals by changing the size, or amplitude, of the radio waves. These AM stations broadcast on frequencies that are measured in thousands of cycles per second, or kilohertz. Other stations broadcast by making small changes in the frequencies of their radio signals. These FM stations are assigned frequencies in millions of cycles per second, or the megahertz range.

In AM broadcasts, sound vibrations in the form of amplified electrical signals are impressed onto electrically generated carrier radio waves by adjusting the amplitude of the carrier waves to keep them aligned with the audio signals. FM gives clearer transmission and reception, but it doesn't affect the amplitude of the carrier wave. Instead, it varies the wave's frequency in accordance with the sound to be transmitted.

When a station is selected on a radio, a tuning circuit picks one and tunes out all the others by permitting current to oscillate at a single frequency. The two conducting plates of a capacitor, or condenser, store energy as electricity, while a coil to which they're linked stores energy as a magnetic field. The magnetic field collapses and sends an electric current to recharge the capacitor, which discharges again through the coil, instigating an oscillating current of one frequency. Essentially, the capacitor blocks the flow of direct current while allowing alternating and pulsating currents to pass. The clear reception of a given radio frequency of course depends on the strength of the signal.

»»*See Also*
RELATED PRINCIPLES: Power and Electricity, 163 • Wave Energy, 58
RELATED APPLICATIONS: Radar, 251

SATELLITE RADIO

Satellite radio, a business venture launched (literally) in 2002, has now attracted millions of customers. In digital radio, analog signals are digitized, or compressed, using formats like mp2. This allows many more stations to broadcast within an allocated frequency band. Also, the broadcast area is significantly increased. With digital radio, other information, such as a song's title and artist, can be broadcast along with the music. Subscribers pay a monthly fee for coast-to-coast, static-free news, music, and entertainment. A satellite broadcast begins at a radio station on the ground. It's transmitted via radio signals to satellites that bounce their digitized signals back to portable radio receivers or to receivers installed in a car. The receivers decode the signals for you.

4G NETWORK

Wireless technology is often referred to in terms of generations. The first generation used analog signals, the second generation used digital signals, and the third generation provided wireless Internet access to mobile devices and networks. The fourth generation (4G), especially the type called 4G LTE, is the latest version. It's faster than 3G and boasts a wide range of features, including the ability to download high-definition video, better voice quality, and high rates of data transmission. The downside: 4G LTE signals eat up your battery faster (that's why 4G phones are usually bigger than 3G ones—they need bigger batteries).

Radio Waves

CELL PHONES

At first, the cell phone was just a way to make phone calls from anywhere. Today, wafer-thin, beautiful, lightweight smartphones transmit e-mail and photographs, download games, connect to the Internet, take pictures, execute spoken commands, and much more. Cellular phones—so-called because they cover compartmentalized, cell-like areas—transmit over radio waves, either via an antenna located in a base station in each cell or by way of satellites. At the center of it all is wireless technology. A cell phone is actually a specialized, sophisticated radio. It's a full-duplex device, meaning that you can interrupt someone else—one frequency is used for talking and a separate frequency is used for listening. The communication occurs on channels, of which the average cell phone can hop among over 1,650.

In a typical cell phone network, a carrier (who provides you with your cell phone service) is assigned 800 frequencies, which it divides into hexagonal units called cells, each containing about 10 square miles. Within each cell are a base station and a tower. Cell phones have low-power transmitters in them, as does the tower. The cell phone and the tower communicate with each other using a special frequency (if this frequency can't be found, an "out of range" or "no service" message is displayed). As you use the phone, moving from one cell to another, the frequency is "passed" from one cell to the next. The carrier maintains the frequency you need to communicate with the person on the other end and continually monitors the signal strength. Sophisticated computer controllers keep you connected on the move.

»»See Also

RELATED PRINCIPLES: Binary Code, 190 • Power and Electricity, 163 • Wave Energy, 58
RELATED APPLICATIONS: Cell Phone Jammer, 63 • Integrated Circuit, 191 • Radio Wave Frequencies, 241

Radio Waves

GPS

The Global Positioning System (GPS) may be built into cars and cell phones, but it started out as a military satellite system. It was actually first launched in 1978 by the Department of Defense (DoD). The satellites are located 12,000 miles above Earth and travel at a speed of 7,000 miles per hour. The manufacturers of GPS equipment quickly recognized its mass-market potential and begged the DoD to let them develop civilian uses. The DoD complied with their request in the 1980s.

The GPS network is 24 satellites (3 of which are backup in case of failure) orbiting about 10,600 miles above Earth. Each satellite makes two complete rotations around the planet every day. The satellites are spaced so that, from any point on Earth, four satellites are always above the horizon. Each satellite has an atomic clock, a computer, and a radio. Using its own orbit and clock, a satellite broadcasts its changing position and time. The computer in your GPS receiver, for example, in your car, figures out where you are to within a few feet by calculating your distance from three of the four satellites in the network with which it is communicating via radio waves.

GPS is already pretty accurate, but developers are working to give it the ability to provide increasingly fine-grained information on a given location. You might have already noticed this when you plug in a route in the GPS in your car and are advised of nearby restaurants in the area or an unexpected detour.

TRACKING AND PRIVACY

Advances in tracking technology have led to some obvious concerns relating to. If you're relying on GPS, can your cell phone maker track your comings and goings?

Whether these concerns are warranted comes down to expectations of privacy. Two recent court cases came to different conclusions regarding the use of GPS to track a crime suspect's moves, and the issue is likely to remain controversial. In the meantime, companies like Microsoft and Google are giving users the ability to opt out of tracking or prevent it.

PROJECT GALILEO. The European Space Agency is in the process of implementing a new advanced global navigation satellite system. On schedule to be fully operational by 2020, the fully deployed system will consist of 30 satellites, will be under civilian control, and will be interoperable with GPS.

»»See Also
RELATED PRINCIPLES: Gravity, 121 • Power and Electricity, 163 • Wave Energy, 58
RELATED APPLICATIONS: Radar, 251

BLOCK III NEXT-GENERATION GPS

The GPS network is managed by the U.S. Air Force, and it contains both old and new satellites, arranged into groups that we might think of as generations but which the military industry calls blocks. Block III, which is currently being developed by Lockheed Martin, is the newest generation of satellites. They're designed to emit more powerful signals, more reliable navigation, enhanced signal integrity, and more.

Radio Waves

WI-FI

Contrary to popular belief, Wi-Fi doesn't stand for wireless fidelity. The term is actually a brand name of sorts for a standard more formally known as 802.11 networking, or just wireless networking. The Institute of Electrical and Electronics Engineers (IEEE) is in charge of setting standards for this and other protocols. So what is Wi-Fi? It's a means of wireless communication that works like a radio. An adapter inside a computer converts data into a radio signal and transmits it via an antenna. On the other end of the transmission, a router receives the signal and converts it back into data. This pathway can be between a device (such as a laptop computer) and the Internet or vice versa.

However, Wi-Fi is not just a fancy radio. It operates at a higher frequency than traditional radio waves, which means more data capacity. A Wi-Fi signal can be split into different streams or sub-channels, which improves capacity, signal strength, and range. And Wi-Fi has the ability to hop from frequency to frequency, which means that multiple devices can use just one router. For the most part, Wi-Fi networks are easy to set up and maintain, are reliable, and are even easier to connect to; that's why nearly every hotel, airport, library, and coffee shop these days boast free Wi-Fi access.

»»See Also
RELATED PRINCIPLES: Binary Code, 190 • Electricity, 163 • Wave Energy, 58
RELATED APPLICATIONS: Personal Computer, 195

FIREWALLS

A firewall is a kind of blockade for your computer. It stops unauthorized data from getting either in or out of your machine.

A firewall can take the form of either hardware or software. Large corporations have used them for years, but even private home users are recognizing the need for them. Firewalls work using one of three techniques: packets of data may be analyzed against a set of filters, and any potentially unsafe information is flagged and not allowed through; a proxy service retrieves the information from the Internet, which is then retrieved by the firewall and sent through the destination server; or a newer method, called stateful inspection, can analyze the data, looking for key characteristics that identify it as being unsafe.

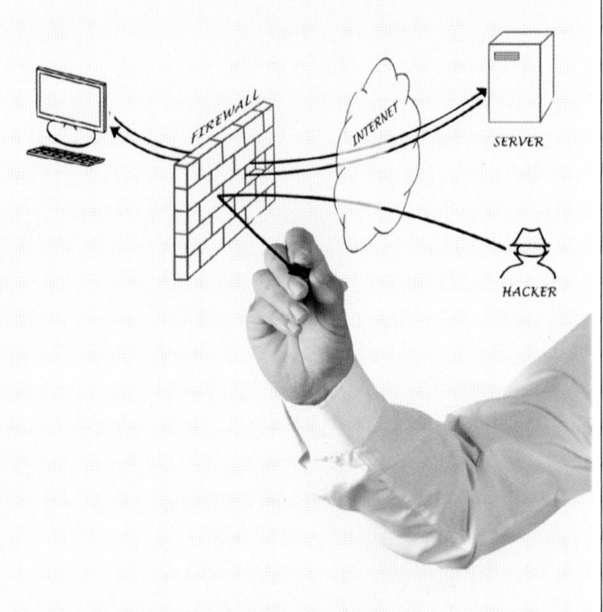

BLUETOOTH

Bluetooth is a radio frequency–based standard that lets nearby devices—gadgets within about 30 feet of each other—communicate wirelessly. Your phone might connect wirelessly to a Bluetooth headset, for example.

This connection is not as easy as it sounds; there must be agreement among the devices on what data to send, how much data to send, how to send it, and where to send it. Bluetooth provides this agreement via low-power radio waves, in an inexpensive, reliable fashion. The name comes from Harald Bluetooth, who was king of Denmark in the late 900s. King Bluetooth's claim to fame lies in the fact that he united Denmark and part of Norway into one kingdom.

Bluetooth was invented by Danish company Ericsson, hence the name.

Radio Waves

NEAR-FIELD COMMUNICATIONS

How we shop, bank, and communicate are all being changed with the use of near field communication (NFC) devices. With one touch, information can be exchanged between NFC devices.

This short-range wireless technology allows communication between devices no more than about 4 inches apart. The devices establish a peer-to-peer (P2P) network to exchange data. After the P2P network is established, another wireless communication technology, like Bluetooth or Wi-Fi, can be used for longer range communication or for transferring larger amounts of information. NFC devices can be either active (battery-powered) or passive (radio energy–powered).

With NFC-enabled devices, you can take a photo with a smartphone, then touch a television or computer and transmit the image for display; you could make a payment by touching your phone to another; you could take advantage of interactive advertising by touching your phone to a coupon, for instance, on a poster. Smartphones with NFC will take the place of plastic credit cards for making purchases.

See Also
RELATED PRINCIPLES: Binary Code, 190 • Power and Electricity, 163 • Wave Energy, 58
RELATED APPLICATIONS: Cell Phones, 243 • Personal Computer, 195

PRINCIPLE

SEEN AND UNSEEN POWER

Electromagnetic Radiation

More light is invisible to us than visible, and the term electromagnetic refers to the complete spectrum of light, which takes the form of alternating waves of electricity and magnetism. The energy that is released as a result is called electromagnetic radiation. This radiation can vary in strength and intensity and can be harmless or beneficial. At high enough amounts or with long enough exposure, electromagnetic radiation can even be deadly. It comes in both natural and human-made forms. Examples of natural electromagnetic radiation include the Sun and rainbows. Human-made examples include radio waves and microwaves, although these can be emitted naturally as well.

Electromagnetic Radiation

SPECTRUM OF WAVELENGTHS

Every segment of the electromagnetic spectrum—visible and invisible—can be reflected, bent, or spread out to separate constituent wavelengths. All waves convey energy. Sometimes the amount of energy in a wave depends on how hard it compresses the medium through which it travels—like in sound, ocean, and seismic waves. But then there's electromagnetic radiation, which is made up of individual waveforms, or energy packets, called photons that can pass through a vacuum as well as air, glass, water, or anything else that is transparent. To pass through denser materials, such as flesh and bone, it takes light waves of smaller size and higher frequency than visible light, such as x-rays. Gamma rays have the smallest wavelengths and the most energy of any other wave in the electromagnetic spectrum. These waves are generated by radioactive atoms and in nuclear explosions. Gamma rays can kill living cells. Medicine harnesses this property, using gamma rays to kill cancerous cells.

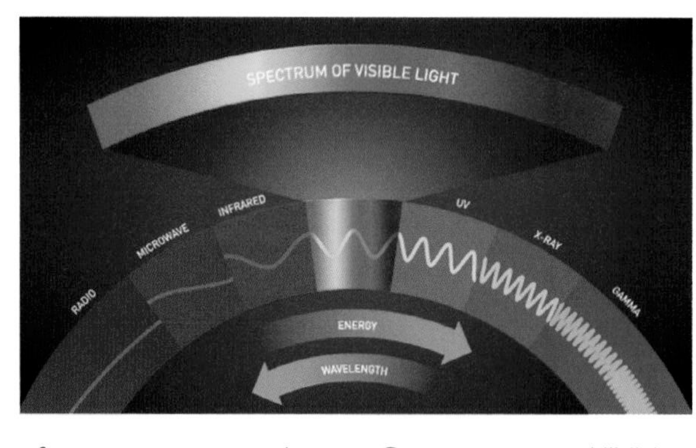

»»See Also
RELATED PRINCIPLES: Absorption and Light, 204
RELATED APPLICATIONS: Sunburn and Sunblock, 217 • UV Light and Eyes, 215 • Visible Spectrum, 205

Electromagnetic Radiation

INCANDESCENT LIGHTBULB

In 1879 Thomas Edison passed electricity through a strand of carbonized cotton sewing thread, causing the filament to glow for more than 13 hours in a glass vacuum tube. His feat was marred, though, by the loss of power through heat and by the short bulb life. The coiled filament made of tungsten, a metal with a high melting point that easily accommodates the heat needed for the best light, made a brighter, longer-lasting bulb. The life of a bulb got even longer when an inactive gas—nitrogen, argon, or krypton—was placed inside to slow the filament's evaporation. Inside the bulb of a halogen lamp are molecules of bromine or iodine, elements that combine with tungsten given off by the filament to form a gas. When this gas comes in contact with the hot filament, the tungsten atoms separate from the halogen and adhere to the filament, essentially rebuilding it.

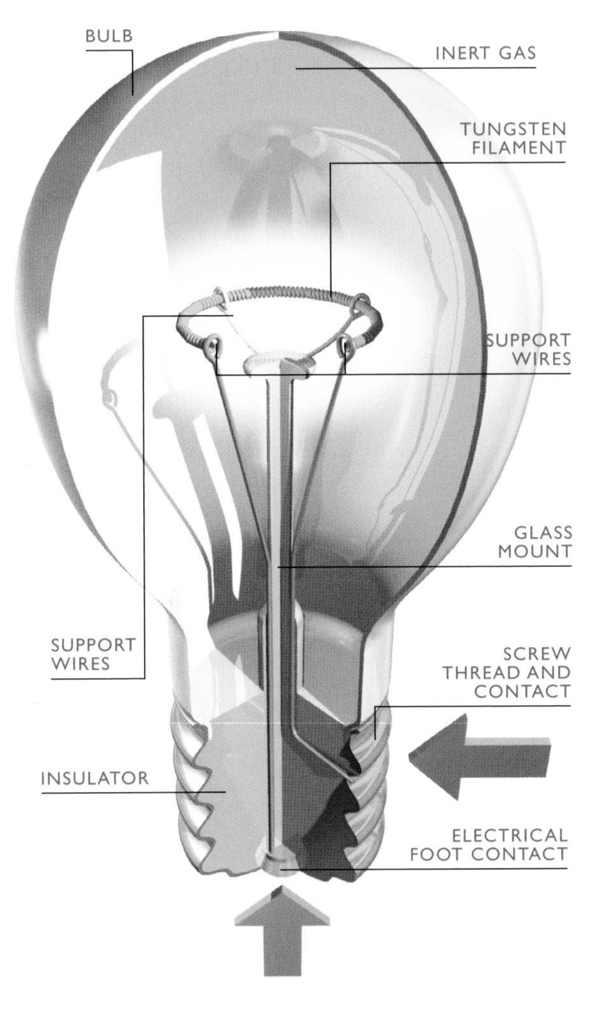

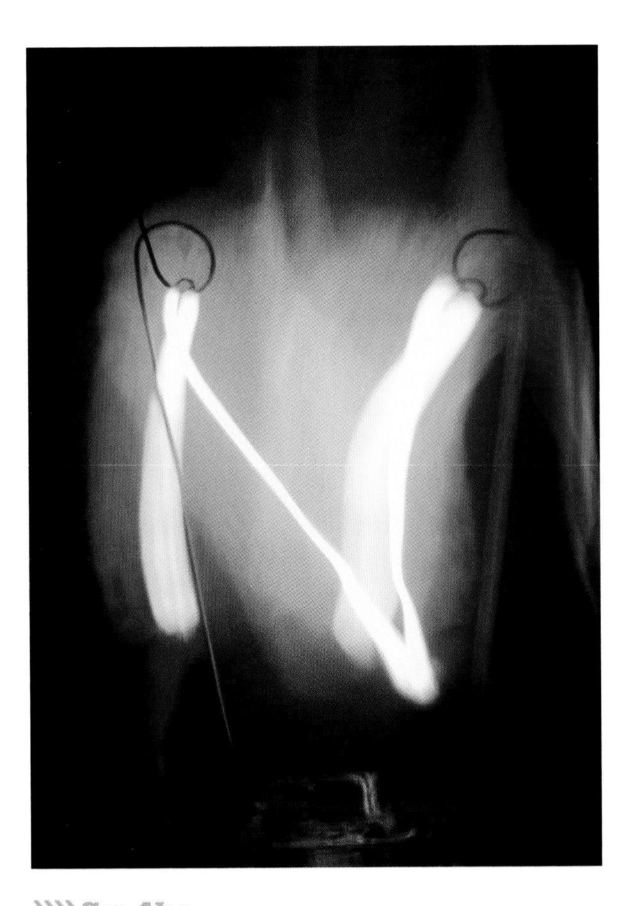

»»See Also
RELATED PRINCIPLES: Elasticity of Air, 78 • Power and Electricity, 163 • Wave Energy, 58
RELATED APPLICATIONS: Argon, 272 • Neon, 264

Electromagnetic Radiation

MICROWAVE OVEN

Microwave ovens produce waves that are just the right frequency to cause the highly polarized molecules in water to flip around faster and faster. Because even "dry" foods often contain moisture in the form of water vapor, the oven can warm them as well. A microwave oven produces high-frequency electromagnetic waves. When these waves pass through food, the waves reverse polarity billions of times a second. The food's water molecules also have polarity, and they react to each change by reversing themselves. Friction results, heating the water and cooking the food.

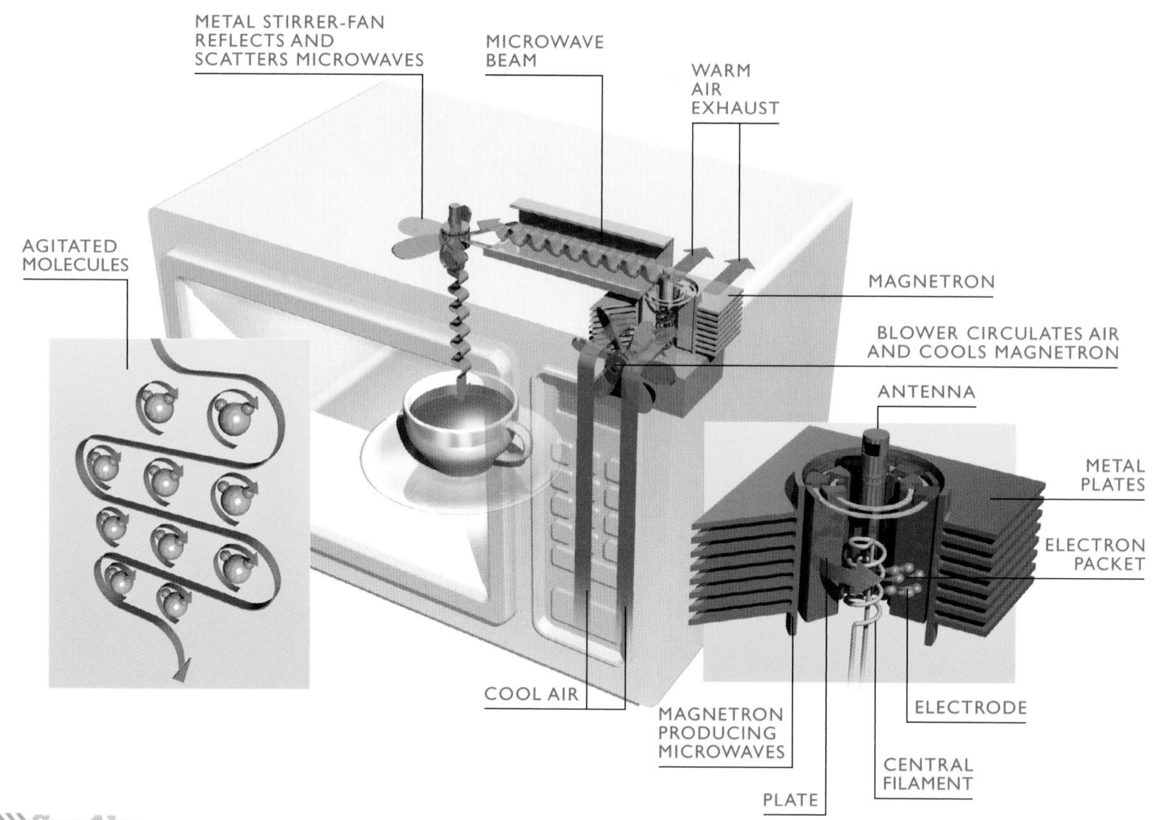

»»See Also

RELATED PRINCIPLES: Power and Electricity, 163 • Wave Energy, 58
RELATED APPLICATIONS: Convection Oven, 103

Electromagnetic Radiation

RADAR

In the form of radar, radio waves help detect distant objects and determine their positions. Like laser, radar is an acronym—it stands for radio detection and ranging. It was introduced shortly before World War II primarily to detect aircraft, but radar is now used to detect more than airplanes. It can detect the speed of an object, and it can warn us of impending storms. Radar works based on two things: echo and the Doppler effect. Everyone is familiar with the concept of an echo. The Doppler effect (also called Doppler shift) refers to the phenomenon in which sound waves either compress or expand in response to changes in distance between the sound itself and the observer. Measure this change in pitch, and you can tell how far away something is. When radar first appeared in the 1940s, the U.S. Navy identified it by the acronym RADAR. The acronym stood for RAdio Detection And Ranging. Since then, the term radar has come to be recognized as a common noun and is no longer capitalized.

»»*See Also*
RELATED PRINCIPLES: Electricity, 163 • Wave Energy, 58
RELATED APPLICATIONS: Radio, 242 • Sonar, 64

DOPPLER RADAR

Radar systems make use of both echo and the Doppler effect, but if a system focuses on just the Doppler effect, essentially disregarding everything else as "noise," a more detailed picture can be created. Weather radar sends out electromagnetic waves of about the same frequency as your microwave oven, with a wavelength of a few inches, as opposed to radio waves. The reflected signal, captured by the receiver, shows the shape of the storm. Doppler radar for weather forecasting had its origins in World War II when radar operators, seeking enemy craft, noticed that weather caused interference in the form of echoes on their screens. This led to postwar experimentation with surplus radars to detect precipitation. In the years since, weather radar evolved and is now routinely used in weather forecasting and analysis.

NEXT-GENERATION AIR TRANSPORTATION SYSTEM

The Next-Generation Air Transportation System (NextGen) is an initiative of the U.S. Federal Aviation Administration (FAA) to overhaul the air traffic control system. It involves replacing the current radar-based system with a satellite-based one, with the goal of making air transportation more efficient, economical, and environmentally friendly. For example, by using GPS technology to guide pilots on a modified landing path that curves as it nears the final approach to a runway instead of coming in straight, flight times are decreased, and less fuel is consumed.

NextGen is still years away, though. It requires that each plane be outfitted with glorified GPS units, for which most airlines are reluctant to pay.

Electromagnetic Radiation

AIR TRAFFIC CONTROL

The job of an air traffic controller is laden with responsibility—and stress. Radar is still a useful tool for controllers, but it's augmented by satellite data and various devices that allow controllers to warn pilots of altitude problems, sequence takeoffs and landings safely, and warn of flights into restricted areas. In the air, pilots are guided and monitored by onboard computers and sensors that control aircraft movements via fiber optic cables. Pilots use the cables to transmit coded digital signals directly to the motors moving the control surfaces. Computers and software also establish safety parameters, check speed and direction, and correct for weather changes and possible pilot miscues. Weather radar in the nose sends information to the plane's data system; sensors detect wind shear and monitor fuel consumption.

See Also
RELATED PRINCIPLES: Electricity, 163 • Wave Energy, 58
RELATED APPLICATIONS: Global Positioning System, 244

Electromagnetic Radiation

X-RAY IMAGING

The discovery in 1895 of the x-ray is credited to physicist Wilhelm Röntgen (for which he won the first Nobel Prize in physics in 1901). X-ray imaging uses a certain spectrum of electromagnetic energy to create waves that enable us to see inside the human body at the doctor's office, to scan luggage for potentially dangerous items at the airport, and to detect electrical wiring in a building before making renovations, for example.

Possessing even shorter wavelengths and higher frequencies than ultraviolet light, x-rays are created by bombarding a metal target with a high-speed electron stream. Electrons smash into the metal and give off radiation. A second metal plate filters the beam, ensuring that the image produced as a result is clear and readable. X-rays have so much energy that they easily penetrate soft tissues and are only slightly slowed or deflected by harder ones, such as bones. This difference in penetration shows up as light and dark areas on a photographic plate or film. In addition, mineral deposits in the body help identify broken bones, tumors, or foreign objects.

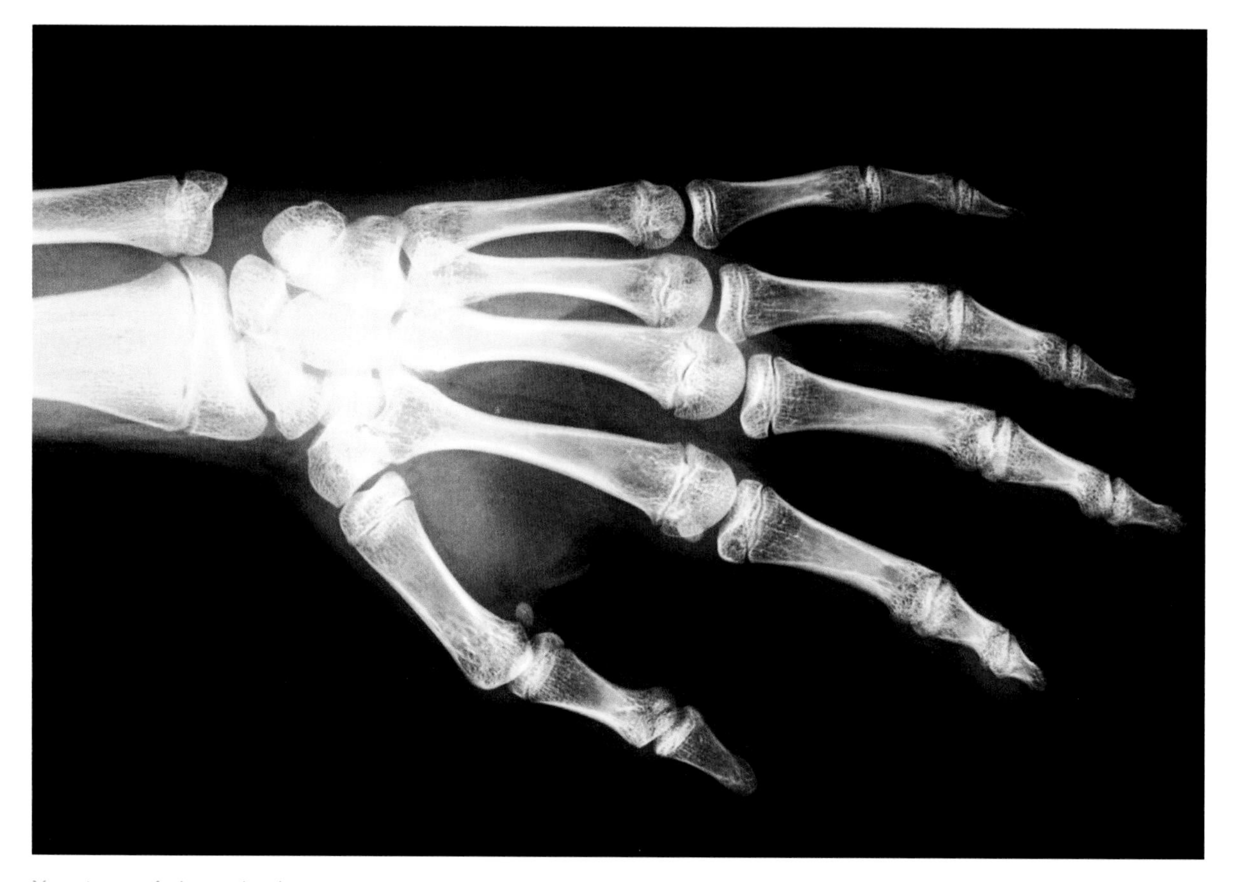

X-ray image of a human hand

»»See Also

RELATED PRINCIPLES: Electricity, 163 • Health and Medicine, 354 • Wave Energy, 58 •
RELATED APPLICATIONS: Electron Microscope, 152 • Magnetic Resonance Imaging, 142

X-RAY CRYSTALLOGRAPHY

X-rays are diffracted by crystals, meaning their directions are changed or bent as they pass through the structure of crystals. Scientists measure the angles and intensities of these diffracted beams to create 3-D pictures of the density of electrons in these crystals. Since many things can form crystals—salts, minerals, metals, even semiconductors, as well as different types of molecules—x-ray crystallography has proven to be very useful in examining things at a molecular level. The most famous discovery using this technology so far has been the structure of DNA. X-ray crystallography has also helped researchers determine the structure of diamonds, table salt, proteins, bacteria, and viruses. Here's how it works: a sample with high purity and a very regular structure is subjected to an intense x-ray beam. The sample is rotated, and as the rays reflect off it, a pattern emerges, corresponding to areas of electron density.

Electromagnetic Radiation

CAT SCAN

Computed tomography (CT), also known as computed axial tomography (CAT), has been hailed as the greatest advance in radiology since Wilhelm Röntgen's discovery of x-rays. It emerged in 1972 largely through the research of Dr. Allan MacLeod Cormack. A South African native, Dr. Cormack was a professor of physics at Tufts University and shared a Nobel Prize for the rationale behind CT. Linking x-ray and digital technology, a CT scan shows the body in cross sections, from which 3-D images are constructed. An x-ray tube revolves about a person's head, for example, converting the images into a digital code. Differences in density between normal and abnormal tissues are revealed as well as bone details and the location of tumors and other signs of disease. Sophisticated computers create 3-D views of tissues by taking multiple x-rays, each at a slightly different angle. Software programs then reassemble and superimpose the data into a series of images so that interior structures can be viewed at various angles and depths.

Patient preparing for a computerized axial tomography (CAT) scan

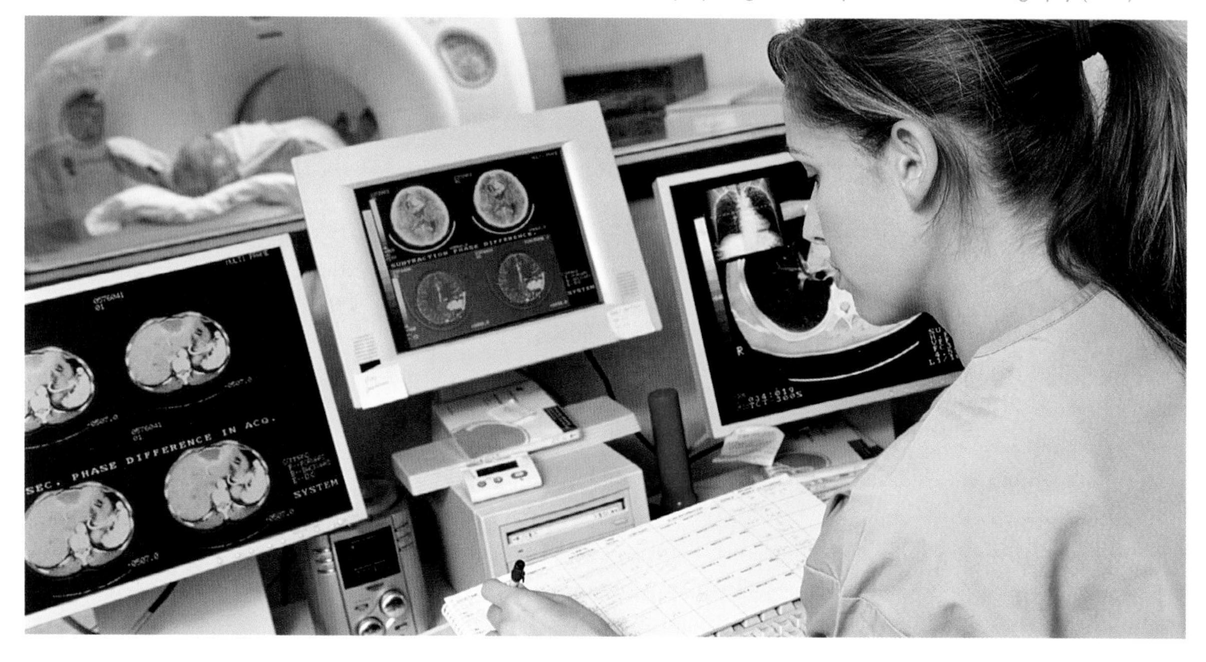

»»See Also

RELATED PRINCIPLES: Health and Medicine, 354 • Magnetism, 135 • Power and Electricity, 163 • Wave Energy, 58
RELATED APPLICATIONS: Magnetic Resonance Imaging, 142

TSA BODY SCANNERS

Controversy has surrounded the decision of the Transportation Security Administration (TSA) to employ body scanners at airports in the wake of the attacks on September 11, 2001. Regardless of which side of the debate you fall on, the technology behind it is pretty cool. A low-intensity beam of radiation is fired at a person, and the reflected signal is processed to create an image. The Rapiscan scanner uses low-level x-rays to create what looks like a naked image of screened passengers to target weapons hidden under the clothes. The TSA terminated the contract with Rapiscan, however, when the company failed to meet a deadline on developing software that would make the controversial scanners less invasive. A second type of TSA scanner, built by L-3 Communications Holdings, uses radio waves and shows hidden objects on an avatar image on a screen—not on an image of a passenger.

Despite the TSA's claim that more than 99 percent of passengers choose to be screened by these scanners and that, according to a CBS poll, four out of five Americans support the use of the body scanners at airports, in January 2013 the TSA said it would remove scanners from airports, and under court order the agency began a process of collecting public comments on such screening procedures.

PART 3
MATERIALS AND CHEMISTRY

CONTENTS

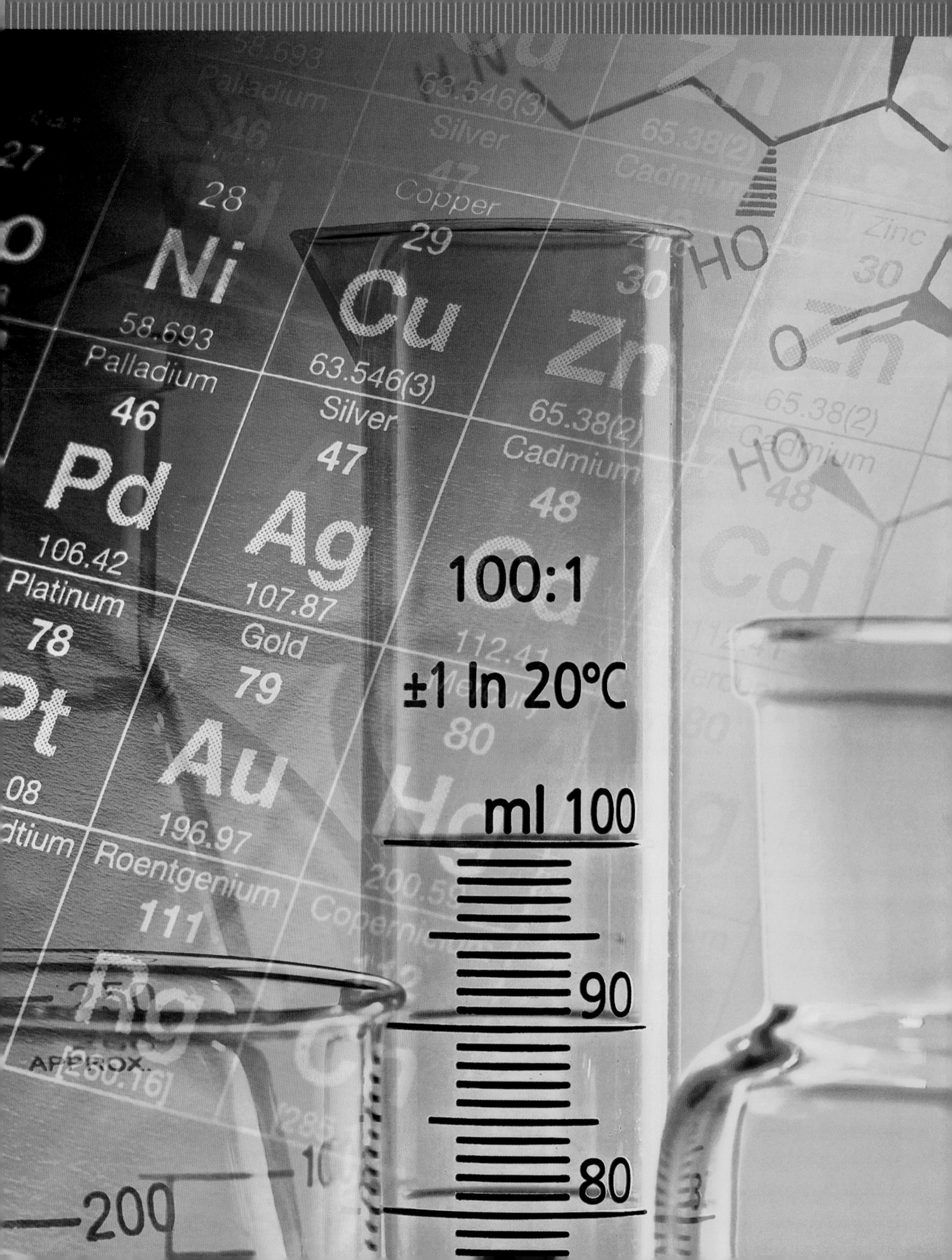

28
Ni
58.693
Palladium
46
Pd
106.42
Platinum
78
Copper
29
Cu
63.546(3)
Silver
47
Ag
107.87
Gold
79
Au
196.97
Roentgenium
111
Zinc
30
Zn
65.38(2)
Cadmium
48
112.41
80
Copernicium
100:1
±1 In 20°C
ml 100
90
80
HO
APPROX.
200

CHAPTER 11

Elements

We are surrounded by elements. They're part of the air we breathe, the food we eat, the clothes we wear, and the sidewalks we walk on. If it weren't for the centuries of study that led to the discovery of these elements and their properties, the world would be a very different place.

The elements are the most basic building blocks of chemistry, and the science of chemistry examines their properties and ability to be combined to create compounds and materials.

Most of the materials, tools, and conveniences we use can be traced to the discovery of an element or its combination with others. Each new element on the periodic table (a table that lists the chemical elements according to the number of protons each has in its nucleus) has meant another step forward for humankind, from the discovery of neon to gold and silver and their hundreds of uses.

The discovery of neon forever changed the world at night, thanks to the creation of neon signs that can be seen miles away and drive customers to a business's door. Thanks to the element lithium, all our electronics stay charged longer, including what has become one of the most vital electronic instruments to humankind, the cell phone.

Without iodine, humans would suffer from thyroid disease, and aluminum's production led to lighter-weight aircraft and space exploration.

Nations' economies have been based on gold since 700 B.C. (beginning in the ancient civilization of Lydia), and our eating utensils and jewelry would be lusterless without silver. Chlorine keeps our drinking water safe. Without silicon the world wouldn't have sand, clay, or electrical steel. Argon keeps the filaments in lightbulbs from corroding, and titanium creates brilliantly white paint.

Without understanding the properties of these and the other elements, life would be more difficult, dimmer, less productive, and, most likely, less enjoyable.

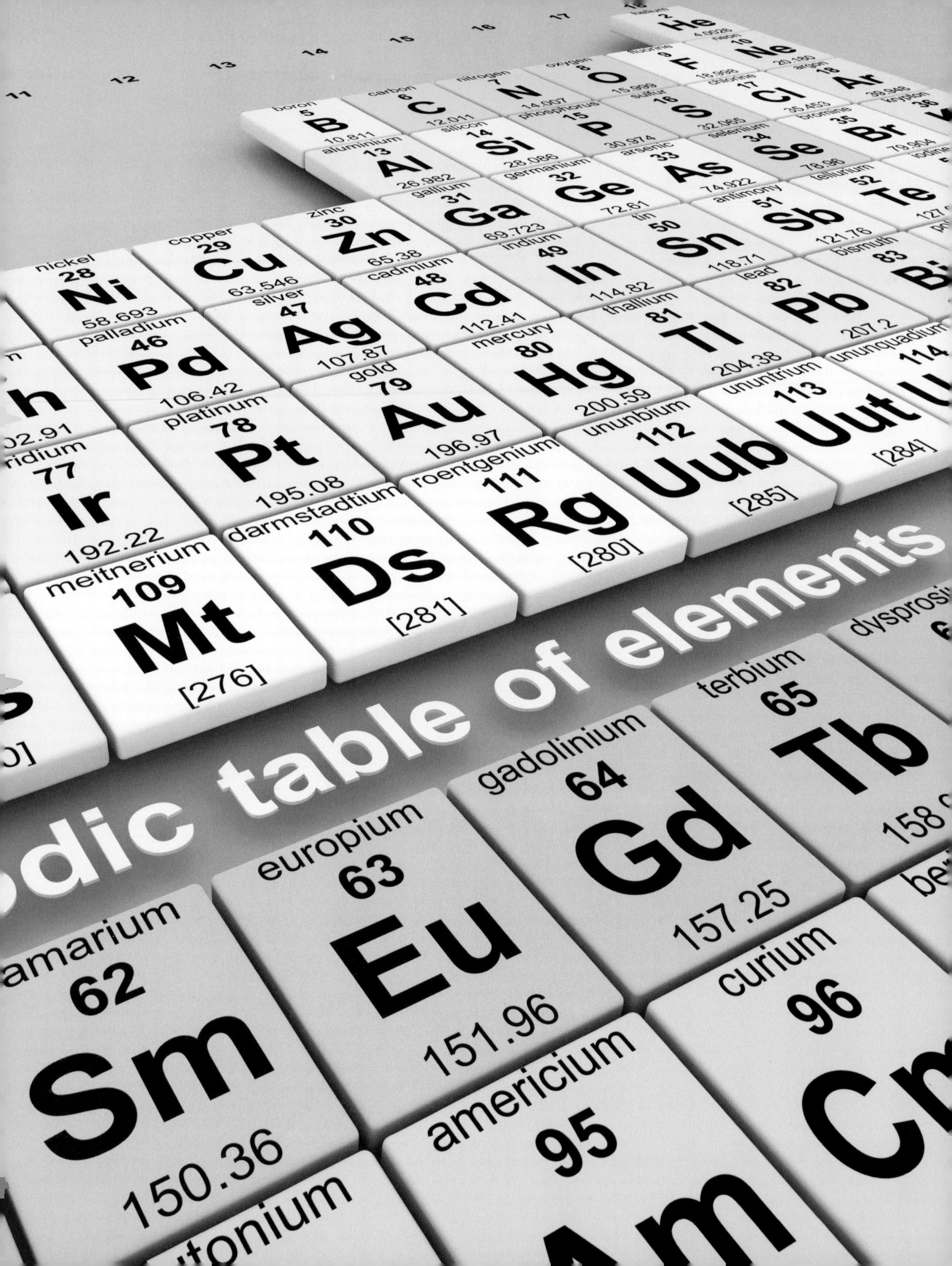

dic table of elements
boron 5 B 10.811
carbon 6 C 12.011
nitrogen 7 N 14.007
oxygen 8 O 15.999
fluorine 9 F 18.998
neon 10 Ne 20.180
helium 2 He
aluminium 13 Al 26.982
silicon 14 Si 28.086
phosphorus 15 P 30.974
sulfur 16 S 32.065
chlorine 17 Cl 35.453
argon 18 Ar 39.948
gallium 31 Ga 69.723
germanium 32 Ge 72.61
arsenic 33 As 74.922
selenium 34 Se 78.96
bromine 35 Br 79.904
indium 49 In 114.82
tin 50 Sn 118.71
antimony 51 Sb 121.76
tellurium 52 Te
thallium 81 Tl 204.38
lead 82 Pb 207.2
bismuth 83
nickel 28 Ni 58.693
copper 29 Cu 63.546
zinc 30 Zn 65.38
palladium 46 Pd 106.42
silver 47 Ag 107.87
cadmium 48 Cd 112.41
iridium 77 Ir 192.22
platinum 78 Pt 195.08
gold 79 Au 196.97
mercury 80 Hg 200.59
meitnerium 109 Mt [276]
darmstadtium 110 Ds [281]
roentgenium 111 Rg [280]
ununbium 112 Uub [285]
ununtrium 113 Uut [284]
ununquadium 114
samarium 62 Sm 150.36
europium 63 Eu 151.96
gadolinium 64 Gd 157.25
terbium 65 Tb
americium 95
curium 96

PRINCIPLE

IT'S ELEMENTAL

Atomic Structure

An element is a substance in its simplest form; it cannot be broken down into other substances. Each element has a specific number of protons in its nucleus. For example, all krypton atoms have 36 protons. If you were to add or subtract a proton, it wouldn't be krypton anymore.

Chemists have been discovering elements for centuries, sometimes by accident, as when Bernard Courtois discovered iodine in 1811 when he was using seaweed ash to create saltpeter (potassium nitrate, mainly an ingredient in explosives). Currently there are 118 known elements; they appear in the periodic table (originally developed in 1817 with 63 elements) in order of the number of protons they have. Hydrogen appears first because hydrogen atoms all have one proton. Only 94 of these elements occur in nature; the rest are synthetic.

After chemists discovered these elements, they began to combine them to form compounds. They learned that some elements didn't react with others—they were stable—while other combinations resulted in chemical reactions. That led to the idea of valences: each atom of an element has an outer electron shell that contains a set number of valence electrons, which is eight for light elements. When an element already contains the full number, it doesn't react when added to others. The stable or noble gases—helium, argon, neon, krypton, xenon, and radon—already have eight valence electrons. The other elements, when combined, form new chemical bonds, losing some valence electrons or gaining some from the other element, so that their total valence electrons becomes that magic stable number, eight.

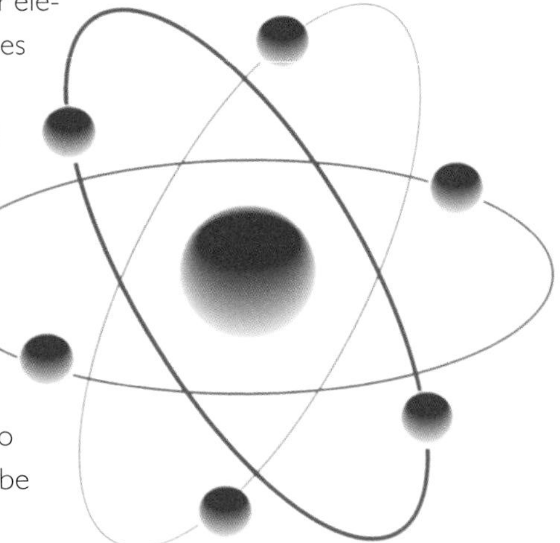

Chemists have discovered these elements over the years by finding ways to isolate them from compounds that contain them, like Courtois discovering iodine in seaweed. Sometimes they were discovered by accident when searching for a different element. Once they were found or synthesized, like technetium, chemists' next step was to understand their properties, which could then be harnessed by technology.

Atomic Structure

NEON

Neon's discovery led to the perfect marriage between two industries: the neon sign industry and car sales. It started with the determination of Scottish chemist William Ramsay. Having discovered argon in 1894, Ramsay knew there had to be an element between argon and helium.

In 1898 he decided to analyze argon for the answer, freezing a sample of it using liquid air (air is liquefied by lowering its temperature to about −374°F [−190°C]). Ramsay and his fellow researcher, Morris Travers, slowly evaporated the argon under reduced pressure, collected the first gas that came off, and then applied a high voltage to that gas in a vacuum tube to obtain the gas's spectrum. The result was dramatic: a flare of bright crimson light—neon.

Since its discovery, neon's primary role has been as the main ingredient in bright lights and signs for businesses. The first neon signs in the United States were sold to a Los Angeles, California–based Packard dealership in 1923. It was the first of a continuing and important tradition of signage in the car sales industry. In 1960 neon was combined with helium to create gas lasers, and today these lasers can be found in academic and optical research labs due to their low cost and low power requirement.

See Also

RELATED PRINCIPLE: Ohm's Law, 164

RELATED APPLICATIONS: Argon, 274 • Barcode Scanners, 230 • Laser, 232

The Salar de Uyuni salt desert in southern Bolivia holds around half the world's reserves of lithium.

Atomic Structure

LITHIUM

Lithium boosts power. Compared with the traditional lead-acid battery, lithium-ion batteries produce six times the charge. Its most prevalent use is in high-energy rechargeable batteries, which revolutionized the electronics industry when they were first commercialized and introduced to the market in the 1990s. Lithium batteries in cell phones and computers have given them a much longer life than alkaline batteries would, making them much more practical. Alkaline batteries are still the main choice for most other applications.

In 1949 psychiatrists began using lithium ions from lithium carbonate as a drug to inhibit the manic periods of patients with manic-depressive (or bipolar) disorder because of its ability to stabilize moods. Lithium may also have neuroprotective effects for individuals at risk of developing Alzheimer's disease by inhibiting the protein responsible for forming amyloid plaques and neurofibrillary tangles.

Lithium is one of only three elements (hydrogen and helium are the other two) believed to have been produced in large quantities by the Big Bang and was present when Earth was created. Its atomic number is 3 (it has three protons), and it was discovered in 1817 by Johan Arfvedson.

In addition to batteries and medical treatment, lithium is used to create light but exceptionally strong alloys used in aircraft and heat-resistant glass and ceramics. Ovenware made with lithium oxides comprises the largest use of lithium worldwide.

See Also

RELATED PRINCIPLES: Electrochemistry, 187 • Ohm's Law, 164
RELATED APPLICATION: Lithium-Ion Battery, 187

Atomic Structure

IODINE

A very tiny amount of iodine is all that stands between humans and thyroid disease. In the human body, iodine is an essential ingredient for cell metabolism and for the thyroid to produce its hormones. Very small amounts of iodine are added to salt so that humans can receive enough to prevent thyroid disease. Lack of iodine can create a goiter (a lump in the neck due to the expansion of the thyroid gland). An isotope of iodine has become a treatment for thyroid cancer, and iodine is also used with other ingredients as an antiseptic. Another commercial use of the element is in printing inks and dyes.

This element, number 53 in the periodic table, occurs naturally in all types of seaweed and in brines and brackish water and was discovered accidentally in 1811 by Bernard Courtois.

Courtois was using seaweed ash to create saltpeter (potassium nitrate, mainly used for explosives) when he accidentally added too much sulfuric acid; a violet-colored cloud was the result. This cloud settled on metal objects in the room and became solid iodine.

»»See Also
RELATED PRINCIPLE: Immunobiology, 357
RELATED APPLICATION: Magnetic Resonance Imaging, 142

Atomic Structure

ALUMINUM

Aluminum is everywhere, from soda cans to airplanes, but it was once considered a rare metal. Even though it's the most abundant metal in Earth's crust and the third most abundant element in Earth's interior, it's not found in a pure form anywhere on Earth's surface. Until the 1880s, it sold for nearly $1,200 per kilogram.

After two chemists, Charles Martin Hall in the United States and Paul L. T. Héroult of France, each discovered a way to extract aluminum from aluminum oxide in 1886, it opened the door for mass production. Karl Joseph Bayer, an Austrian chemist, also developed a method for extracting it cheaply from bauxite, an abundant ore. Hall began manufacturing aluminum in 1888, and by the early 1900s, he was producing such great quantities of it that the price fell to $0.60 per kilogram.

Aluminum's light weight and strength make it the perfect material for plane and automobile manufacturing, and it can be found in several other commercial products, such as drink cans and window frames. A great conductor of electricity, it's part of electric transmission lines around the world and used to produce cooking utensils. It's also popular because it can be easily recycled.

»»*See Also*

RELATED PRINCIPLE: Faraday's Law and Electromagnetic Induction, 157
RELATED APPLICATION: Airplane, 70

Atomic Structure

GOLD

Humankind's fascination with gold is unequivocal, and hundreds of myths, legends, and stories abound of people who were willing to do anything to get their hands on it. Let's not forget the sad myth of King Midas and how he turned his daughter into gold.

Even though historians can't agree on when gold was first discovered, it's been found in ancient Egyptian tombs and was first mined by the Egyptians as early as 2000 B.C. It was first exchanged as money in 700 B.C.

Perhaps one reason humans have held gold in such awe is its amazing properties. It's the most malleable of all metals as well as the most ductile, meaning that it can be hammered into shape and stretched into wire without breaking. It's a great conductor of electricity, and it never tarnishes or decomposes under normal conditions. That means your pure gold necklace could've been worn by a woman in ancient Egypt.

Ancient philosophers conducted an inexhaustible search for a method of turning common metals into gold. These alchemists, as they were known, were ultimately unsuccessful. It was from their tinkerings, however, that modern chemistry emerged.

Gold is rare. Incredibly, it's estimated that all the gold in the entire world would fit into a 60-foot cube. Nearly 85 percent of gold mined today is made into jewelry and measured in carats (the proportion of gold to the alloy in pieces of jewelry—24 carat gold is pure gold).

The combination of nanoparticles of a radioactive isotope of gold (a form of gold with an unstable nucleus that stabilizes itself by emitting ionizing radiation) and a compound found in tea leaves is showing promise as a treatment for prostate cancer. Treatment targets radioactivity to the site of the tumor and causes far less damage to healthy tissues than chemotherapy does. And while some people are allergic to gold, therapy using gold sodium thiosulfate, or gold salts, reduces joint inflammation for rheumatoid arthritis.

Gold is also used in electrical parts such as circuit boards and connectors due to their high conductivity and low susceptibility to corrosion, but don't tear your computer apart to find them; the very minute amount they contain isn't worth it.

»»See Also
RELATED PRINCIPLES: Electromagnetism, 147 • Faraday's Law and Electromagnetic Induction, 157
RELATED APPLICATIONS: Loudspeakers and Headphones, 148 • Silver, 269

Atomic Structure

SILVER

Second only to gold when it comes to humans' fascination, silver has a long history. It was first mined around 3000 B.C. and subsequently graced people's necks and dinner tables. When the first Europeans arrived in the Americas, they found silver already a part of the native culture.

Sterling silver is the purest form, at about 92 percent silver, and even after hundreds of years, its market is jewelry and silver tableware such as eating utensils, pitchers, and trays.

Silver's light weight and high reflectivity make it an ideal choice for creating mirrors, and its high conductivity has made it a popular choice for solder, electrical contacts, and in batteries. Painted circuit boards also contain the precious metal. Unlike gold, silver tarnishes (turns black) over time due to the sulfates in the air.

»»See Also

RELATED PRINCIPLE: Absorption of Light, 204
RELATED APPLICATIONS: AC/DC Edison Versus Tesla, 165 • Gold, 268 • Reflecting Telescopes, 211

Atomic Structure

CHLORINE

Next time you take a refreshing dip in a swimming pool, remember that its water contains an element that was part of a biological weapon during World War I.

Chlorine, a yellowish green gas first discovered by the German-Swedish chemist Carl Wilhelm Scheele in 1774, is a genuine good news/bad news story. The good news is its value as a bacteria killer. Used in the right proportion in drinking water and swimming pools, it's a good thing. But in its gaseous form or

in larger quantities, it's toxic and can corrode skin. Its attributes are the product of its instability—it has just seven valence electrons.

The United States is a leading producer of chlorine, despite its potential hazards. Chlorine starts its life as salt. The chlorine is extracted from salt. Manufactured paper, textiles, petroleum products, medicines, antiseptics, insecticides, solvents, paints, and plastics all contain chlorine. Fifty million tons of chlorine are produced every year.

»»See Also
RELATED PRINCIPLE: Natural Resins, 283
RELATED APPLICATION: Cellulose, 276

Atomic Structure

SILICON

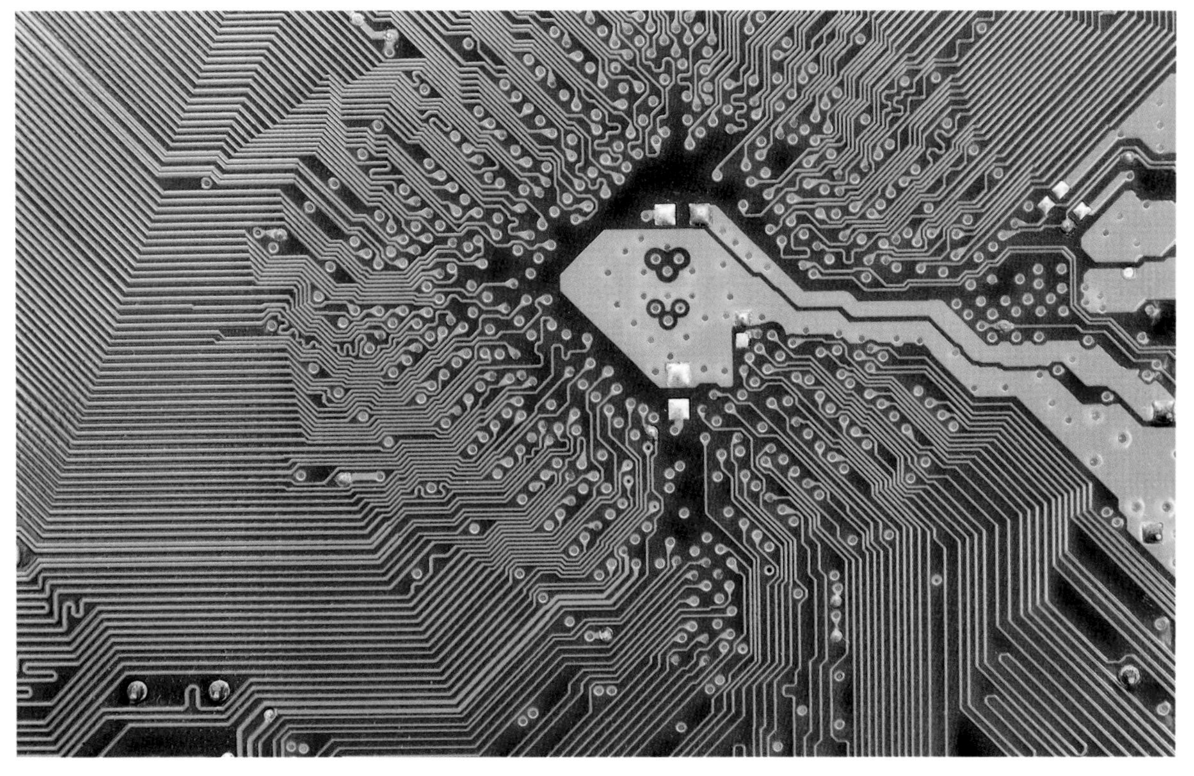

Silicon, a highly effective semiconductor, has been at the center of the microelectronics industry.

You may not realize it when you're out for a hike, but you're surrounded by silicon. Silicon is the second most abundant element in Earth's crust. It's contained in rock such as quartz and in amethyst, agate, and opals. There's a lot of it on the beach.

Even though it's so abundant, silicon wasn't isolated and discovered until 1824 by Swedish chemist Jöns Jacob Berzelius of Stockholm.

The silicon derived from sand is one of the most widely used elements today. This sand is used almost everywhere you look, from the sidewalks you walk on (concrete) to the glasses you drink from. Silicon is also a component of electrical steel (such as that used in motors and transformers), pottery, and enamels.

Sand dunes in the Sahara

The microelectronics industry owes a great deal to its discovery (think Silicon Valley). Because of its high melting point and effectiveness as a semiconductor, it's used in electronics in the form of transistors, solar cells, and similar instruments.

»»See Also
RELATED PRINCIPLE: Electrochemistry, 187
RELATED APPLICATIONS: Aluminum, 267 • Graphene, 290 • Integrated Circuit, 191

Atomic Structure

ARGON

The noble gas argon is valued not for what it does, but what it doesn't do. Argon is present in less than 1 percent of the air, but it's considered the third most abundant gas on Earth. And its most valuable characteristic is that it's almost completely inert, meaning that it doesn't react or combine with other gases. That makes argon the perfect element to fill light bulbs, for example. This marvelous stability is thanks to argon's eight valence electrons. If it weren't for argon, our light bulbs would give off too much heat, and the filament would break.

A nice, inert blanket of argon is used in manufacturing processes, in arc welding, and in making semiconductors. It's also an important element in the medical industry as part of ophthalmology lasers that treat blood vessel leakage, macular degeneration, and glaucoma.

Bubbles of argon gas trapped in polystyrene gel

Argon can be used as a shielding gas in metal arc welding.

See Also

RELATED PRINCIPLES: Absorption of Light, 204 • Electromagnetic Radiation, 248 • Ohm's Law, 164
RELATED APPLICATIONS: Incandescent Light Bulb, 249 • Neon, 264

Atomic Structure

TITANIUM

Titanium's special properties justify its being named after the mythological Greek gods, the Titans. For one thing, titanium is as strong as steel but much less dense. That, along with its ability to withstand very high temperatures, makes it the ideal ingredient to combine with other alloys in the aerospace industry. It's also malleable and ductile at high temperatures, allowing it to be shaped.

Titanium is highly unreactive to seawater. It takes 4,000 years for seawater to corrode its surface. Unsurprisingly, it's a vital component of the hulls of ships.

One of titanium's famous compounds is titanium oxide, which is so pure white in color that it's used in paint, another of titanium's primary uses.

Titanium is also the only element that will burn in pure nitrogen.

See Also
RELATED PRINCIPLE: Macromolecular Chemistry, 288
RELATED APPLICATIONS: Aluminum, 267
• Gold, 268 • How We See, 227 • Jet Engine, 26

STRONG STUFF. Titanium is used to manufacture sports equipment, including golf clubs and tennis rackets.

CHAPTER 12
Polymers

and Resins

When you're holding a plastic bowl or trash bag in your hand, what you're actually holding is a series of polymers. Polymers are very long chains of molecules. While some occur in nature, such as starch and cellulose, there are many synthetic polymers, from plastic bags to nylon.

Natural polymers are found in the human body, animals, plants, and minerals. In animals and plants, polymers—for example, cellulose—play very important roles. They provide structure and energy.

Synthetic polymers have replaced many natural polymers because chemists have learned how to engineer them with specifically helpful characteristics. For instance, synthetic rubber is less sensitive to temperature differences and sunlight than its natural counterpart and is often less expensive to manufacture (unless natural rubber prices decrease). Polymers, whether natural or synthetic, are everywhere, from tires to paper and fabric.

From jewelry to perfume, humankind has found many great purposes for resins. Plants secrete resins—usually a clear liquid such as pine trees' sap. Because resins can be hardened using chemicals and made indissoluble in water, they are excellent sealants. Plus, they are often aromatic, which has led humans to make perfume and incense from them. Amber, a fossilized tree resin, is valued for its beauty and made into jewelry. In some ancient cultures amber was prized as highly as gold.

Although resins do not have as many applications as polymers, humans continue to find new purposes for resins, from varnishes to food glazes.

PRINCIPLE

A CHAIN REACTION

Natural Polymers

Molecules bound together in long chains have strength and other properties that make them important, desirable, and very helpful. Take starch, for example. Starch is a natural polymer found in potatoes of all varieties and other plants. Its hundreds of glucose monomers, or molecules that repeat in a chain, are what provide the "starchiness" humans love to consume and use for other applications. Found in plants and animals, natural polymers become building material and storage units, and they can initiate important biochemical reactions. Over the years, humankind has taken advantage of their best characteristics. For example, silk is a natural polymer that is 78 percent protein, giving it its highly prized strength.

You can visualize polymers like chain necklaces, many single monomers, or small molecules linked together, with at least 1,000 atoms in a chain. It's this huge number of bonds that gives polymers their desirable characteristics, from softness to brittleness to pliancy and bounce.

Natural Polymers

CELLULOSE

Humans and cows have something in common: we both like to eat cellulose. One of the most abundant organic compounds on Earth, cellulose is a natural polymer contained in most plant's cell walls. Cotton is an example of cellulose in its most natural form. Cellulose was first isolated by the French chemist Anselme Payen in 1834.

Cellulose fibers in trees contain a polymer called lignin, which is pliant, but strong and durable because the position of its glucose molecules is opposite that of other carbohydrates. These fibers are the main ingredient of paper. Film, explosives, and plastics may all contain cellulose. It's also an important part of the human diet. Humans get cellulose by eating lettuce and vegetables.

Recycled pressed wood

»»*See Also*
RELATED PRINCIPLE: Macromolecular Chemistry, 288
RELATED APPLICATIONS: Chlorine, 270 • Polyethylene, 294

Natural Polymers

SHELLAC

For us, shellac is a safe and popular sealant. But if you're a lac bug, it's a lifesaver. Lac bugs infest trees in Asian forests and feast on tree sap, which then is secreted through their pores and forms a tough shell, protecting them and their young. The adults die while the young bugs break free from the shell and seek new feasting trees. The lac is harvested by breaking off millions of twigs covered in the stuff and taking them to a factory, where they are scraped off and processed into shellac. According to Ramesh Singh at the Department of Zoology at Udai Pratap Autonomous College in India, it takes 300,000 lac insects to make a kilogram of lac resin.

A natural polymer, shellac is a combination of hydroxyl acids (the acids that define alcohols) and carboxyl groups found in organic acids. It's this combination that causes shellac to harden when exposed to air and form elaborate constructions.

Shellac was once an extremely popular sealant used in construction, but its use has declined since the invention of synthetic resin compounds in the mid-20th century. Shellac is a clear, thick resin that dries quickly.

The first manufacturer of shellac was William Zinsser & Company of New York, which began producing the sticky stuff in 1849.

Today natural shellac is appealing to manufacturers because a glaze made from shellac is safe to use as a coating on candy (including Milk Duds and Raisinets), pills, and fruit, and for sealing furniture for babies and children. Fruit, like apples, are coated with edible shellac to preserve them, replace natural wax lost in processing, and make them look more appealing.

See Also
RELATED PRINCIPLE: Molecular Chemistry, 288
RELATED APPLICATION: Superglue, 293

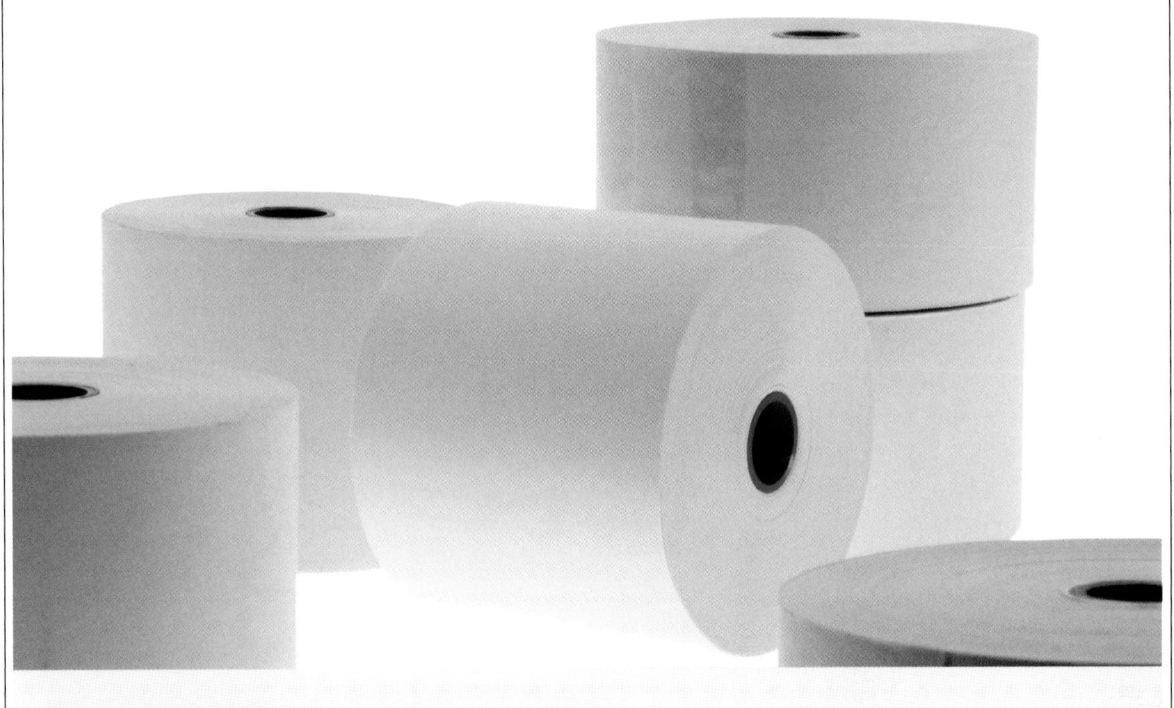

PAPERMAKING

The reports of paper's death have been greatly exaggerated. Despite the fact that the world is steadily becoming an e-world, paper is still everywhere, from toilet paper in the bathroom to currency, menus, and coffee filters—the list is virtually endless.

Human beings weren't the first creatures on Earth to make paper. Paper wasps have been making it for much longer. They chew tiny pieces of wood until a paste is created, which they then spit out. When the paste dries, it's paper. The wasps use this paper to build their nests.

As with so many other important advancements, the Egyptians were among the first people to create a papermaking process using papyrus. The papyrus was woven into a mat and then pounded flat into a sheet. Paper—in its current form and the actual word—was derived from this process.

Paper starts its life as pulp, which comes mainly from trees, paper waste, cotton, flax, and sugarcane. Wood becomes pulp as it's held against a huge grindstone. The pulp moves through a machine that removes the water, leaving the cellulose fibers that become matted together. These sheets are fed through heated rollers that flatten and dry them.

Paper can be recycled or repulped five times before the fibers become too short to reuse. Paper production has slowly declined since 2000 as more people try to reduce their consumption. Newspapers, books, and other forms of reading material are becoming more popular in their e-formats, but other uses of paper, such as in packaging and food uses (paper plates, napkins, etc.) are still prevalent. While the industry has been producing less, it's also been producing more sustainably, pulping from hemp and other nonwood sources. Paper-pulping mills also have managed to dramatically reduce the amount of dioxin that is released into water sources. Dioxin is the name for dozens of toxic chemical compounds that are used to whiten paper. The common ingredient in all of them is chlorine, which makes for a highly toxic by-product. Manufacturers achieved this great improvement in the amount of dioxin in the water by replacing chlorine with chlorine dioxide, which degrades far more rapidly than chlorine in water and blood and poses significantly reduced health risks.

Natural Polymers

RUBBER

It all started with an excellent bounce. Natives in Central America, where there were plenty of rubber trees, had developed rubber balls and used them in games amazingly similar to volleyball and basketball. The first European explorers were very impressed by these balls. After all, they'd been using balls made of leather, and the bounce didn't really compare.

Once a man named MacIntosh in England realized rubber was also a great repellent of water and created boots out of the material, it wasn't long before many other applications for rubber were developed. Rubber coats, balls, and medical devices were just a few of them. Rubber's long and flexible chain of carbon atoms, with the occasional hydrogen atom thrown in, is what creates its amazing elasticity. Rubber's history turned a corner in 1844 when Charles Goodyear patented a processing involving heating rubber and sulfur (vulcanizing process) and later, in 1888, when John Boyd Dunlop invented the pneumatic tire (a rubber tire filled with air). Once rubber was heated, it wouldn't become sticky in high temperatures or brittle in cold ones.

After World War II, the world relied more on synthetic rubber, but natural rubber still has many uses. Rubber trees were planted in Asian colonies of England in the 19th century, and by the 21st century, more than 90 percent of the world's natural rubber supply came from Asian countries such as Thailand, Indonesia, Malaysia, India, Vietnam, China, Sri Lanka, the Philippines, and Cambodia. More than 50 percent of all rubber, both natural and synthetic, goes into manufacturing tires. Bouncy rubber balls, such as basketballs, are still made from both natural and synthetic rubber.

»»See Also
RELATED PRINCIPLE: Macromolecular Chemistry, 288
RELATED APPLICATION: Neoprene, 291

SYNTHETIC RUBBER

Although forms of synthetic rubber had been known since the 1890s, it was not until the early 1930s that Wallace Carothers at DuPont developed the first commercially viable artificial rubber. Neoprene hit the market in 1932, and it was a success due to its ability to resist sunlight, abrasion, and extreme temperatures better than natural rubber or any of the earlier synthetics.

Natural rubber comes from plantations in Asia and Africa, and the synthetic version is manufactured across the globe due to its lower cost and easier-to-acquire materials. Its widespread use has outpaced natural rubber since 1960, and you'll find it in tires, shoes, tools, textiles, and adhesives. It is also used in the aviation and space industries.

Dr. Wallace Carothers

PRINCIPLE

A STICKY SITUATION

Natural Resins

Cover your skin with a layer of resin, and you'd feel like you were encased in plastic. Resin, a clear substance that can be hardened or altered with chemicals, is an essential ingredient in adhesives, varnishes, perfumes, and for food glazing.

One major function of resin in plants is protection. Some resins are toxic to harmful insects or pests, while others attract predators to eat the pests. The resin produced by most plants is composed primarily of terpenes (for example, citrus and cinnamon), which give it its thick, sticky texture that has inspired so many of its applications. What gives terpenes their unique characteristics is the fact that they are composed of five-carbon building blocks called isoprenes, which are volatile—they evaporate easily. This volatility is also why they can be quite fragrant.

Natural Resins

INCENSE AND PERFUMES

From spirituality to protecting the skin, incense and perfumes have been part of our culture for thousands of years.

Incense was the forerunner of all perfumes, and our ancestors in Mesopotamia were the first to develop it 4,000 years ago. The ancient people of Mesopotamia burned tree resin and wood during religious ceremonies, soaked them in oil and water, and then applied the liquid to their bodies.

Soon the ancient Egyptians were also using oils and incense in their religious ceremonies, to bury their dead, and just to make their bodies smell good. In modern times, the scents and perfume industry is worth billions of dollars. We have a complete wardrobe of perfumes or scents for purposes from freshening ourselves and our homes to improving our moods. Today, the uses for incense range from the religious—preparing us to commune with God or warding off evil spirits—to simply improving the smell of our homes.

»»See Also
RELATED PRINCIPLE: Natural Polymers, 276
RELATED APPLICATION: Antimicrobials, 306

Natural Resins

VARNISH

Preservation has been important to humans for centuries. Historians in the ninth century were the first to record the use of varnish, a protective liquid composed of resins, natural oils, and alcohol (among other ingredients). Furniture makers learned that if they painted coats of the stuff on a piece of furniture or floor, it made that item impermeable to water and water damage, as well as normal wear and tear. Plus, they just liked the glossy shine the furniture had after it was varnished.

Varnishes come in many formulations. Some are oil-based, using linseed or tung oil, while others are water-based. The varnish cures, or dries, onto the item, creating a glossy, clear film on the surface. Some varnishes can turn the wood a yellowish color. Water-based varnish does not. The wood must be sanded and cleaned before the varnish is applied. With some varnishes, these steps must be repeated several times. All varnishes contain resins, terpenes with five-carbon molecules called isoprenes, but they also include a drying oil or solvent to reduce drying time.

While furniture and floors are still the most often varnished items, hulls of wood boats are also varnished to protect them from long exposure to seawater. And after centuries, no one has developed a method for applying it that speeds up the process. It's applied the same way it has been for hundreds of years, one coat (of many) at a time. Another thing about varnish that hasn't changed: people still like the way their pieces look after the varnish is applied.

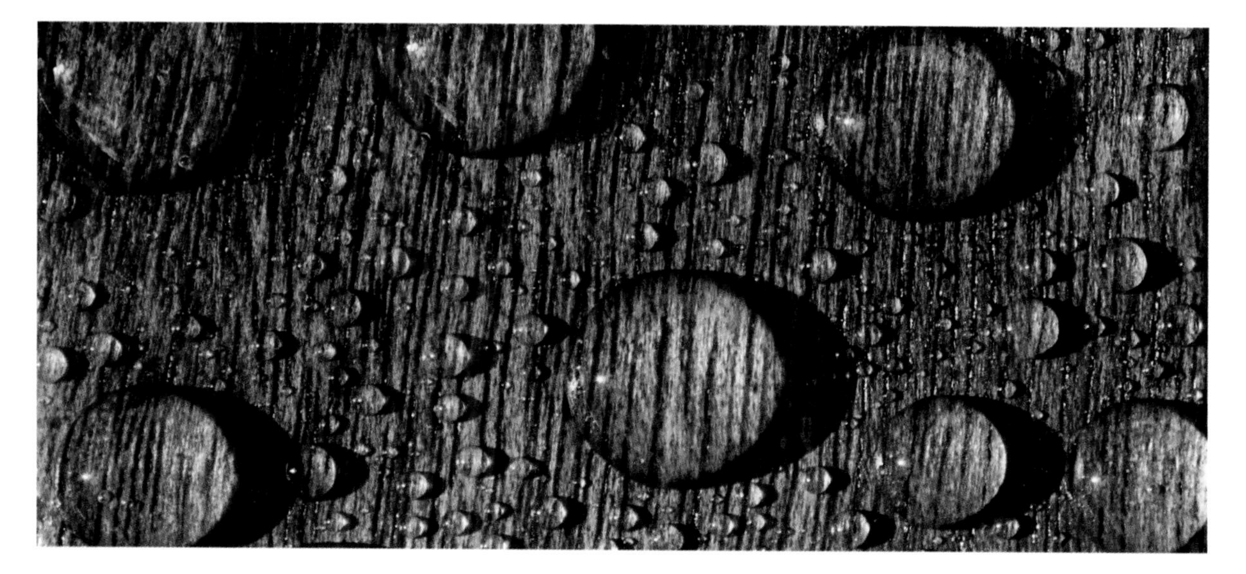

»»*See Also*

RELATED PRINCIPLE: Natural Polymers, 276
RELATED APPLICATION: Shellac, 278

ROSIN

When you need to get a grip on something, nothing works better than rosin. Rosin is also known as colophony, and it's made from the resin of more than 100 types of pine and spruce trees. Manufacturers heat resin to separate out the essential oil, and what is left is solid rosin. Rosin is most commonly known by musicians who apply it to their stringed instruments, such as violins. They put the rosin on their bows or the bridges of their instruments to prevent them from slipping.

But many other people use rosin for the same reason. Bowlers use powdered rosin to keep their fingers from slipping out of the holes of their bowling balls, and gymnasts, ballet dancers, bull riders, and baseball pitchers have all been known to use rosin when a firm grip is required.

Natural Resins

AMBER

Amber can be confusing—is it a fossil or a precious stone? It's actually both. Amber is tree resin that drips down the bark, entrapping anything in its path, from flowers to insects. Over the years, it falls off the trees and ends up in rivers or on the ground, where it continues to fossilize.

Despite the fact that some ancient cultures associated it with death, its gorgeous, shiny appearance was prized as highly as gold in some countries, where it was made into jewelry.

Over the years, its applications have increased, and it has become part of smoking accessories, teething rings, writing utensils, chess sets, and chandeliers. Although it's most famous for its amber color, it can also be milky white, green, and reddish.

In 1996 one of the most famous pieces of amber was discovered by entomologist David Grimaldi, who was then with the American Museum of Natural History. It was a piece that contained three tiny flowers, probably from an oak tree, that dated to the age of the dinosaurs, some 90 million years ago.

As of the 21st century, amber's major market is the jewelry industry, and it's valued highly as a fossil. While it's not considered as valuable as gold, pieces that contain intact "debris" can be expensive—bad news for people born under the astrological sign of Taurus, whose signature stone is amber.

Dragonfly fossilized in amber

»»See Also
RELATED PRINCIPLES: Friction, 52 • Macromolecular Chemistry, 288
RELATED APPLICATION: Stringed Instruments, 54

PRINCIPLE

MODERN ALCHEMY

Macromolecular Chemistry

If molecules are not simply long chains of atoms, and their chemical bonds can be understood and manipulated, then the world really becomes humankind's oyster. Macromolecular chemistry is modern alchemy. When more than 100 atoms get together in repeated building blocks to form such things as RNA, plastics, or proteins, they follow rules that may be used to understand their composition, properties, and mechanisms. Prior to 1917, when the German chemist Hermann Staudinger introduced the idea that true polymers (molecules composed of a chain of smaller units) existed, chemists held fast to the belief that they were simply aggregates of smaller molecules.

Staudinger upheld his belief in macromolecular chemistry alone until an American scientist, Wallace Carothers, used organic condensation reactions to create polymers in the 1920s.

In the 21st century, macromolecular chemistry has revolutionized industries and life and continues to do so. From what have become essential materials such as polymers, plastics, and nylon to materials under development, such as graphene, macromolecular chemistry has expanded humankind's horizons in a way these early scientists never thought possible.

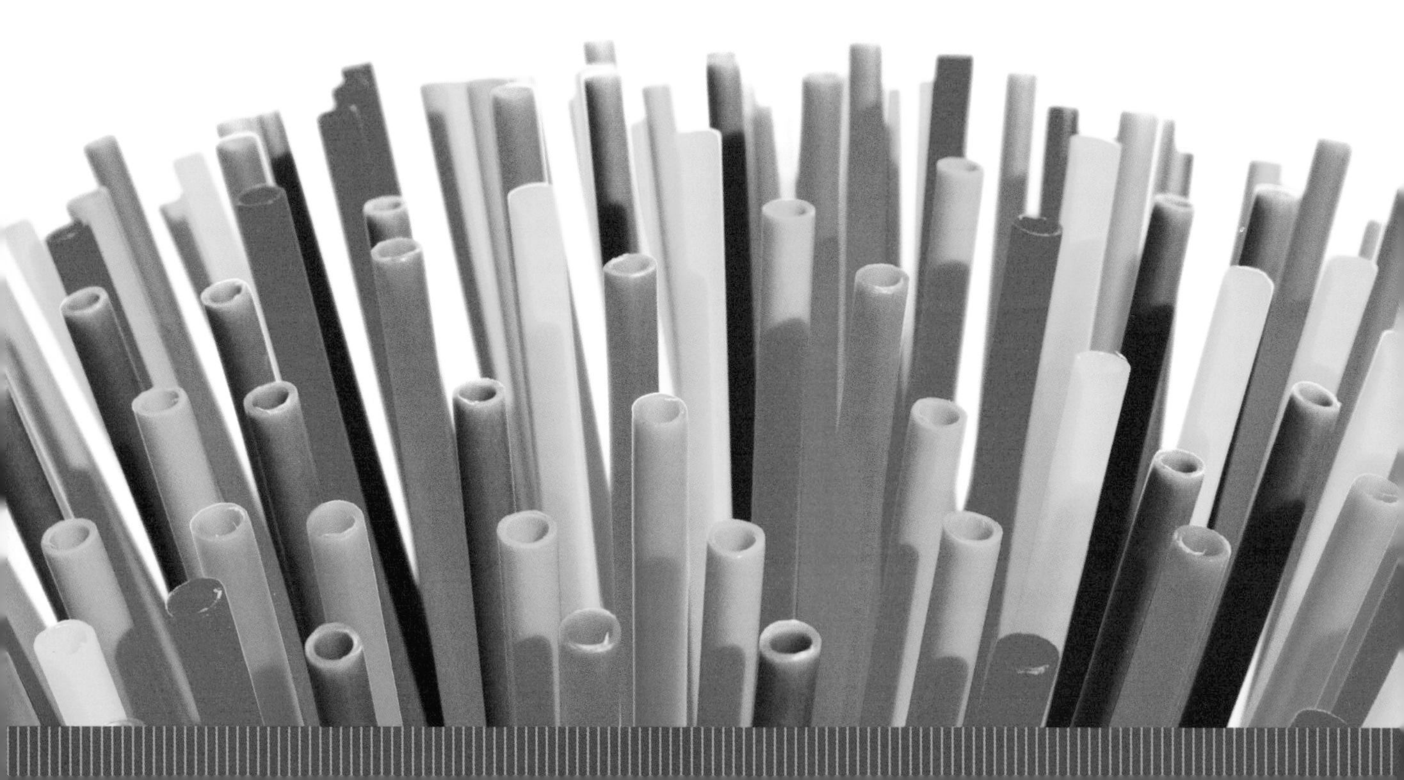

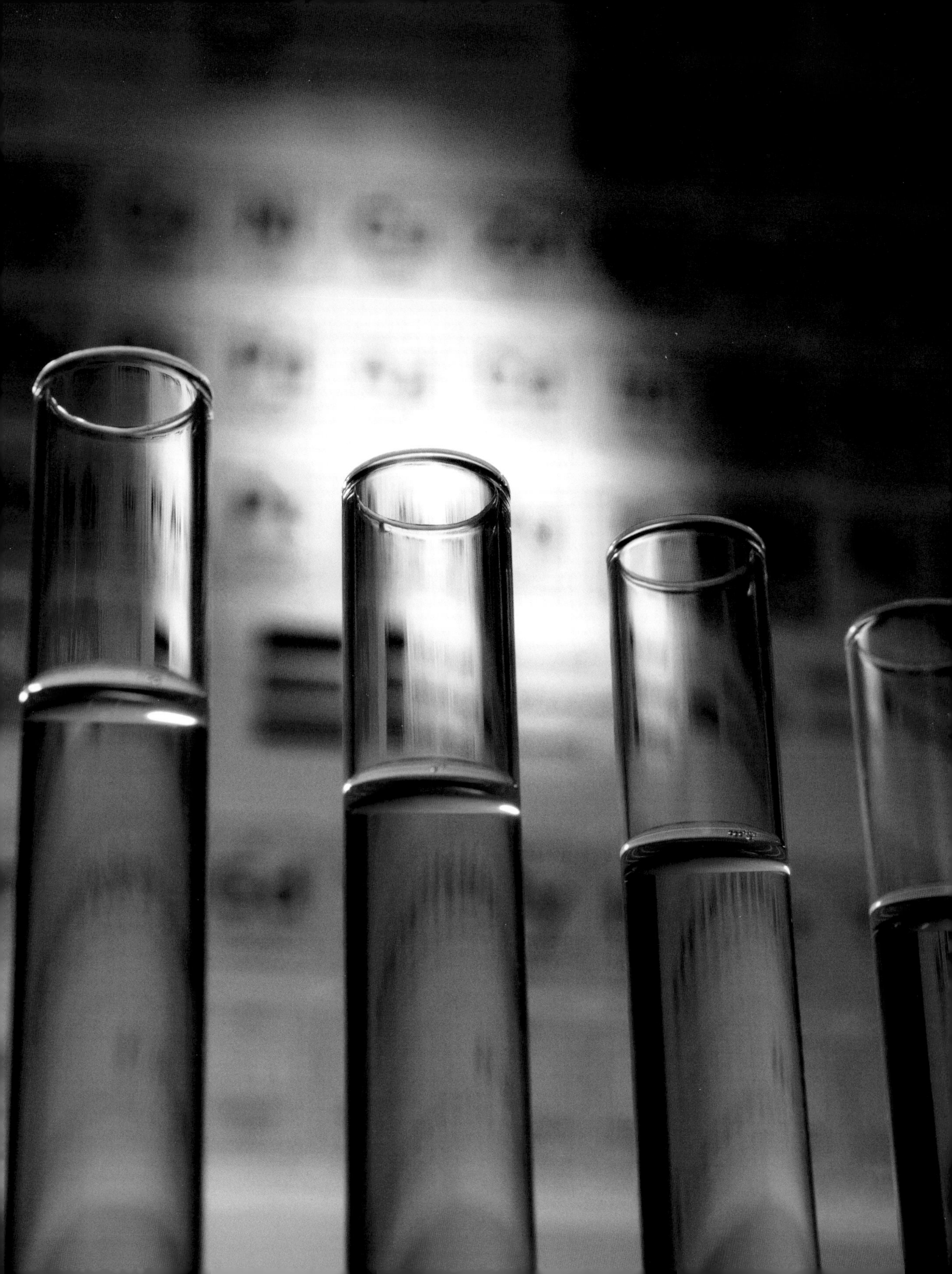

Macromolecular Chemistry

GRAPHENE

Forty times stronger than steel and with electrical conductivity 1,000 times better than silicon, it's not a super hero, it's graphene. Graphene, a one-atom-thick carbon structure that resembles a net, is also an excellent conductor of heat. Scientists have studied it since the 1940s but didn't take it very seriously until two University of Manchester scientists, Andre Geim and Konstantin Novoselov, created graphene and found it to be stable. Their work earned them the 2010 Nobel Prize in physics.

Since then, they and other researchers have continued to learn and explore the potential applications of this new "super" material. It has become the building block for nanotubes and three-dimensional graphite, all of which have improved properties over traditional or previously used materials in fields such as computing and solar energy. Its full of commercial potential.

What many people find exciting about the one-atom-thick material is that it is not only stronger and a better conductor of electricity than current materials but also easier to recycle and potentially much less costly to produce. In 2012 scientists from the University of Manchester discovered that graphene can repair holes in itself spontaneously. This new property makes it an even more attractive material for electronics and medical and solar energy applications. In January 2013 Rice University researchers created a two-dimensional graphene device that combined a conductor with an insulator. It might result in even smaller electronic devices in the future.

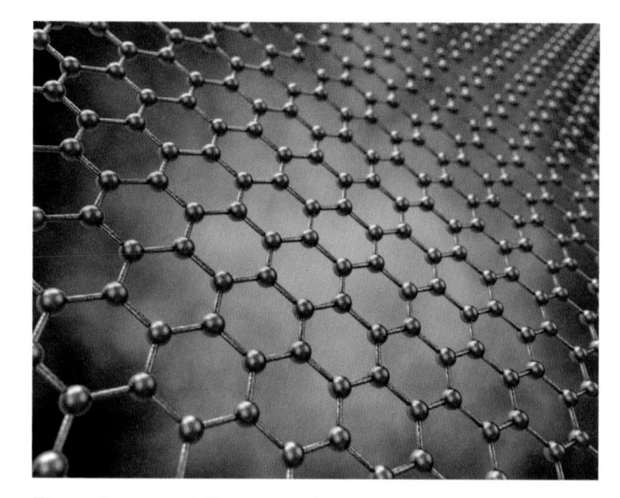

Three-dimensional illustration of a graphene sheet

»»See Also
RELATED PRINCIPLE: Nanotechnology, 296
RELATED APPLICATIONS: Integrated Circuit, 191 • Silicon, 271 • Water and Air Purification, 302

DESALINATION AND NANOPOROUS GRAPHENE

Imagine having a limitless supply of drinking water at your fingertips. That dream is becoming a possibility, thanks to the amazing filtration property of nanoporous graphene.

Researchers at the University of Manchester in England discovered that graphene is impermeable to everything but water. This led scientists at the Massachusetts Institute of Technology (MIT) to create simulations that verify that nanoporous graphene, a one-atom-thick form of carbon, can desalinate seawater more efficiently, and potentially less expensively, than current methods—making the world's oceans one massive drinking fountain.

Macromolecular Chemistry

NEOPRENE

Humankind has often sought ways of creating its own form of natural substances, and rubber is a good example. Neoprene, invented by scientists at DuPont, was developed as a substitute for natural rubber, which was not abundant. Our dependence on it, for example in tires for vehicles, prompted this discovery.

Both natural rubber, which is a natural polymer derived from latex (that comes from tree sap), and neoprene are formed by polymerization (the bonding of two or more monomers to form polymers), but there is a subtle difference in their chemistries. Natural rubber is 95 percent hydrocarbons, while neoprene is primarily chloroprene (the combination of acetylene and hydrochloric acid).

Neoprene was not only a great substitute for rubber, but it could withstand much higher temperatures than its natural counterpart as well. For this reason, it has become part of many insulators and insulating products. These include gaskets and hoses and weather stripping for fire doors. It also weighs much less, is inert to chemicals, and is quite water resistant. This combination of properties makes it ideal for many applications. Soon divers were donning neoprene wet suits, and scientists were wearing neoprene gloves to prevent chemical burn. By the 21st century, our laptops and cell phones were wearing neoprene protectors to prevent damage.

»»*See Also*
RELATED PRINCIPLE: Natural Polymers, 276
RELATED APPLICATIONS: Competitive Swimwear, 55 • Rubber, 280

MEDICAL APPLICATIONS FOR POLYMERS

Although most people aren't aware of it, DNA, RNA, and polypeptides are considered polymers, just like plastics and other synthetic material. These biopolymers have led to experimental treatments, such as gene therapy, and new methods of drug delivery. For example, in gene therapy, new genes are added to a patient's cells—usually introduced using a virus—to replace that patient's defective or missing gene.

Macromolecular Chemistry

NYLON

When a naturally occurring resource is in short supply, humankind is on the spot to create its own version. And in most cases, we have improved upon the natural version and found hundreds more purposes for it. This is true of nylons. First invented as a substitute for silk stockings by Wallace Carothers of DuPont, its improved tensile strength and durability also made it the perfect material for manufacturing parachutes. Silk's high protein content, coupled with the fact that it's a natural polymer consisting of thousands of atoms bonded together, made it a desirable material to efficiently replicate.

Although American women were fond of their nylons by the late 1930s, they sacrificed them to the war effort. During World War II, all nylon manufactured was purchased by the military. Nylon stockings on the black market sold for as much as $10 a pair.

Nylon is a popular material in hundreds of products from backpacks and umbrellas to jackets, swimwear, and sportswear. It's a bit too durable, though—because it does not naturally decompose, it has become a bane of landfills. One company, Interface, has found a solution to this problem. The company developed an efficient method for recovering nylon fibers and recycling them into high-end carpets, which it has been doing since 2007.

See Also

RELATED PRINCIPLE: Natural Polymers, 276

RELATED APPLICATION: Parachute, 53

Macromolecular Chemistry

SUPERGLUE

It might surprise you to know that your mouth may contain superglue, which has become an ingredient in dental fillings. Since its invention in 1942, the amazing adhesive that "sticks to anything" has become a component of many formulas and has found its way into several less commonly known products. Its properties can be explained in part by the fact that it's a product of covalent bonding, or shared electrons.

Superglue was discovered accidentally when scientists were working to create clear plastic gun sights during World War II. They rejected the material because it stuck to everything. Then Eastman Kodak researchers "rediscovered" it in 1951, and by 1958 it hit the market as an extra-sticky glue. The Vietnam War introduced the first medical usage of super glue to stop soldiers' bleeding in the field until they could be transported for conventional treatment. Its success under emergency conditions eventually led to approval by the Food and Drug Administration.

By the early 21st century, a medical-grade version of superglue had been created, and doctors used it for suturing and laceration repair.

A scientist is glued upside down by his boots to demonstrate the strength of a new superglue.

»»See Also
RELATED PRINCIPLES: Immunobiology, 357 • Natural Polymers, 276
RELATED APPLICATION: Shellac, 278

PLASTIC BAG

Before the question "Paper or plastic?" came into existence, plastic bags appeared on the scene in the late 1950s as sandwich bags and to protect dry cleaning. Soon they were popping up everywhere, proliferating for everything from bread packaging to trash. Today they are blown on sheets of melted petroleum, or ethylene, pellets. Though most are used only once, waste-reduction approaches include Japanese inventor Akinori Ito's carbon-negative machine that converts plastic bags and other trash back into petroleum.

Macromolecular Chemistry

POLYETHYLENE

What would the world be like without Hula-Hoops and Tupperware? If it weren't for polyethylene, we would know the answer. Discovered accidentally by English chemists E. W. Fawcett and R. O. Gibson of Imperial Chemical Industries in 1933, polyethylene was first utilized as an insulator in British radar cables. Polyethylene is basically a long chain of two hydrogen atoms on either side of a carbon atom.

After the war, developed nations soon saw the many benefits that a durable plastic like polyethylene could give us. Two types, high density and low density, are manufactured as packaging, containers, and in even more important applications such as cold water pipes, insulation, and water vapor barriers in construction.

»»See Also
RELATED PRINCIPLE: Natural Polymers, 276
RELATED APPLICATION: Cellulose, 276

Macromolecular Chemistry

3-D PRINTING

Ever think how wonderful it would be if you could design a three-dimensional object via computer and then "print" it into being? That's what 3-D printing, a technology that began as early as 1984, is all about. It first started with Charles Hull's invention of stereolithography, which gave engineers and designers the ability to create 3-D models of their designs by printing them. The printer consisted of a laser and materials such as plastic and polymers to create the models.

A MakerBot Replicator 2X 3D desktop printer

Thermoplastic, a material that becomes liquid when heated and solid when cooled, became a leading material in 3-D printing. It can create virtually any shape, and after it's cooled it's strong and durable.

Three-dimensional printing can create some surprising products. Researchers at Wake Forest University, for example, used 3-D printing to synthetize a scaffold to repair the bladder. The scaffold is covered with the patient's own cells, so there is a very low risk of rejection. By 2002 these researchers had employed a similar technique to create a functioning kidney for transplant, and by 2008 customized prosthetics had been created with the method.

It seems no object is too large to be created using a 3-D printer. In 2011 the world's first 3-D-printed car was unveiled by Kor EcoLogic. It gets 200 miles per gallon on the highway and 100 miles per gallon in the city.

»»» *See Also*
RELATED PRINCIPLES: Bioengineering, 364 • Quantum Mechanics, 298
RELATED APPLICATIONS: Electric Car, 160 • Inkjet Printer, 171 • Prosthetic Limbs 370

PROGRAMMABLE MATTER

The ability to give commands—in essence to run code through material—lies at the heart of programmable matter. Working with millimeter-sized parts, scientists at the MIT's Center for Bits and Atoms have created a robot that can fold into complex shapes. The so-called milli-motein, named for its behavior resembling that of proteins, uses electromagnetism to change states by driving a steel ring. Reconfigurable robotics means that one machine will function in many tasks. What one researcher called "turning data into things" is being explored on scales from the nano to the person-sized.

CHAPTER 13

Nanotec

hnology

There is tremendous power and potential in a technology that is too small to see with the naked eye. Nanotechnology, which first began as an idea in 1959 and didn't become a reality until the 1980s after the discovery of the scanning tunneling microscope, is about manipulating materials that are less than 100 nanometers in size. To put that in perspective for you, according to the National Nanotechnology Initiative, there are 25.4 million nanometers in 1 inch. If a marble were one nanometer wide, then one meter would be the diameter of Earth.

So what makes these extremely tiny materials so powerful? Scientists are finding ways to build nanotechnology materials at the atomic level that have amazing properties. By filling nanotubes with titanium oxide and applying them to cotton, for example, you can create a T-shirt that blocks the sun's ultraviolet rays. Fill nanorods with gold, and you can develop a therapy that only affects cancer cells, leaving healthy cells alone and drastically reducing side effects. Nanotube carbon membranes can dramatically improve water purification systems.

Researchers at the University of Buffalo have developed a nanomaterial-based process to produce hydrogen, and the process doesn't require any light, heat, or electricity. This brings humankind one step closer to affordable, hydrogen-powered cars. Researchers at Stanford University have developed a nanoparticle-based battery that is able to store five times more energy than batteries in use today. It's made up of an inner core of sulfur surrounded by an outer layer of porous titanium oxide, and its architecture resembles the yolk and shell of an egg.

These are just a few examples of some of the current and potential applications of this revolutionary technology. As nanotechnology and its materials become more widely produced, people will be using cell phones whose batteries last three times longer, driving cars whose windshields won't fog and that repel water, and wearing socks that never get smelly. And no one likes smelly socks.

Nanotechnology products can potentially save lives as well as give us hundreds of new conveniences from enhanced drug delivery and disease detection to fabric that won't stain. It has the potential to touch nearly every aspect of our lives.

PRINCIPLE

SMALL MATTER

Quantum Mechanics

Before scientists could work with nanomaterials, they had to understand them. Quantum mechanics employs mathematics to predict the behavior of very small, discrete units of energy and matter.

The two scientists given the most credit for quantum mechanics are Max Planck, who conducted his famous "black box" experiment and learned that light was made up of specific "quanta" of energy in 1900, and Niels Bohr, who used quantum mechanics as part of his description of how an atom works in 1922. These were very good theories but tough to prove except in experiments that tested behaviors of materials. In fact, Albert Einstein had proved the existence of atoms in 1905, but the atom itself was just too small to see.

About 60 years later, in 1981, after obtaining a huge amount of experimental data about atoms, the curtain over them lifted when researchers Gerd Binnig and Heinrich Rohrer developed the first working scanning tunneling microscope while working at IBM Zurich Research Laboratories in Switzerland. It was the first microscope that gave scientists the ability to see atoms. By the end of the 20th century, the first commercial applications of nanotechnology hit the market.

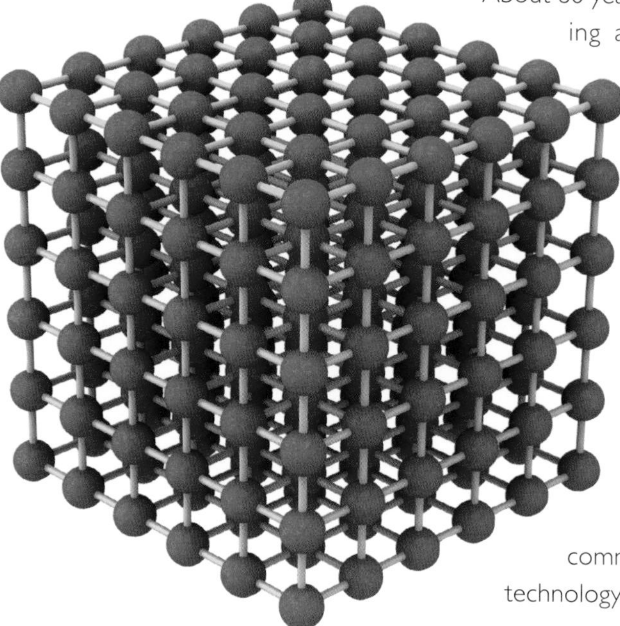

Three-dimensional representation of a crystal lattice structure

NANOTECH LAB. A development engineer works on a precision testing device for the analysis of applications in the field of nanotechnology.

Quantum Mechanics

TARGETED DRUG DELIVERY

Once the stuff of science fiction, nanotechnology is heralding a potential revolution when it comes to drug delivery. Like the doctor in an old episode of *Star Trek* who could cure a patient with practically the push of a button, physicians today can inject therapies that attack only the cells that carry disease. Researchers involved in the development of nanoparticle drug delivery vehicles say nanoscale-level drugs can provide patients with much fewer side effects and a greater degree of therapeutic benefit.

Imagine being injected with nanorods containing gold that target only cancer cells. This is just one example of several of nanotechnology's impact on health care. In the process being developed by researchers at MIT and the University of California at San Diego and Santa Barbara, the gold nanorods are absorbed by healthy blood cells.

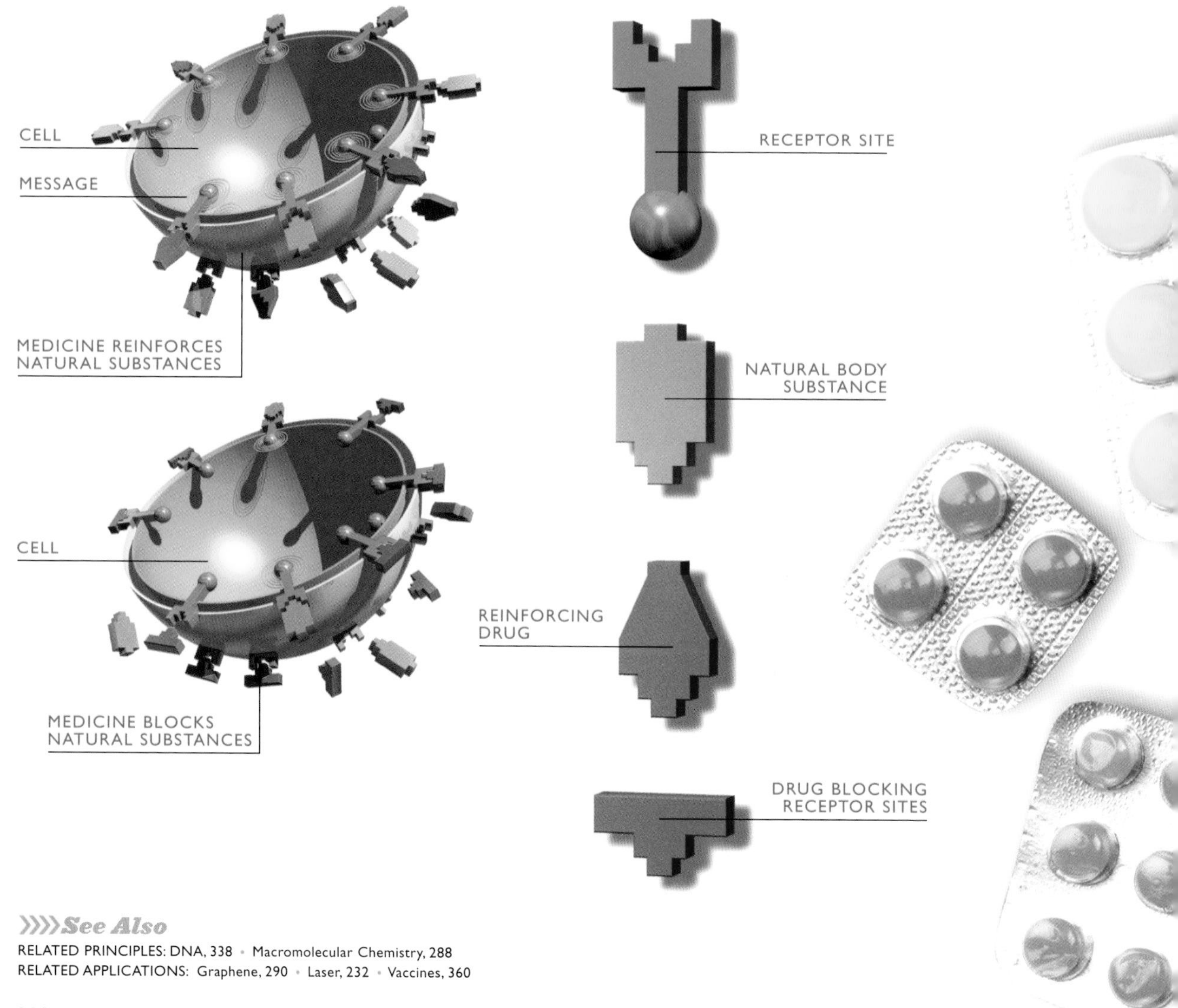

»»*See Also*
RELATED PRINCIPLES: DNA, 338 • Macromolecular Chemistry, 288
RELATED APPLICATIONS: Graphene, 290 • Laser, 232 • Vaccines, 360

But blood cells within a tumor are leaky and allow the gold nanorods to enter the tumor, where they build up. An infrared laser heats up the tumor cells, and this heat causes them to release a protein. After other steps, molecules of chemotherapy are inserted and attach directly to the tumor cells. Instead of every cell receiving the therapy, only cancer cells receive it. This means cancer cells targeted and side effects reduced.

A slew of products using this technology are currently in the development pipeline. Some that may appear on the market are bismuth nanoparticles that can greatly improve imaging of tumors, specific nanoparticles that absorb the free radicals (an atom with an unpaired electron that is unstable and can cause damage to the body) during radiation therapy, and targeted therapy to treat breast cancer tumors.

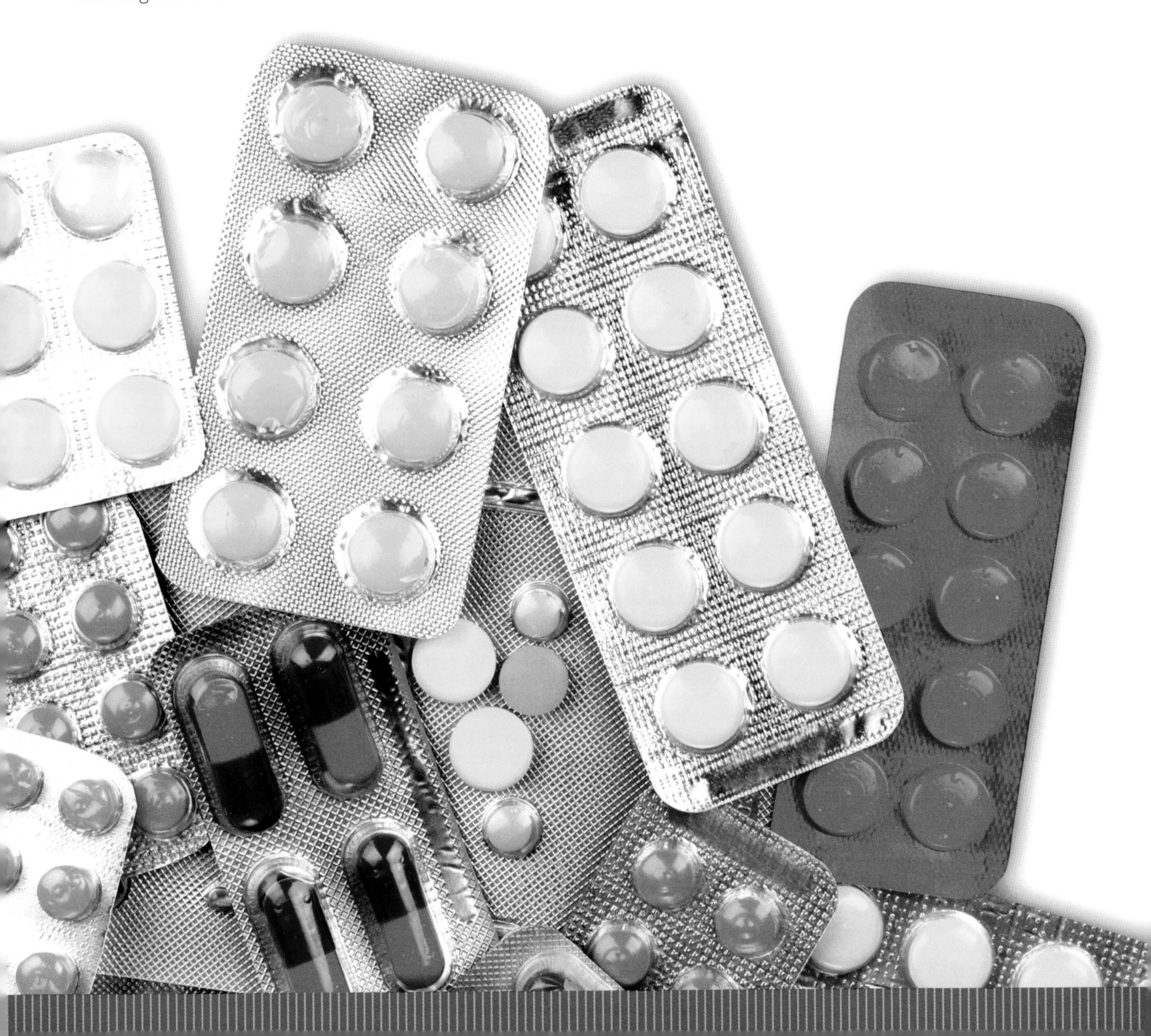

Quantum Mechanics

WATER AND AIR PURIFICATION

It seems that the phrase "the bigger, the better" will soon need to be changed to "the smaller, the better," as nanotechnologies are proving. One of their many applications is as a purifier of water and air. Researchers are excited by the potential uses of these systems, which include preventing and detecting pollution and remediating pollution sites.

Nanotube carbon membranes are already in use in some countries, where scientists have discovered that they have an equal or faster flow rate in purifying water than traditional filters, despite the fact that their pores are so much smaller. This is likely because of their smooth interiors. Nanotube carbon membranes can remove almost all kinds of water contaminants, from oil and bacteria to viruses and organic contaminants. Nanoscale water purifiers are also often easier to clean and cost less. Despite the fact that it sounds like a television commercial version of a paper towel, they are woven from tiny wires of potassium manganese oxide and absorb 20 times their weight in oil for cleanup applications.

The increasingly versatile element gold can also play a big part in nanotechnology focused on air purification. Associate Professor Zhu Huai Yong from Queensland University of Technology's School of Physical and Chemical Sciences says that gold nanoparticles that are energized can destroy airborne pollutants such as volatile organic compounds. Yong's team is developing a technology that combines gold with manganese oxide to harness this ability.

»»See Also
RELATED PRINCIPLE: Macromolecular Chemistry, 288
RELATED APPLICATIONS: Gold, 268 • Graphene, 290 • Vacuum Cleaner, 79

CHEMICAL AND BIOLOGICAL SENSORS

What's better than a highly trained, sensitive canine nose for detecting explosives? It's not a trick question. Since the first nanotechnology-based company began commercializing products in 1997 (a robot using nanotech systems), scientists have worked to create ultra-sensitive chemical and biological sensors. These sensors include nanoscale devices that are able to detect trace amounts of the target chemical, and by 2012 many were under development or becoming a reality. One example is Oak Ridge National Laboratory's sensor that will detect explosives, biological agents, and narcotics. Functionalized gold nanoparticles invented by laboratory researchers are modified and positively charged so that in as little as one minute they interact with and identify specific pollutants such as perchlorate or uranium. Current methods require laborious preparation of samples and lengthy laboratory analysis.

OIL AND TOXIC WASTE CLEANUP

Trichloroethylene, or TCE, is a great degreaser and notoriously stable carcinogen when it enters groundwater. Its bond of two carbon and three chlorine atoms breaks down a billion times faster when treated with gold-palladium nanoparticles than when remediated with iron filings. Also, the gold-palladium nanoparticles break all the bonds, whereas iron misses some, resulting in even worse by-products, such as vinyl chloride. Because of these by-products, air stripping and carbon adsorption are commonly used for TCE, meaning you dry it out or remove it—without changing it. If gold-palladium nanoparticles pan out, they could be put to work at more than half the nation's Superfund hazardous waste sites and save billions of dollars in the process.

Quantum Mechanics

FLEXIBLE DISPLAYS

Drop your cell phone and crack the screen? That won't happen if flexible OLED displays become the gold standard for display screens. OLED stands for organic light-emitting diode, and researchers are investigating methods of creating OLED displays that could result in flexible screens that don't crack or break when flexed. Not only that, these screens would be much lighter in weight, be able to provide high-definition displays, and would require less energy, making your cell phone battery last nearly twice as long. Flexible displays will allow users to roll up or fold their electronic devices or drop them without breaking the screen.

To compete with current displays, the OLEDs will need to be highly conductive, transparent, flexible, and less expensive. One method being considered uses silver nanowires that are deposited on plastic sheets. A method that uses sheets of metallic carbon nanotubes is also a contender. These sheets are extremely flexible, highly conductive, and transparent. Graphene, a one-atom-thick sheet of carbon, is also being considered as a material for creating OLED displays. The University of Arizona has manufactured a flexible OLED using advanced mixed oxide thin-film transistors. Samsung is developing a version of flexible OLED using plastic, and Sony is developing a version that uses an organic transistor.

Not only can this OLED display technology be incorporated into laptops, cell phones, and TVs, but it can also be employed as a filter on glasses and as a bendable, foldable display that can be used as a brochure or for marketing purposes.

Korean electronics company Samsung has been the leading manufacturer of OLED displays since 2004. It is also the largest holder of patents in the world for active-matrix OLED technology.

»»See Also
RELATED PRINCIPLES: Absorption of Light, 204 • Macromolecular Chemistry, 288 • Magnetism, 135
RELATED APPLICATIONS: Graphene, 290 • Plasma, LCD, and LED Screens, 290 • Visible Spectrum, 145

An organic
light-emitting
device in
the research
laboratory

Quantum Mechanics

ANTIMICROBIALS

Scientists have discovered that bacteria are smarter than they first thought. For decades, researchers thought they were mostly unorganized, free-floating organisms. By 2005, however, many forms of bacteria had become resistant to antibiotics, and scientists had discovered that some had bonded together, forming biofilms. Once together, they chemically communicated with each other and were effective in fighting a "host's" defenses.

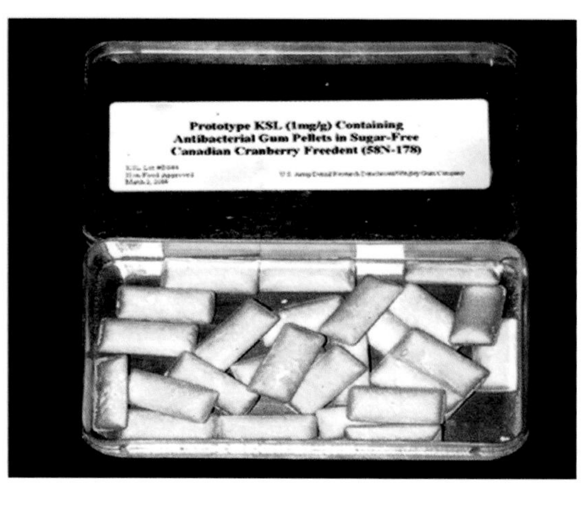

Thankfully, nanotechnology, in the form of specific nanoparticles, such as zinc and silver, may be able to break down these biofilms. Silver and zinc atoms hold a unique electrical charge called zeta potential. It's this charge that neutralizes the bacteria's communication without creating corrosive oxidation. Not only that, the nanoparticles can work as a barrier to prevent bacterial contamination in the ground, in fabrics, and in water.

Nanosized-antimicrobial silver is currently under investigation for a broad spectrum of medically relevant microorganisms, including bacteria, fungi, and yeasts. As little as one part per billion of silver may be effective in preventing biofilm cell growth.

One example of the application of this antimicrobial is as a treatment for dental plaque, which is actually a biofilm. Other potential applications include protective coverings on artificial hearts and stents and controlling or preventing biofilm production on contact lenses, urinary catheters, and joint implants.

»»» *See Also*
RELATED PRINCIPLES: DNA, 388 • Macromolecular Chemistry, 288
RELATED APPLICATIONS: Antibiotics, 359 • Silver, 269

ANTIMICROBIAL BANDAGES

Smart technology is everywhere, and now nanotechnology is responsible for the development of the world's first smart bandage, which will release antibiotics when it's triggered to by detecting "bad" bacteria. Its intended primary application will be on burn victims, half of whom die from bacterial infections.

Researcher Toby Jenkins of the University of Bath in the United Kingdom got the idea of turning forms of toxic bacteria on themselves. They use their toxins to rupture vesicles containing an antimicrobial agent.

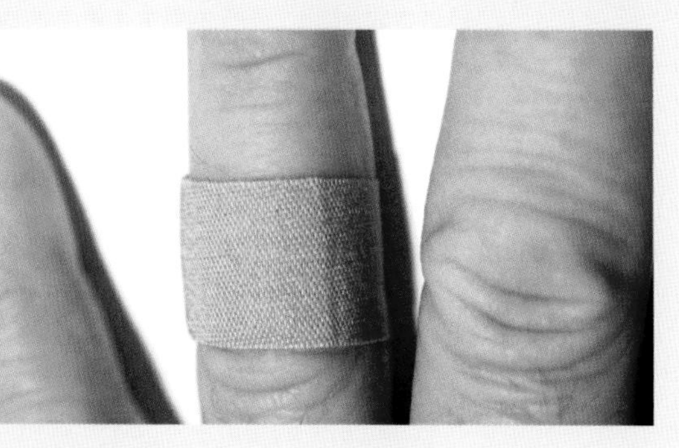

ODOR-REDUCING FABRIC

Embarrassed by excessively smelly socks? Nanoparticles of silver can take care of that embarrassing odor. Imagine cotton socks and other garments that prevent odor. Researchers at the Southern Regional Research Center operated by the U.S. Department of Agriculture's Agricultural Research Service are working on methods for inhibiting microbial growth in cotton. Using silver particles ranging from two to six nanometers in size, the method can be applied on all cotton fabrics, including socks and T-shirts. The silver acts as a catalyst and has germicidal effects. It kills harmful bacteria without affecting other animals.

Quantum Mechanics

UV-BLOCKING FABRIC

It turns out those lifeguards wearing zinc oxide on their nose to prevent sunburn were on to something. Titanium oxide is a chalky white substance, works like zinc oxide by reflecting the Sun's rays, and is a major ingredient in fabric that blocks ultraviolet (UV) light. The titanium oxide is contained in nanoparticles and applied to the fabric, enhancing its ability to protect the wearer's skin from the UV rays of the sun. It's now widely recommended that in addition to sunblock, people should limit exposure to harmful UV rays by covering up.

»»See Also
RELATED PRINCIPLE: Ultraviolet Light, 215
RELATED APPLICATION: Sunburn and Sunblock, 217

Quantum Mechanics

SELF-CLEANING GLASS

Most modern commercial buildings are faced in glass on four sides and that's a lot of windows to clean. Fortunately, nanotechnology has come up with a solution in the form of self-cleaning glass. Pilkington Glass in 2001 brought the first self-cleaning glass to market. They have since been joined by several others, but they all use the same science. The glass is coated with a thin film of titanium dioxide, that reacts with sunlight to break down the dirt on the glass. Then, it allows rain to wash away the dirt with almost no streaking because the surface coating causes the water to spread evenly across it. Because the self-cleaning is a two-step process, in order to work properly, it must be installed in a location that receives both sunlight and rain.

Self-cleaning glass roof based on TiO_2-nanoparticles

»»See Also
RELATED PRINCIPLES: Absorption of Light, 204 • Macromolecular Chemistry, 288
RELATED APPLICATIONS: Graphene, 290 • Rubber, 280

Quantum Mechanics

WATER- AND STAIN-RESISTANT FABRIC

Will a washer and dryer be obsolete someday? It's always a possibility, with nanotechnology. Researchers and companies are developing fabrics with greatly improved water and stain resistance, thanks to nanotreatments.

Two companies have created products based on this technology, drawing their inspirations from objects in nature. Shoeller Technologies noticed that tiny spheres on a lotus leaf repel water and dirt, so they created nanosphere technology that can be applied to fabric to mimic this ability.

Nano-Tex, of Oakland, California, took a peach as the inspiration for its nanowhisker technology. Its nanoscale fabric is coated with tiny whiskers that lift liquids right off the fabric.

Applications for these water- and stain-resistant fabrics aren't limited to clothing. That's only the beginning. Companies already have their sights on a host of other applications. Imagine your furniture, window coverings, and entire wardrobe could all be made from the stuff. Dusty drapes and grimy car upholstery could all become a thing of the past. Good-bye, frequent washings.

»»See Also
RELATED PRINCIPLES: Absorption of Light, 204 • Macromolecular Chemistry, 288
RELATED APPLICATIONS: Color-Boosting Detergent, 206 • Graphene, 290

PART 4
BIOLOGY AND MEDICINE

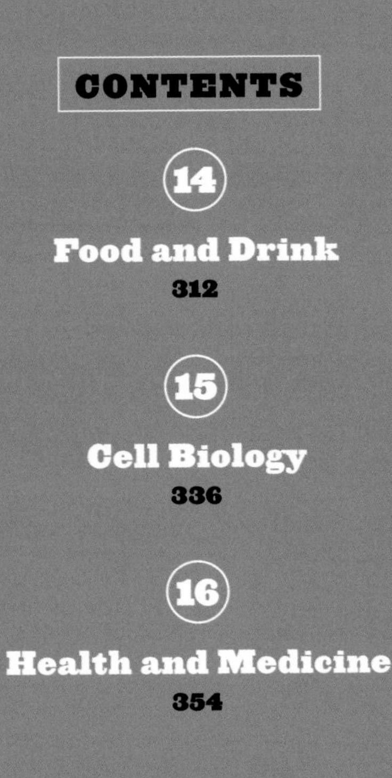

CONTENTS

CHAPTER 14

Food and

Drink

Consuming nutrients—eating and drinking—is, on one level, the ultimate no-brainer. Appetite is a basic instinct evolved over millions of years. Babies come into the world with a rooting reflex that causes them to turn and latch onto the breast. The tongue with its thousands of taste buds tells us what's especially good to eat—sugars and fats that help us store energy to protect against the possibility of famine—and what might be poisonous, such as the toxins in certain plants.

Food supplies us with energy to keep our bodies functioning and healthy. If we don't eat, we die after about five weeks. Water, which makes up more than half the human body, is an even more urgent necessity; we constantly replenish ourselves with it through food and drink. We cannot survive without water for more than three days.

Because it's surprisingly easy to make poor choices in what we eat, nutrition has become an important area of study. It wasn't until the early 20th century that scientists began to describe the nutritional requirements of infants—and to see that inadequate and often-contaminated weaning foods were a significant cause of infant mortality. Today, public health specialists attribute about a third of all cancer diagnoses and an even higher proportion of heart disease cases to poor diet.

As we understand more about our nutritional needs, we also are discovering new ways to meet them—to grow, prepare, and preserve foods—that would have amazed farmers and food workers of even the World War II generation. Through this process we are also coming to understand the impact our food production systems have on the environment. Developing the resources that fuel our own bodies is perhaps the most elemental way we interact with our environment. Because of this, food and drink are at the center of every culture on Earth.

PRINCIPLE

ADVANCED CULTURE

Fermentation

Fermentation is enjoying a revival—just look at all the bottles of kombucha, or fermented sweetened tea, on store shelves these days—but it's also one of the oldest methods of food preparation, indispensable to such ancient staples as bread, wine, and soy sauce.

Fermentation uses the natural process of decomposition—rotting—to preserve food, make it easier to digest, or lend it a stronger, tangier (sometimes literally intoxicating) flavor.

When yeasts, molds, or bacteria consume foods for the energy they need, these tiny organisms convert carbohydrates, such as starches and sugars, to alcohol or an acid. Each organism produces a particular waste product that imparts some desirable characteristic—the lactic acid that preserves and gives pungency to cheese, for example, or the ethanol that gives beer its mild kick and contributes to its aroma.

The human role in all this is to carefully manage the conditions necessary to promote the right kind of rotting. Salt may be used to control the growth rate of yeast or curtail contamination with undesirable bacteria. Temperature is also important; some organisms like it warm, while others prefer it cool.

Fermented vegetables in a traditional food market in South Korea

Checking the fermentation process at an industrial brewery

Fermentation

BEER

Etched into a 4,000-year-old Mesopotamian clay tablet is the oldest known recipe for making beer. The process hasn't changed much since then. First, a grain, most commonly barley, is malted—that is, steeped in water, allowed to germinate, and dried. During malting, enzymes convert starches into simpler sugars that can be digested by yeast. The malt product is mashed and mixed with water to form a sweet liquid, which after filtering and boiling is ready to be fermented with brewer's yeast.

It takes about ten days for the yeast to consume its sugary repast, producing as waste products two of the most valued ingredients of beer—alcohol, of course, and carbon dioxide, the source of its fizz.

Brewers choose particular starting grains and yeasts to achieve subtle variations in flavor. Though barley is the usual choice, wheat, corn, rice, rye, and oats are also used. There are two main categories of yeast strains. Top-fermenting yeasts ferment quickly and tend to produce thick surface foam; they're used for brewing robust ales. Bottom-fermenting yeasts, used to make the relatively light, clean lager beers, grow more slowly and tend to settle. Bitter hops are also added to cut the sweetness of barley malts.

Hops

»»**See Also**

RELATED PRINCIPLES: Macromolecular Chemistry, 228 • Food Microbiology, 329

RELATED APPLICATIONS: Bread, 317 • Cheese, 321 • Vinegar, 319 • Wine, 318 • Yogurt, 321

Fermentation

BREAD

The next time you enjoy a slice of toast for breakfast, be sure to thank yeast. These single-celled fungi ferment the bread dough, converting its sugar into alcohol and carbon dioxide. The carbon dioxide fills the dough with bubbles and makes what might have been a hard cracker into a chewy, airy loaf. The alcohol, though burned off in baking, lends the bread its rich yet subtle flavor.

Of course, the yeast is only pursuing its own ends, consuming the sugar it needs to grow. These sugars are produced when water breaks down the starch in flour; a wetter, softer dough will ferment more quickly. Yeast also needs warmth to thrive, favoring temperatures in the range of 85°F (about 30°C). Place a bowl of dough in a drafty window, and you'll wait a long time for it to rise. The high temperatures used for baking, on the other hand, kill yeast, so the loaf stops rising.

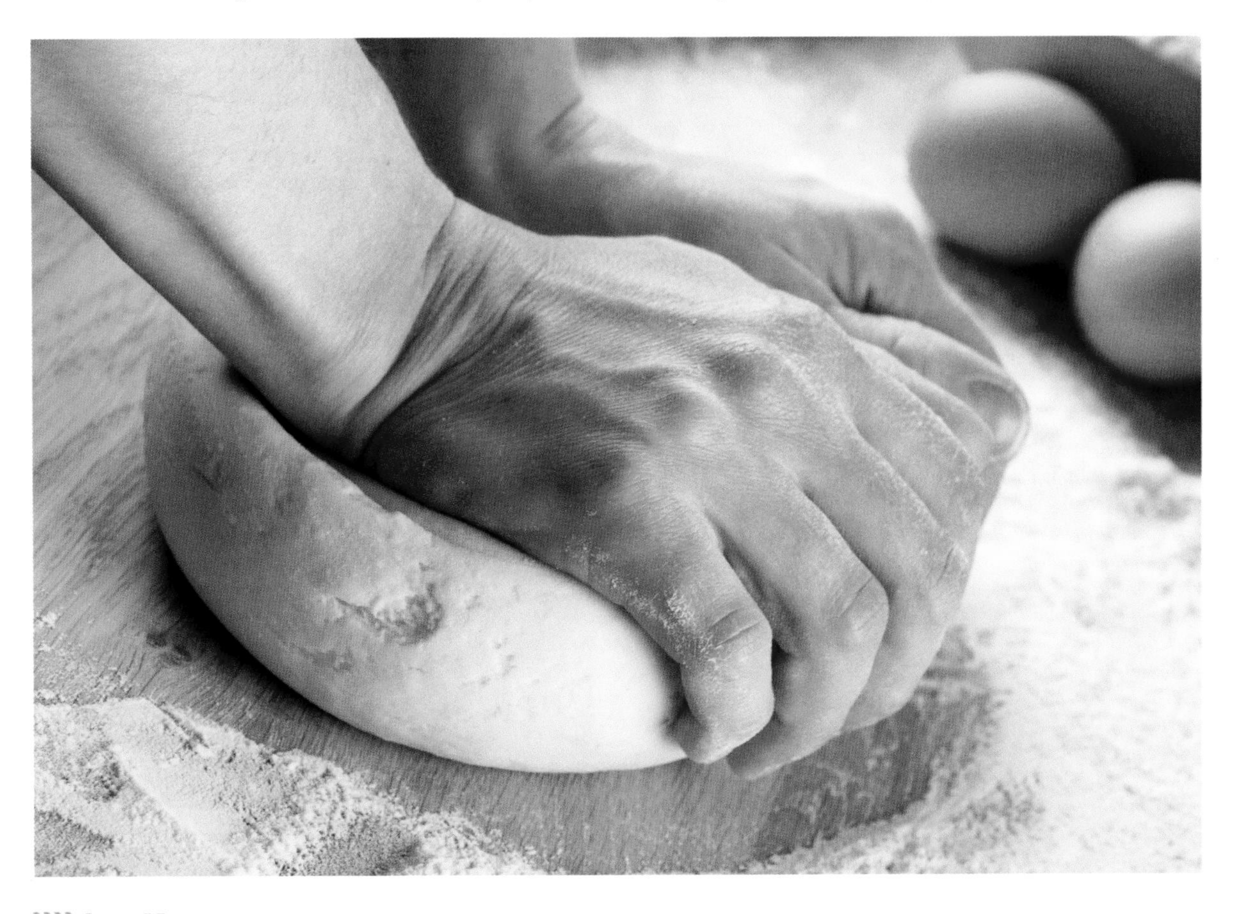

›››› *See Also*
RELATED PRINCIPLES: Food Microbiology, 329 • Macromolecular Chemistry, 288
RELATED APPLICATIONS: Beer, 316 • Cheese, 321 • Vinegar, 319 • Wine, 318 • Yogurt, 321

Fermentation

WINE

The yeast used to turn grapes into wine may be a cultured type chosen for its reliable hardiness under certain conditions, or the wild yeast present on the grapes themselves, which some winemakers believe gives their product complex qualities unique to their winery or locale. The yeast converts the fruit's sugars into alcohol and carbon dioxide, which in still wines (as opposed to sparkling) is allowed to escape through valves or the surface of open vats.

To make sweet dessert wines, winemakers stop the fermentation process before all the sugar is consumed. Red wines sometimes undergo a secondary fermentation by lactic acid bacteria, which metabolize sour-tasting malic acid into a different acid with a softer flavor. Sparkling wines also require a second fermentation process; it takes place in the bottle, where yeast consumes added sugars, producing both alcohol and carbon dioxide.

What's desirable in vinegar is decidedly unwelcome in wine; the sharp tang of spoilage is often the result of contamination with Acetobacter.

»»See Also

RELATED PRINCIPLES: Food Microbiology, 329 • Macromolecular Chemistry, 288
RELATED APPLICATIONS: Beer, 316 • Bread, 317 • Cheese, 321 • Vinegar, 319 • Yogurt, 321

Fermentation

VINEGAR

Vinegar can be made from just about anything containing sugar—from apple juice to grape juice to coconut water. Preparing it involves a two-step fermentation process. The first step uses yeast to convert sugars into alcohol. The second uses bacteria from a family called Acetobacter; with the help of oxygen from the air, the bacteria metabolize the alcohol and produce, as a waste product, acetic acid. This mild acid gives vinegar its sour flavor makes it a natural disinfectant.

Acetobacter also produce cellulose, a fibrous substance that in vinegar combines with the acetic acid to form what's called the mother, the goop that you'll sometimes see at the bottom of vinegar bottles, especially unpasteurized types. Most commercial producers pasteurize their vinegar to kill the bacteria and prevent formation of the mother. It's harmless; herbalists even attribute healing properties to the mother. Nonetheless, many find it unappetizing.

»»» ***See Also***

RELATED PRINCIPLES: Food Microbiology, 329 • Macromolecular Chemistry, 288
RELATED APPLICATIONS: Beer, 316 • Bread, 317 • Cheese, 321 • Wine, 318 • Yogurt, 321

PROBIOTICS

The idea that consuming certain live bacteria promotes digestive health is all the rage, driven in part by marketers pitching an assortment of probiotic drinks, supplements, and foods. The theory began with Russian biologist Ilya Ilyich Mechnikov, the 1908 winner of the Nobel Prize in medicine, who drank sour milk regularly in the belief that its lactic acid bacteria would make him live longer. There's some substance in the claims. Research has shown that certain products help ease digestive troubles associated with irritable bowel syndrome and reduce the diarrhea that can result from taking antibiotics.

The principle behind probiotics is that they seed the intestinal tract with bacteria that are considered helpful, not least because they check the overgrowth of disease-causing germs by competing with them for space and nutrients. But the scientific work needed to prove specific health benefits from eating particular bacterial strains is young. In general, a person would probably have to take a supplement or eat a probiotic yogurt (for example) routinely to realize any benefit.

Lactobacillus

Fermentation

YOGURT

It took decades for Americans to embrace yogurt, the sour, fermented milk product long enjoyed in Europe and elsewhere in the world. It probably didn't help that it was marketed under the slogan "Doctors recommend it." Today, it's a lunchbox staple, prized not only for its fresh, tangy flavor but also for the live bacteria most yogurts contain.

It's not just any bacteria that are used to make yogurt. Milk is first pasteurized to kill any bad bugs and then cultured with two organisms approved by the U.S. Food and Drug Administration (FDA), Lactobacillus bulgaris and Streptococcus thermophiles, that convert milk sugars to lactic acid. The acid imparts flavor and curdles the milk, thickening it to a pudding-like consistency. Sometimes other bacteria, such as Lactobacillus acidophilus, are added for their believed role in balancing bacterial populations in the human gut.

»»**See Also**

RELATED PRINCIPLES: Food Microbiology, 329 • Macromolecular Chemistry, 288
RELATED APPLICATIONS: Beer, 316 • Bread, 317 • Cheese, 321 • Vinegar, 319 • Wine, 318

Fermentation

CHEESE

Enterococcus faecalis doesn't sound very tasty. But these bacteria contribute to the flavor of certain cheeses produced in southern Europe, including Portuguese Picante da Beira Baixa, Spanish Cebreiro, and Greek feta. In fact, the craft of cheesemaking makes use of a menagerie of fermenting microorganisms—including a variety of bacteria, molds, and yeasts. The choice of organism is one of the most important elements in determining taste and consistency.

The most basic step in making cheese is fermenting milk with bacteria that convert its sugars to lactic acid. The milk curdles. Next comes rennet, a substance found in the stomach linings of young mammals that helps them digest their mother's milk. The enzymes in rennet separate the solid curds from the watery whey. The product is salted, and other organisms such as mold may be introduced.

Cheese is easier to store and, thanks to acid and salt, keeps much longer than milk—reasons people have been making it for millennia.

»»**See Also**

RELATED PRINCIPLES: Food Microbiology, 329 • Macromolecular Chemistry, 288
RELATED APPLICATIONS: Beer, 316 • Bread, 317 • Vinegar, 319 • Wine, 318 • Yogurt, 321

PRINCIPLE

HARVESTING THE EARTH

Ecology

Just as a bear learns where to find berries in the summer and some birds journey thousands of miles to their seasonal feeding grounds, people in every region of the world have studied the natural rhythms of their local environments. What time of year do fish swim into the mouths of rivers? Where do nut-bearing trees grow best? What's the best way to harvest raspberries?

Agriculture is a uniquely human way of controlling the environment to not just gather but also produce food. Since agriculture first took root some 10,000 years ago in the eastern Mediterranean, people have invented increasingly sophisticated agricultural methods, from modifying the DNA of seeds to forcing plant growth under artificial lights.

At times, people have had to learn from their mistakes—mistakes that led to unforeseen destruction of natural resources. For example, the terrible dust storms in the American plains during the 1930s taught farmers to protect and replace nutrients in the soil by rotating crops and applying fertilizer.

Tilapia raised in a fish farm

Ecology

HYDROPONICS

Many North Americans first encountered the words hydroponically grown on a tomato sticker. All but unknown to U.S. shoppers as recently as the 1980s, tomatoes raised without soil now account for some 40 percent of the fresh-tomato market. Growers are also increasingly producing salad greens, herbs, and other produce hydroponically.

The term derives from the Greek words hydro (water) and ponos (labor). This soilless cultivation method lets water do the work of distributing food to the plant.

Fertilized or naturally fertile soil holds mineral nutrients such as nitrogen, phosphorus, and potassium that are essential for plant growth. But the dirt is merely a medium. The nutrients dissolve in water, and plants feed themselves by drawing this nutrient-rich water into their roots. In hydroponic growing, the plant root is suspended in an oxygenated nutrient solution or packed in a medium such as stones or perlite that's saturated with the solution. The roots get their nutrient-rich water without soil.

Because growers can carefully measure out the precise amount of nutrients required and recycle the water, hydroponic systems conserve both. Also, pesticides and fertilizers don't get into local soil and groundwater.

Perhaps the biggest attraction of hydroponic growing is its high, year-round yields. Though hydroponically grown produce is shipped from around the world, in some cases the method lets consumers get out-of-season greens or strawberries locally. That means the food arrives on the table fresher and faster. Of course, these operations often entail not just soilless cultivation but also protecting the plants from the weather and perhaps adding grow lights.

Terraced rice fields in Vietnam

»»» *See Also*

RELATED PRINCIPLES: Absorption of Light, 204 • Ohm's Law, 164 • Second Law of Thermodynamics, 98 • Ultraviolet Light, 215

RELATED APPLICATIONS: Plants and Turgor Pressure, 91 • Visible Spectrum, 205

Ecology

AQUACULTURE

People are eating more fish than ever. Fish accounts for 18.5 percent of the total animal protein consumed around the world. This growing demand for seafood has placed enormous strain on fisheries. Some fish populations, such as cod and Atlantic bluefish tuna, have been harvested nearly to the point of collapse.

The characteristically human solution: make more fish. The breeding of fish and shellfish in captivity—aquaculture—has grown dramatically over the last half-century. Production is more than 50 times what it was in the 1950s. The Asia-Pacific region dominates the industry, with China as the leader.

There are two basic but overlapping approaches to raising fish. Intensive aquaculture manages everything

»»See Also

RELATED PRINCIPLES: Convection, 103 • Evaporation, 113 • Geothermal Energy, 175 • Pascal's Law, 80 • Surface Tension, 88 • Ultraviolet Light, 215
RELATED APPLICATIONS: Tides, 130 • Water and Air Purification, 302

about the fish's environment. They're reared in aerated, circulating tanks or ponds at high density and given commercial feed. In extensive aquaculture, farmers stock a natural ecosystem—a river, coastal waters, or even the open ocean—with fingerlings, or very young fish. Penned in by netting, the fish eat phytoplankton and other food.

Though often marketed as a way to ease pressure on wild fish populations, certain kinds of fish farming, like factory farming of livestock and poultry, have a serious impact on aquatic environments. Mangrove swamps in Southeast Asia and Africa are disappearing to make way for shrimp ponds. And in intensive operations, large quantities of uneaten food and waste may decompose in the water, leading to oxygen depletion and algal blooms that can decimate aquatic life. Raising valuable favorites like shrimp and salmon often places heavy demands on marine resources; by the time they're full-grown, those kinds of seafood-to-be consume twice their weight in fishmeal, which is made of smaller fish like sardines and anchovies. The environmental group Sierra Club Canada has partnered with others in preparing a seafood sustainability guide for interested consumers at seachoice.org.

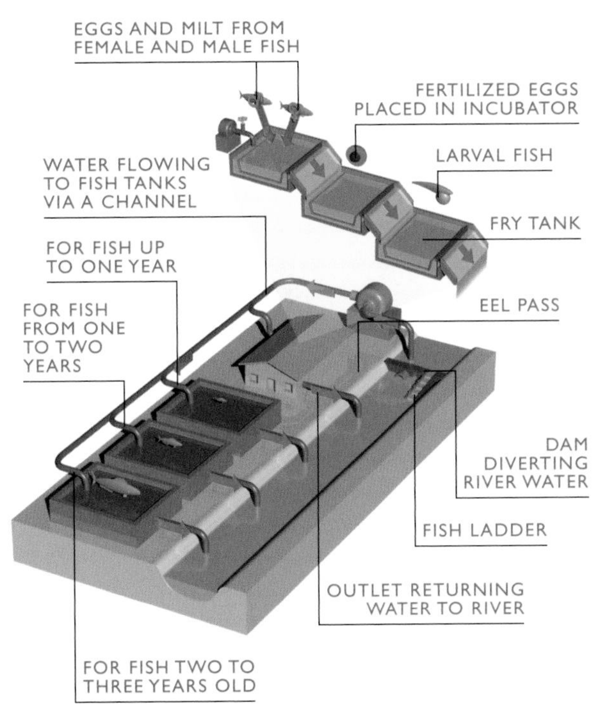

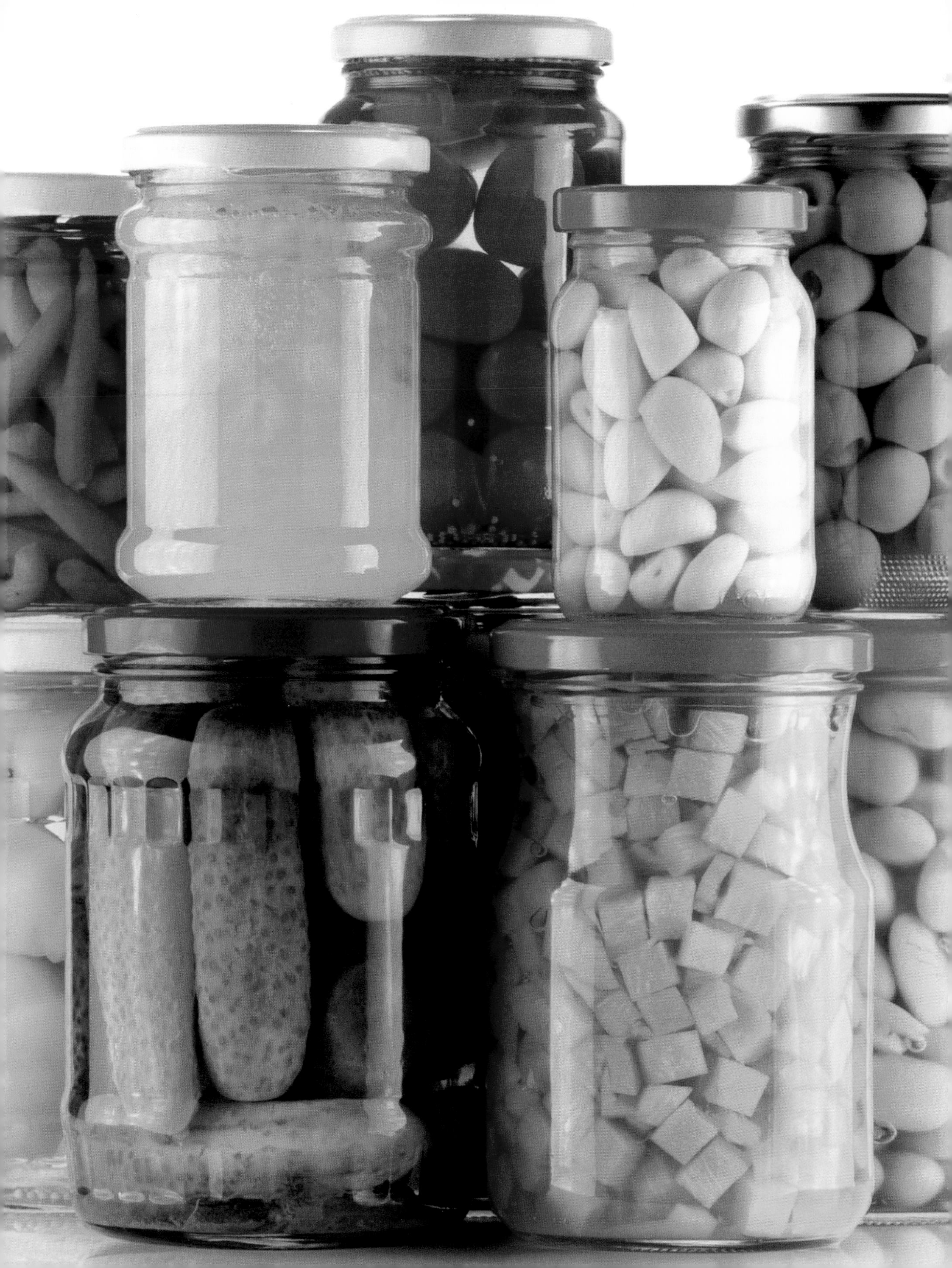

PRINCIPLE

WASTE NOT, WANT NOT

Food Microbiology

Making food is one thing; keeping it is another. In their natural state all the organic materials we eat are perishable. After a certain period of time they go bad—smell bad, taste bad, look bad. Even if they don't become downright unappetizing, foods lose nutritional value as vitamins degrade due to heat, strong light, air, or water.

In preserving foods, the first order of business is to prevent yeasts, molds, bacteria, and insects from devouring them before we do. Yeasts and molds thrive on acidic fruits and convert their sugars to an acidic fizz. Bacteria prefer meat, dairy products, and low-acid vegetables; if allowed to proliferate, they create acid and other waste products that putrefy food. Certain bacteria can overrun a person's gastrointestinal tract or release dangerous toxins, causing the misery of acute food poisoning.

People have tried to prevent food spoilage since they first started cooking meat over a fire—smoking, salting, drying, and pickling have been around for millennia. But the science of food preservation advanced considerably in the 19th century with the advent of such processes as pasteurization.

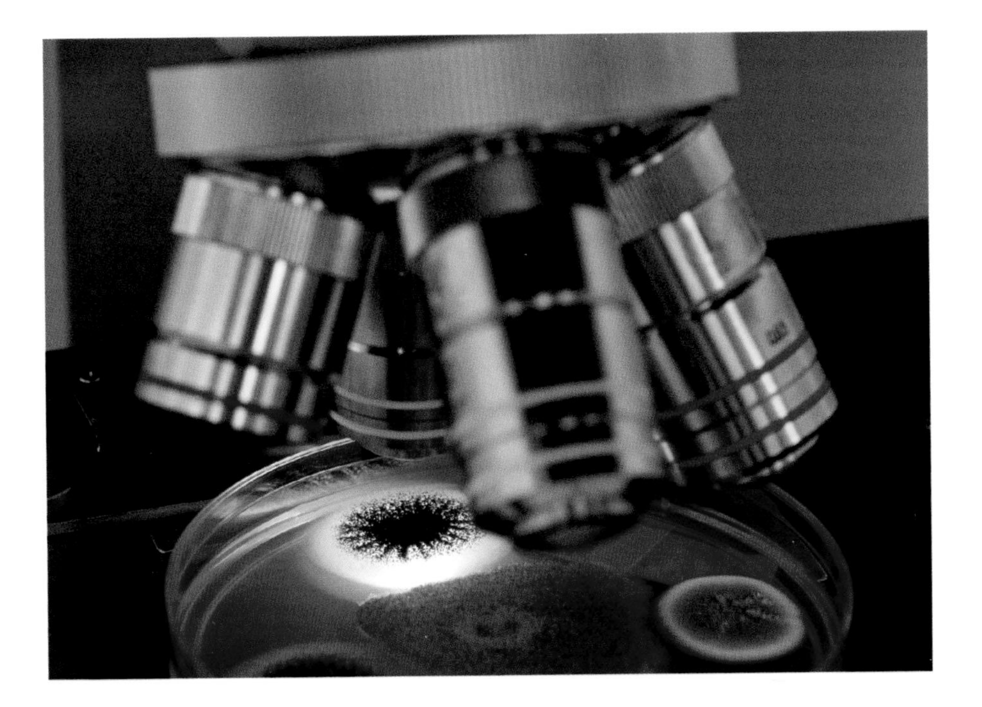

Microbiology

FROZEN FOOD

The bacteria and molds that are quick to colonize and digest fresh food thrive in warm, moist conditions. Refrigeration slows their growth. Freezing does one better; it kills many microorganisms and locks the rest in a kind of suspended animation. Food that might have been edible for a few days or a week can, when encased in ice, last for months.

Water is a major part of most organic materials. When you freeze a bushel of peas or a pound of hamburger, you're turning the water in these foods into ice crystals. It sounds simple enough. And in fact people in cold climates have always taken advantage of frigid weather to stash game or fish outdoors and extend its freshness. But in the United States, freezing food at home didn't become commonplace until the invention of commercial flash frozen foods—and along with them, the first home freezers—in the mid-20th century.

The flash freezing process offers important advantages over home freezing even today. First, it applies subzero cold that's more lethal to microorganisms than the temperatures in most home freezers. Second, it freezes the food very quickly—in a matter of minutes—by circulating the air all around the food with a fan. Slowly freezing foods forms larger ice crystals, which can deform and burst open cells. The result is a mushy consistency and, perhaps, a loss of color and flavor.

Before they're frozen, many fruits and vegetables are first blanched—quickly heated by boiling or steam. This halts the activity of the enzymes that would otherwise overripen the food.

»»See Also
RELATED PRINCIPLES: Electromagnetic Induction, 157 • Ohm's Law, 164 • Pascal's Law, 80 • Second Law of Thermodynamics, 98 • Surface Tension, 88
RELATED APPLICATIONS: Turgor Pressure in Plants, 91 • Refrigerator, 101

CLARENCE BIRDSEYE

The name Birdseye instantly brings to mind the rectangular boxes of frozen vegetables available in every American grocery. If it weren't for Clarence Birdseye, in fact, groceries might not have a frozen food section at all.

A Brooklyn taxidermist, Birdseye got his inspiration to freeze foods with a blast of ultra-low temperature while living in Labrador (now in northern Canada), where he noticed fish caught by native Inuit froze quickly in the Arctic air and tasted good for a surprisingly long amount of time.

Flash freezing was the basis of the company he started in 1923. It was a true advance. Though the freezing process can degrade certain water-soluble vitamins, vegetables harvested at peak ripeness and frozen quickly may be more nutritious than out-of-season fresh produce that travels far and then sits on a shelf for several days.

Microbiology

CANNED AND BOTTLED FOODS

In canning and bottling, the key ingredient is the container itself. Sterilized and airtight, the vessel seals out the germs that would make a meal of its contents. What's on the inside—the deviled ham or lima beans—remains inviolable, ready to eat for up to five years.

First, the contents undergo some form of preservation, usually cooking at high temperature; bacteria like extreme heat even less than they like cold. Sometimes, the canner adds acidic ingredients as preservatives.

The prepared food is then sealed in the container, which is sterilized by heat—a blitzkrieg on remaining microorganisms. The temperature required for sterilization depends on the food. Traditional home-canning recipes often are for high-acid pickled foods and fruity preserves that can be safely sterilized by simple boiling. But to kill off dangerous germs such as the bacteria that cause botulism, low-acid meats have to get very hot—in the range of 215°F (about 102°C)—on the inside. This requires cooking under pressure.

If canned goods spoil (and they rarely do), it's always because bacteria slipped into the food during preparation or canning. The live bacteria inside produce carbon dioxide as waste, which bloats the can in that telltale sign of spoilage.

Canning is an old form of food storage whose invention is attributed to the Parisian confectioner Nicolas Appert. After more than a decade of experiments, in 1810 Appert published *The Art of Preserving Animal and Vegetable Substances*. With chemistry in its infancy as a science and scant knowledge of bacteriology, Appert proceeded largely by trial and error.

»»*See Also*

RELATED PRINCIPLES: Elasticity of Air, 78 • Electromagnetic Induction, 157 • Ohm's Law, 157 • Pascal's Law, 80 • Second Law of Thermodynamics, 98
RELATED APPLICATIONS: Freeze-Dried Foods, 334 • Frozen Foods, 330

Microbiology

FREEZE-DRIED FOODS

The bacteria and molds that make food foul or downright toxic can't live without water—another Achilles' heel people have exploited to keep their eatables edible for longer.

A NASA food scientist shows a finished packet of freeze-dried shrimp fried rice.

Freeze-drying first subjects a food or drink to temperatures in the range of −60 to −110°F (−16 to −43°C), turning its water content to ice and stunning germs in the bargain. Then, in a process called sublimation, the food is heated in a low-pressure chamber, which forces the solid ice molecules to break apart into a gaseous state, skipping the liquid phase. The water vapor is condensed and pumped from the chamber. The process can take hours.

What's left is the intact structure of the food, with airy pores where the water was. It's dry as a bone and wholly inhospitable to microorganisms. The lack of water even stops enzymes from rotting produce. To prevent the food from absorbing more moisture from the air, it's packaged in an airtight container.

As long as 30 years later, just add water and stir!

Freeze-drying is the technology behind instant coffee, dried soup mixes, and some baking mixes, powdered drinks, and dried-fruit snacks. Though they have to be cut into small chunks, most foods can be freeze-dried. Emptied of water, freeze-dried foods are compact and extremely lightweight, making them especially useful to soldiers in the field and campers on long treks.

»»See Also

RELATED PRINCIPLES: Elasticity of Air, 78 • Electromagnetic Induction, 157 • Ohm's Law, 164 • Pascal's Law, 80 • Second Law of Thermodynamics, 98
RELATED APPLICATIONS: Canned and Bottled Foods, 332 • Frozen Foods, 330

FREEZE-DRYING PHARMACEUTICALS

Freeze-drying was one of many innovations stimulated by the urgent necessities of battle during World War II. It proved an excellent way to preserve blood plasma to transfuse wounded Allied soldiers in the field. Today not only blood plasma but also biologic supplies such as vaccines, antibodies, and hormones are often freeze-dried, eliminating the need for costly refrigeration during transport and storage.

FOOD IN SPACE

Before landing humans on the moon in the Apollo missions of the 1960s, NASA conducted a great deal of research on suitable foods for space flight; freeze-dried ice cream was one result. NASA commissioned the Whirlpool Corporation to create it.

But if it isn't cold, smooth, and creamy, can it really be ice cream? Unlike other freeze-dried foods, astronauts eat ice cream without added water; it's crumbly and dry. The invention made it to space only once but has enjoyed a busy afterlife as a novelty product.

Eating in space can be tricky, requiring care and coordination. The astronauts shown here are eating in their sleep racks. A tray of food is attached to the rack's sliding door. They simultaneously take a bite of their meal and wipe their mouths.

CHAPTER 15

Cell Biology

In the beginning, there was a cell. All life on Earth evolved from single-celled organisms that lived about 3.5 billion years ago. More than 500 million years ago, these single cells began to form clusters. The cell is the fundamental unit of all life on the planet. It's the key to understanding animals and plants as well as our own human bodies—how they work, how they fail, and what can be done to fix them. Nearly every cell in the body contains the genetic material that tells the tissues and organs what to be and do.

There are hundreds of different cell types in the body, each with its special function. A brain cell can transmit electrical impulses. White blood cells rush to a site of infection and surround the invader or produce antibodies to fight it.

People first glimpsed the building blocks of life through a microscope in the 17th century. In the 21st century, the cell is still a frontier, as scientists seek novel ways to improve and save lives—and even create new ones.

PRINCIPLE

UNLOCKING THE CODE

DNA

Deoxyribonucleic acid: this elegant molecule is the very stuff of life, holding the instructions that make each person biologically human and, at the same time, unique.

Packed into every cell's nucleus, DNA is made up of strands of nucleotides, which in turn consist of alternating sugar and phosphate components and one of four nitrogen bases: adenine, thymine, cytosine, or guanine. With the nitrogen-base pairs on the inside, the strands are twisted together in the double helix discovered by Francis Crick and James Watson in 1953.

Because the nitrogen bases are always paired in a specific way—cytosine goes with guanine, adenine with thymine—a single rung of the twisted ladder serves as a pattern for the whole. Before cell division, the helix unzips, and each daughter cell builds a full copy using one strand as a template.

The order of these base pairs along the strands is the simple language that—through the messenger molecule ribonucleic acid (RNA)—instructs amino acids how to arrange themselves into proteins that regulate just about every process in the body, from muscle contraction to growth.

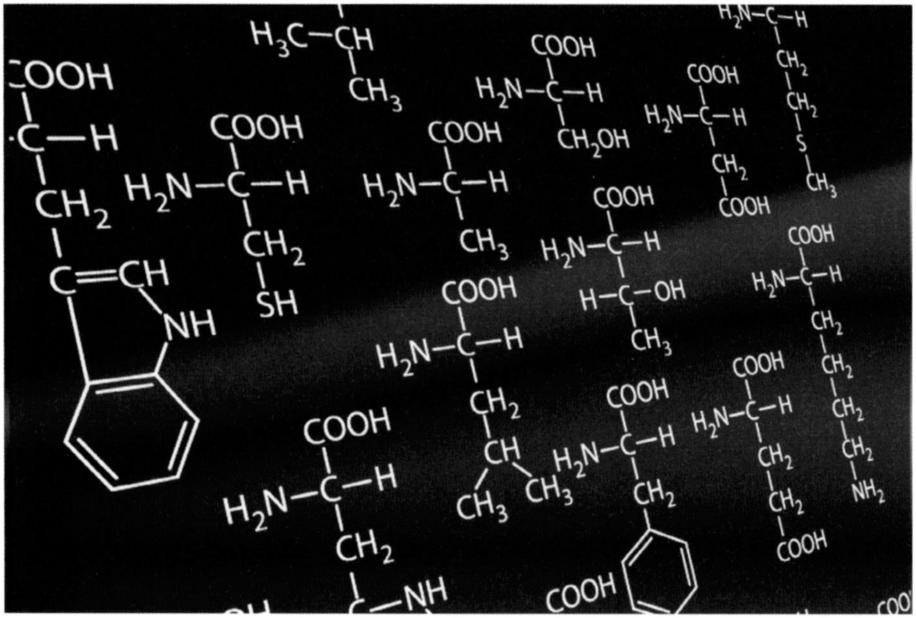

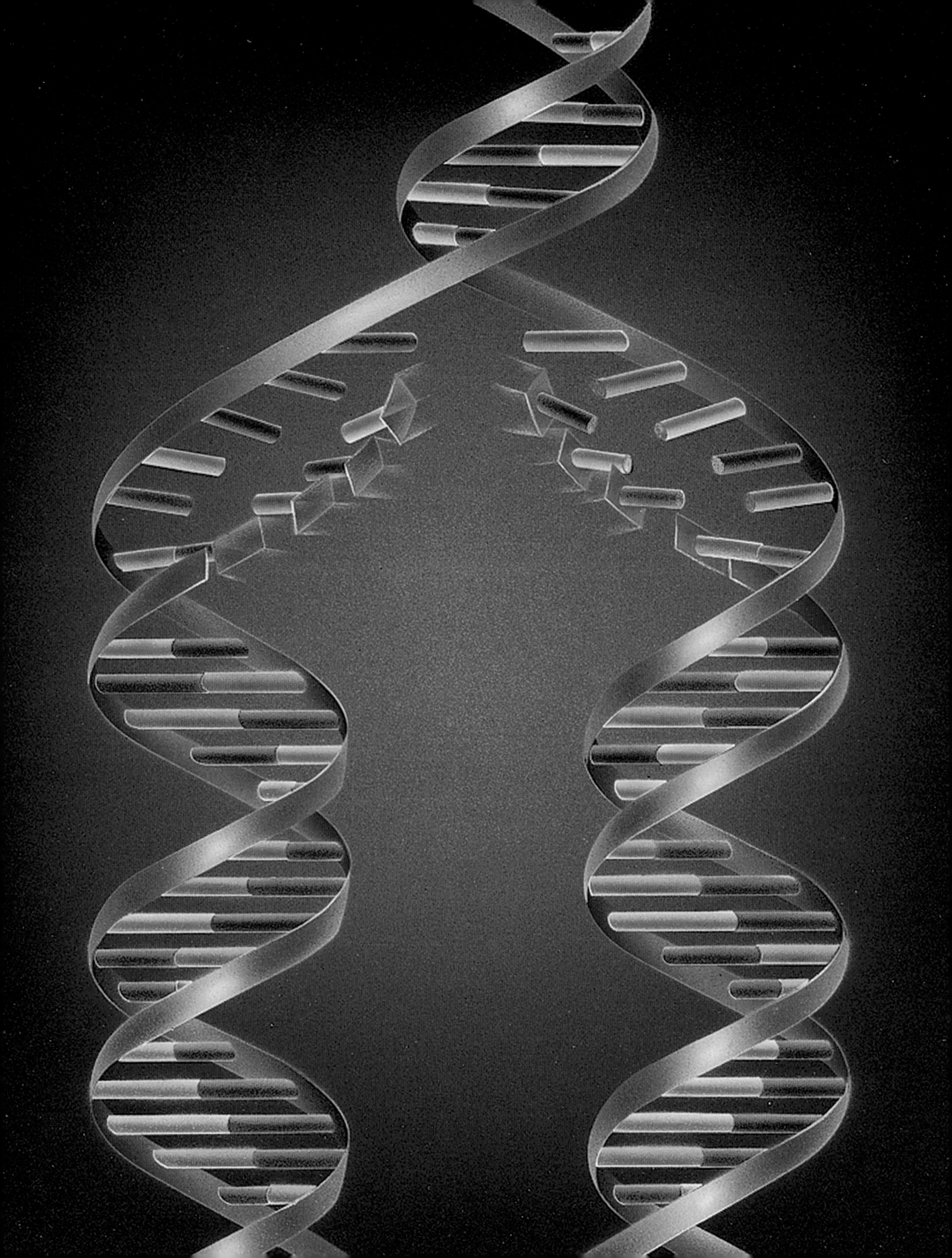

GENETICALLY MODIFIED CROPS

For thousands of years, people have bred plants for favorable traits—resistance to drought, a tastier fruit—by selecting and saving the best plants or by cross-pollinating plants with different characteristics.

Since the 1980s, though, a new method has come on the scene: creating a better specimen by directly inserting genetic material from one plant (or in some cases from a bacterium or virus) into the genes of another. Among the most common genetically modified crops are corn, soybeans, canola, cotton, and potatoes. Though they may not know it, most Americans eat these genetically tweaked plants, which are present in many processed foods.

Typically, genetic modifications aim to boost a crop's resistance to disease, insects, pesticides, or weed killers.

The first step is locating and isolating the desirable gene sequence of an organism, made possible only by sophisticated gene-mapping techniques developed in the last decades of the 20th century. There are many methods for delivering new genes to the host. One of the most common is to insert the genes into a kind of bacteria called Agrobacterium tumefaciens, the natural cause of crown gall in plants, then infect the target plant with the altered bacteria. Another is to attach the desired DNA to microscopic particles of gold or tungsten, then shoot these into the target cells using high-pressure gas.

To some people, genetically modified plants are the hope of an increasingly populous and hungry world, permitting farmers to produce more food at less expense and with fewer bug- and weed-killing chemicals. The World Health Organization has said these crops likely pose no health risk to people. But they are controversial. Many people are concerned the plants may have unknown health effects, introduce allergens into the food supply, or spread genetic traits to other plants.

See Also
RELATED PRINCIPLE: Ecology, 322
RELATED APPLICATION: Electron Microscope, 152

DNA

LIVESTOCK CLONING

COUNTING SHEEP. Dolly, right, the first cloned sheep produced through nuclear transfer from differentiated adult sheep cells.

Livestock breeders have been using artificial insemination and other assisted-reproduction technologies for decades. These tricks allow breeders to choose a mother and father, but the uncertainties of sexual reproduction remain. Which traits will be passed on? Will the offspring be male or female?

Cloning, on the other hand, yields an exact genetic copy of an animal chosen for its desirable characteristics—the ability to thrive in a certain climate, or produce lean meat, or produce milk.

Since the birth of the famous cloned sheep Dolly in 1996, scientists have also successfully cloned cattle, pigs, and goats, among other animals.

Like Dolly, most are created by a process called somatic cell nuclear transfer. Scientists remove the DNA-containing nucleus from an immature egg and replace it with a nucleus from the cell of a donor animal. If the fused cell survives and divides, the resultant embryo is implanted in a surrogate mother.

It's an expensive process used mainly to create high-quality breeding stock rather than food animals. In 2008 the U.S. Department of Agriculture estimated there were 600 animal clones in the United States and asked their makers to maintain a voluntary moratorium on sending the animals' milk or meat into the food supply. The agency planned to "ensure a smooth and seamless transition into the marketplace for these products."

That same year the U.S. Food and Drug Administration issued a review of research on cloned-animal food, saying it was just as safe for people as ordinary animal products. The report did say that compared with conventionally bred livestock, cloned animals have more health problems and are more likely to die early in life, and their surrogate mothers are more likely to suffer complications.

»»See Also

RELATED PRINCIPLES: Ecology, 322 • Food Microbiology, 329

RELATED APPLICATIONS: Gene Therapy, 352 • Human In Vitro Fertilization, 344 • Stem Cell Therapy, 350

DNA

IN VITRO MEAT

Is it possible to grow meat in a lab, without having to kill a living, breathing creature? That's the hope of a small group of bioengineers, along with environmentalists and animal-rights activists who worry about the suffering of animals, and the environment, in large-scale livestock operations. Livestock farming generates an even higher percentage of human-generated greenhouse gases than transportation does.

Efforts to produce test-tube meat started with stem cells—cells that can continuously divide and develop into any number of specialized cell types, like skin, blood, bone, and brain cells. In 2012 a scientist from Maastricht University in the Netherlands announced he had made strips of muscle using cow myosatellite cells, stem cells that multiply rapidly to repair damaged muscle. He cultured the cells in calf serum and introduced them to a synthetic scaffold made from chemical compounds.

Scaling up such an experiment poses a major challenge; the taste test, of course, lies far in the future.

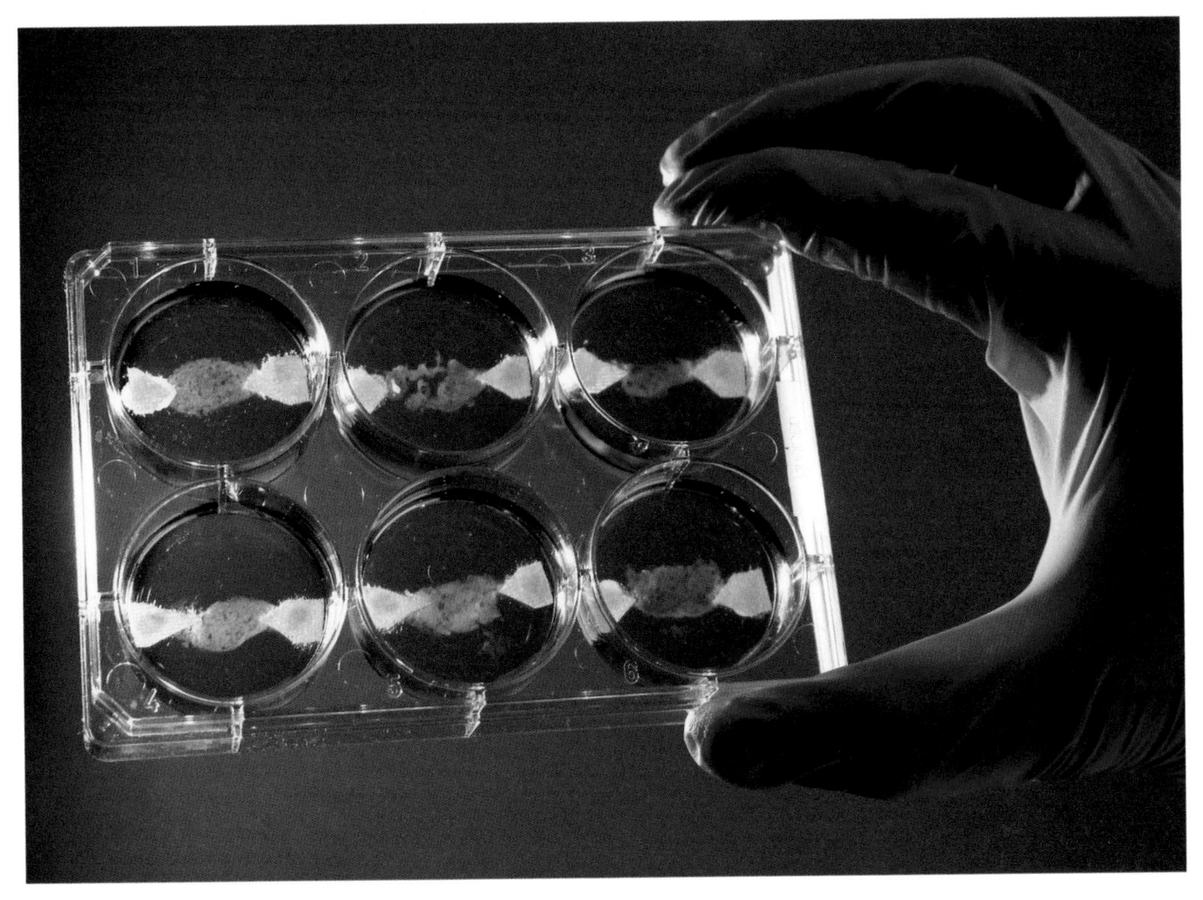

Meat cells cultured in the laboratory

»»*See Also*
RELATED PRINCIPLES: Ecology, 322 • Food Microbiology, 329
RELATED APPLICATIONS: Gene Therapy, 352 • Human In Vitro Fertilization, 344 • Human Tissue Cultivation, 349 • In Vitro Meat, 343 • Livestock Cloning, 342 • Stem Cell Therapy, 350

DNA

HUMAN IN VITRO FERTILIZATION

It took British fertility researcher Robert Edwards decades to unravel the mysteries of conception. First, he worked with animals; then he struggled for years to create just the right hormonal conditions to fertilize and grow human eggs in a petri dish. At last a successful implantation resulted in the 1978 birth of "test-tube baby" Louise Brown. Thirty years later, more than four million people had been born using that method.

Now a common, if costly, solution to infertility, in vitro fertilization (IVF) involves first giving a woman fertility drugs to induce ovulation of multiple eggs, then retrieving the eggs from her ovaries with a hollow needle in a minor surgical procedure. Each egg is placed in an incubator with thousands of sperm. The next day, doctors look to see if fertilization has taken place. By day three, an embryo will be six to eight cells and may be transferred to the woman's uterus via a thin plastic tube.

Sometimes multiple embryos are introduced to improve the chances of a live birth, although the American Society of Reproductive Medicine suggests no more than two should be implanted in women under 35 to prevent high-risk multiple births.

IVF technology is well developed, but its success depends on the quality of the egg and sperm. A woman's ova degenerate with age, a problem that's sometimes addressed by using a donor egg. If sperm are few in number or don't swim well, a single sperm may be injected directly into the egg.

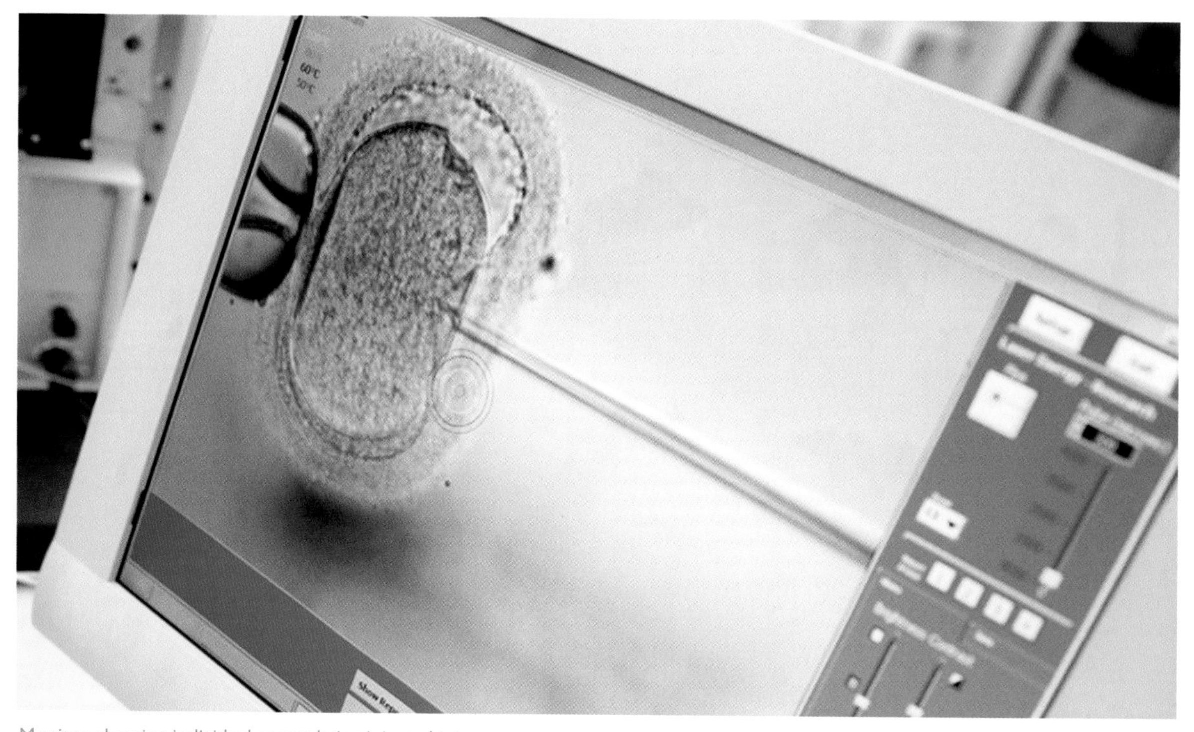

Monitor showing individual sperm being injected into an egg

»»See Also

RELATED PRINCIPLES: Bioengineering, 364 • Immunobiology, 357 • Reflection of Sound Waves, 64

RELATED APPLICATIONS: Gene Therapy, 352 • Human Tissue Cultivation, 349 • Livestock Cloning, 342 • Stem Cell Therapy, 350

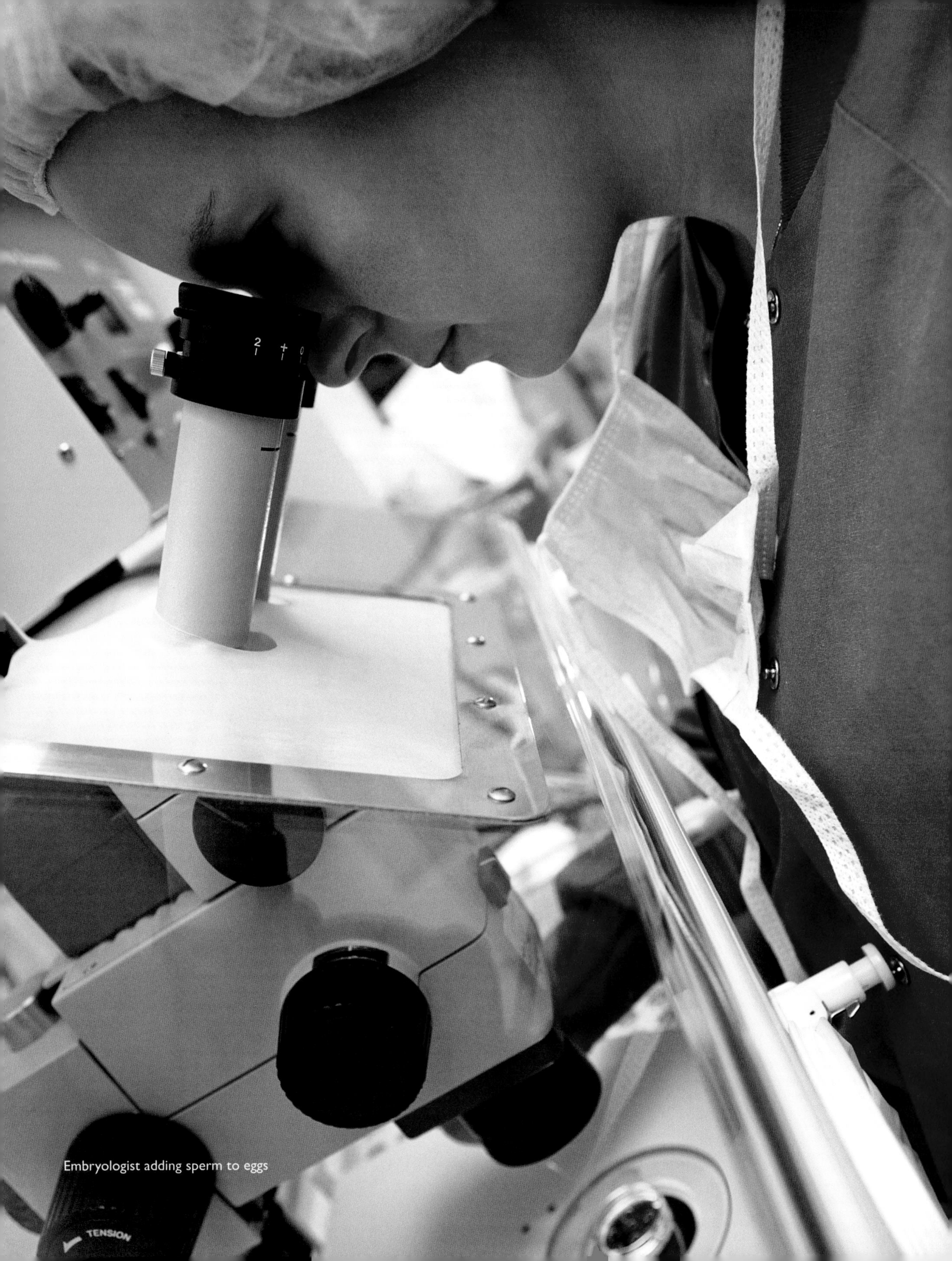

Embryologist adding sperm to eggs

DNA

CANCER

In cancer, cells run amok. Sometimes, something goes wrong with the intricate genetic programming that tells them when to proliferate in order to grow new tissue or replace lost cells and when to die because they're old or damaged. Usually, we can't identify the cause of the genetic damage, although risk factors for cancer are well known. Many genes may be affected. Whatever the cause, the effect is the same: wild, unregulated growth of abnormal cells.

These cells multiply until they form a tumor. Cancer cells are distinguished by their insidious ability to spread, either by direct invasion of adjacent tissue or by slipping through the bloodstream and lymphatic system to distant body parts.

Cancer can emerge just about anywhere in the body. Malignancies are named for their site of origin and the type of cells from which they arise. So, for example, melanoma begins in skin cells called melanocytes, which are responsible for producing pigment. Genetic changes, perhaps spurred by sun exposure, cause the melanocytes to grow out of control. If they're not caught early, they spread through the blood vessels to distant regions such as the lungs or brain; that's a metastatic melanoma.

To make a prognosis and decide how to treat a cancer, doctors look at the type of cancer involved—some types are highly lethal, while others are almost always curable—and the extent to which the cells have spread. They also consider how the cancer cells look under a microscope. More aggressive cancer cells tend to be undifferentiated; they look primitive, immature, and unlike the particular cell type from which they originated.

Few cancers are entirely genetic in origin. Studies suggest as many as two in three are related to environmental factors, which may include smoking, a high-fat diet low in fruits and vegetables, certain infections, and exposure to radiation or pollutants. These interact with a unique pattern of random genetic mutations, some of which may be inherited.

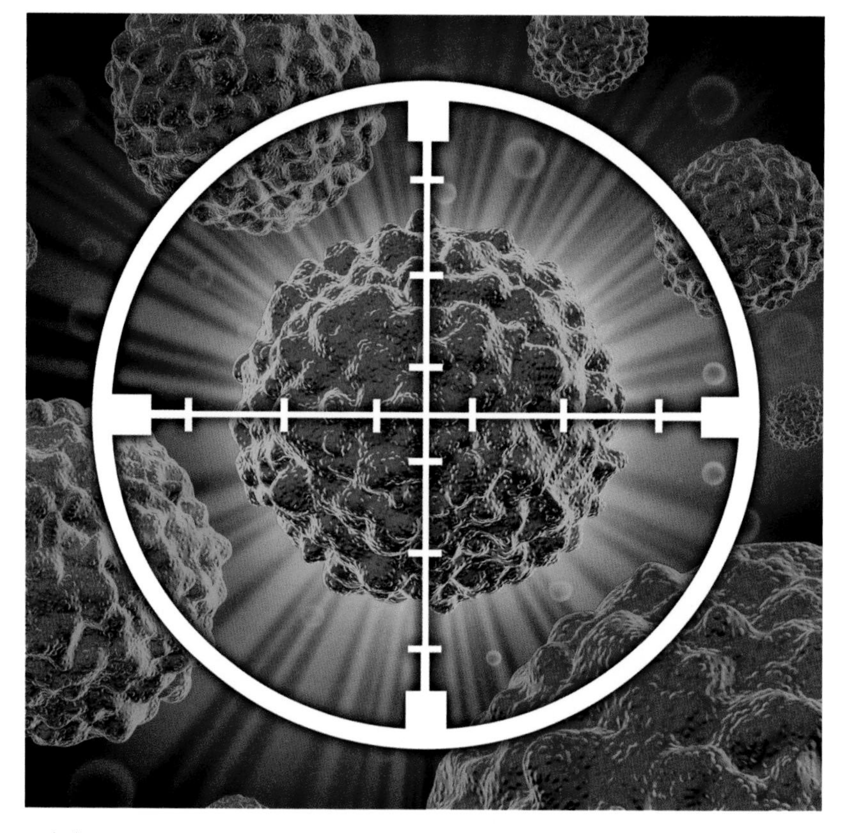

TAKING AIM AT CANCER. Targeting cancerous cells without harming healthy cells is the goal of cancer research.

»»See Also
RELATED PRINCIPLES: Ecology, 322 • Immunobiology, 357
RELATED APPLICATIONS: Gene Therapy, 352 • Stem Cell Therapy, 350

Cancer cells (green) spreading and growing through the body via red blood cells

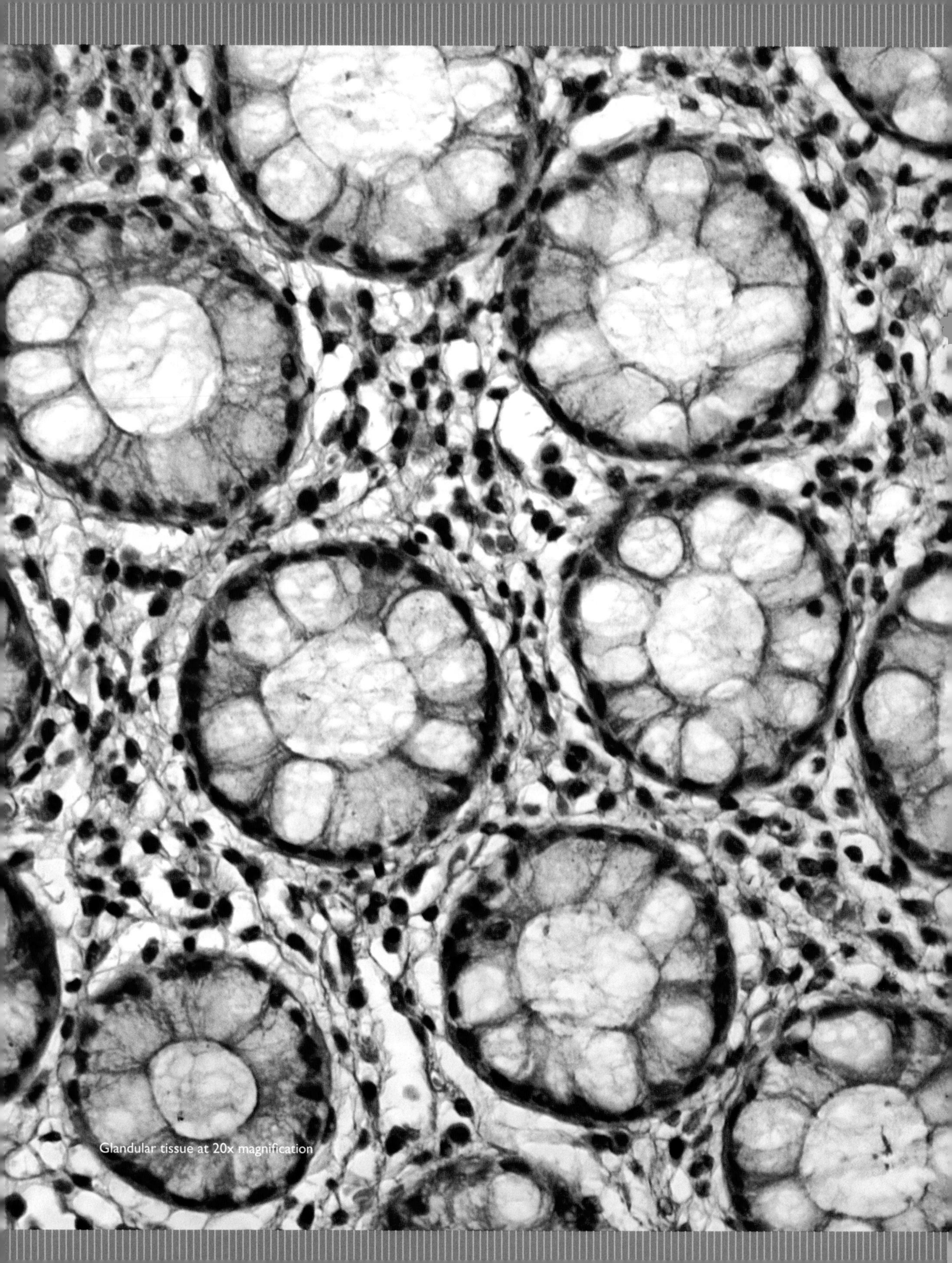

Glandular tissue at 20x magnification

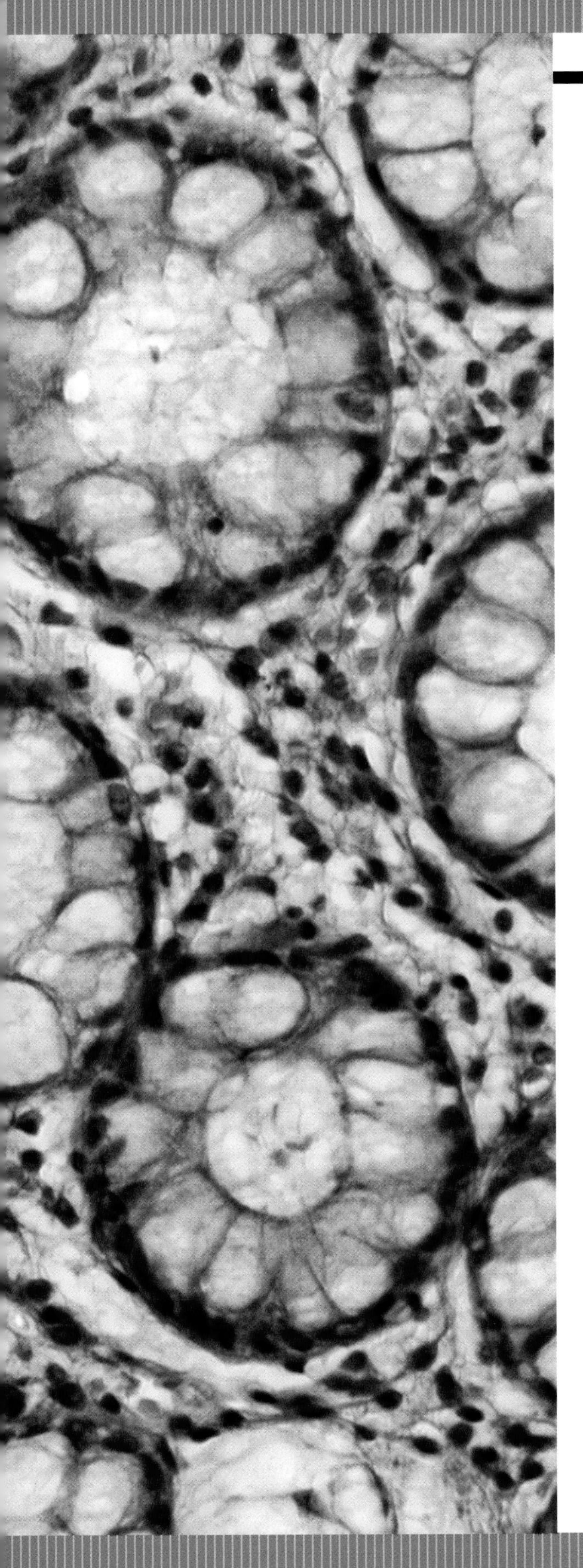

DNA

HUMAN TISSUE CULTIVATION

The developing field of regenerative medicine seeks nothing less than to provide patients with replacement body parts. Here, the parts are not steel pins and silicone. They are the real thing: living cells, tissue, and even organs. Though still a mostly experimental enterprise, the field has produced some astounding successes.

In 2006 researchers at Wake Forest Institute for Regenerative Medicine in North Carolina announced they had successfully implanted the first working, lab-grown organs—new bladders—into people. The scientists collected bladder and muscle cells from several patients, grew them in the lab, then used the cells to line a bladder-shaped scaffold of biodegradable material with muscle cells on the outside and bladder cells inside. The cells grew together, and the scaffold eventually dissolved.

More recently, scientists at the McGowan Institute for Regenerative Medicine at the University of Pittsburgh have stimulated the growth of new muscle using just the scaffold—a sheet of what's called extracellular matrix from a pig bladder—sewn into the injury site.

Extracellular matrix is the material, made mostly of collagen, that holds cells and guides their growth in connective tissue. Stripped of its own cells, the donor pig matrix actually stimulated patients' stem cells—the growth and repair cells that multiply fast and can specialize into particular cell types—to migrate to the damaged area and muscle cells.

Scientists are trying to grow liver, kidney, and heart tissue. Regenerative medicine promises to sweep aside two huge problems with traditional organ transplant medicine: the shortage of donor organs and the problem of transplanted foreign tissue being rejected by the body's immune system.

»»See Also

RELATED PRINCIPLE: Bioengineering, 364

RELATED APPLICATIONS: Gene Therapy, 352 • Human In Vitro Fertilization, 344 • In Vitro Meat, 343 • Stem Cell Therapy, 350

DNA

STEM CELL THERAPY

Stem cells are the master cells of the human body. They can proliferate rapidly and take on a variety of forms to grow new organs and tissues. Stem cells from an embryo a few days old can turn into any of the body's hundreds of specialized cell types, from brain cells to those that line the inner surfaces of the body. Stem cells in bone marrow and blood from an infant's umbilical cord can make new blood and immune cells, while those in skin, muscle, and various organs can produce new cells to repair those tissues.

For decades, scientists have sought ways to use these special engines of growth and repair as medicine to restore diseased parts of the body. The first successful bone marrow transplant, performed in 1956 in Cooperstown, New York, on a three-year-old girl with end-stage leukemia, was a mixed success. The child's irradiated bone marrow was rescued by the stem cell–laden marrow of her identical twin, but after six months her cancer returned, and she died.

The procedure was an early demonstration of the potential of stem cell therapy. Today transplantation of bone marrow or umbilical cord blood is an established (though difficult) treatment for people whose bone marrow is not functioning to make blood cells due to disease or the destructive effects of chemotherapy. About 15,000 Americans receive such transplants each year. Though still experimental, broader applications of stem cell therapy hold promise in repairing damage to the spinal cord, brain, and heart.

Research in stem cell medicine was stymied in its early days by controversy over the use of cell lines originally collected from very early embryos. But in the 21st century, much investigation has focused on adult stem cells. A breakthrough came in 2006 when scientists at Japan's Kyoto University discovered a way to coax ordinary stem cells—those in connective tissue, for example—to become pluripotent (meaning many potentials) and grow into all cells in the body.

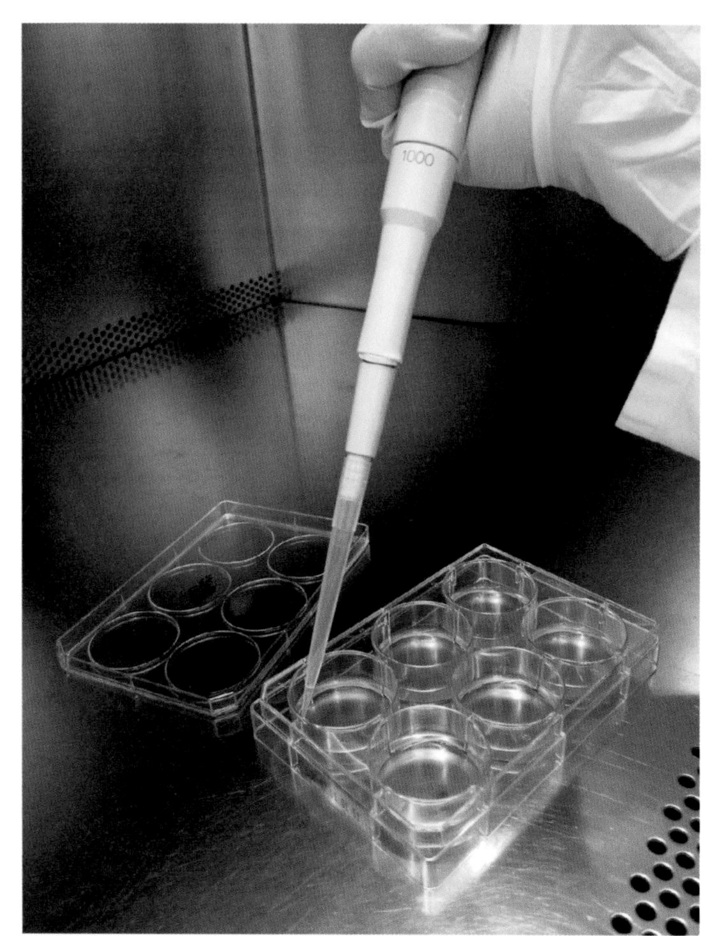

CANCER RESEARCH. A clean bench in a laboratory is essential for avoiding contamination of cell culture experiments.

»»See Also
RELATED PRINCIPLES: Bioengineering, 364 • Ecology, 322 • Immunobiology, 357
RELATED APPLICATIONS: Cancer, 346 • Gene Therapy, 352 • Human In Vitro Fertilization, 344 • Human Tissue Cultivation, 349

TYPES OF STEM CELLS

There are different types of stem cells. All can divide and create exact copies of themselves.

Embryonic stem cells (from embryos) are found in the very earliest stage of embryonic development, up to about day five. They can develop into all the cells of an adult body. Embryonic stem cells can be grown in laboratories, which allows them to continue to grow and divide indefinitely, maintaining the ability to form any of the more than 200 adult cell types. Because of this, they're referred to as pluripotent (many potentials).

Adult stem cells are specialized cells found in the tissues of adults, as well as in children and fetuses, for example, in the skin, brain, liver, intestines, and blood. Bone marrow stem cells can be used to generate red or white blood cells. Stem cells from the brain can develop neurons and brain cells but not tissue outside the brain. Umbilical cord blood cells that are collected and stored after birth are a form of adult stem cells. Unlike embryonic stem cells, adult stem cells have not yet been grown in laboratories.

Induced pluripotent stem cells, which are taken from adults and children, are genetically modified to behave like embryonic cells. Like embryonic cells, then, they can develop into any adult cell type.

Exciting research is being done with cancer stem cells, which can develop into any of the cells found in certain tumors. The hope is that, by isolating stem cells, treatment can be directed to destroying them, and destroying the potential of recurring tumors.

Illustration showing DNA being injected into a stem cell

DNA

GENE THERAPY

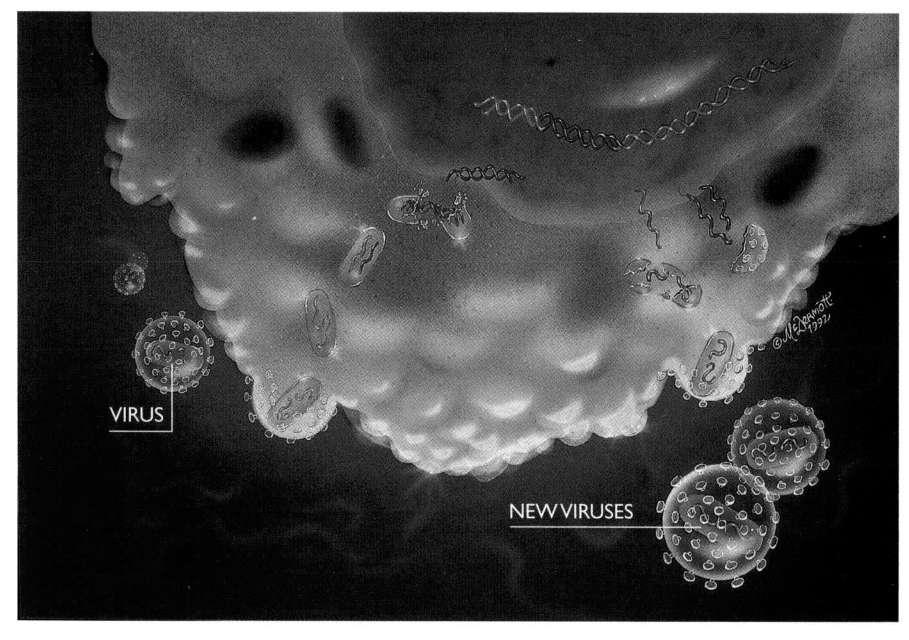

GENE THERAPY SUCCESS. The microscopy image (opposite) shows immune system T-cells, center, binding to beads that cause the cells to divide. The beads, depicted in yellow, are later removed, leaving pure T-cells, which are then infused to cancer patients. Using this method, scientists at the University of Pennsylvania reported in 2011 the first clear success with gene therapy to treat leukemia, using the patients' own blood cells to hunt down and wipe out their cancer. Scientists are preparing to try the approach in other kinds of cancer.

Genes are like an instruction manual for building and running the body. So why not tackle chronic faults in the body's functioning by correcting the instructions themselves? It's a straightforward idea. But given the complexity of human genetics, it's no easy task.

In 1990 scientists at the U.S. National Institutes of Health launched gene therapy's first human trial, treating two girls with combined immunodeficiency (SCID), also called "bubble baby" syndrome. They collected the children's defective white blood cells, inserted normal genes using viruses as delivery agents, and then returned the cells to the patients. Results were promising.

Then, in 2000, a team from Paris reported results of a gene therapy trial in ten boys with a different form of SCID; the boys had received transplants of their own corrected bone marrow cells. Following the procedure, nine of the boys gained functioning immunity. Tragically, two years later the team reported that one boy, then another, developed a leukemia-like condition, apparently because the virus used to deliver the boys' corrected DNA had lodged near a gene that regulates cell growth. This alarming news led to the halting of not only the French trial but also studies in Germany, Italy, and the United States.

For years the SCID case served as both gene therapy's cautionary tale and one of its few success stories. Since the first trials, the technique has been used experimentally—and successfully—in a few dozen people worldwide.

Recent gene therapy experiments have also produced promising results in the treatment of leukemia, congenital blindness, and a degenerative brain disease affecting the nerves' fatty coating. In 2012 a University of Pennsylvania team reported that, after more than a decade, 43 patients who tested positive for the human immunodeficiency virus (HIV) and were treated with gene therapy remained healthy. In 41 of them, immune cells genetically altered to resist the virus were still alive, suggesting a possible alternative to lifelong drug treatment.

»»» *See Also*

RELATED PRINCIPLES: Bioengineering, 364 • Ecology, 322 • Immunobiology, 357
RELATED APPLICATIONS: Cancer, 346 • Human In Vitro Fertilization, 344 • Human Tissue Cultivation, 349 • Stem Cell Therapy, 350

Genetically altered viruses have proven an excellent delivery vehicle for gene therapy.

CHAPTER 16

Health and

Medicine

Many scholars trace the origins of Western medicine to a radical moment in ancient Greek society, when Hippocrates and his followers first took the art of doctoring away from the magicians and priests. The Hippocratic physicians of the fifth and fourth centuries B.C. insisted that disease be attributed to natural or material causes and counseled a gentle, conservative approach to treatment that strove to assist the body's own restorative powers. First, do no harm was the crux of the oath taken by physicians. Doctors still take the Hippocratic oath today.

Yet alongside it there is a tradition of bold experimentation and aggressive intervention, of trying new approaches to protect human health—from inoculating people with infectious agents to stimulate immunity to fitting them with synthetic limbs so they can walk again. When they work, such innovations seem almost like magic.

Little more than a century ago, death in childhood from infectious disease was commonplace even in wealthy nations. Today, cancer, heart disease, and autoimmune disorders—the so-called diseases of civilization—are the more common maladies. So, in pursuing an old goal—the preservation of human health—medicine must forever chart a new course. It must observe and test results rather than rest on assumptions of the past. Organized scientific inquiry is what joins medical traditions based on care of the patient and audacious striving for new cures.

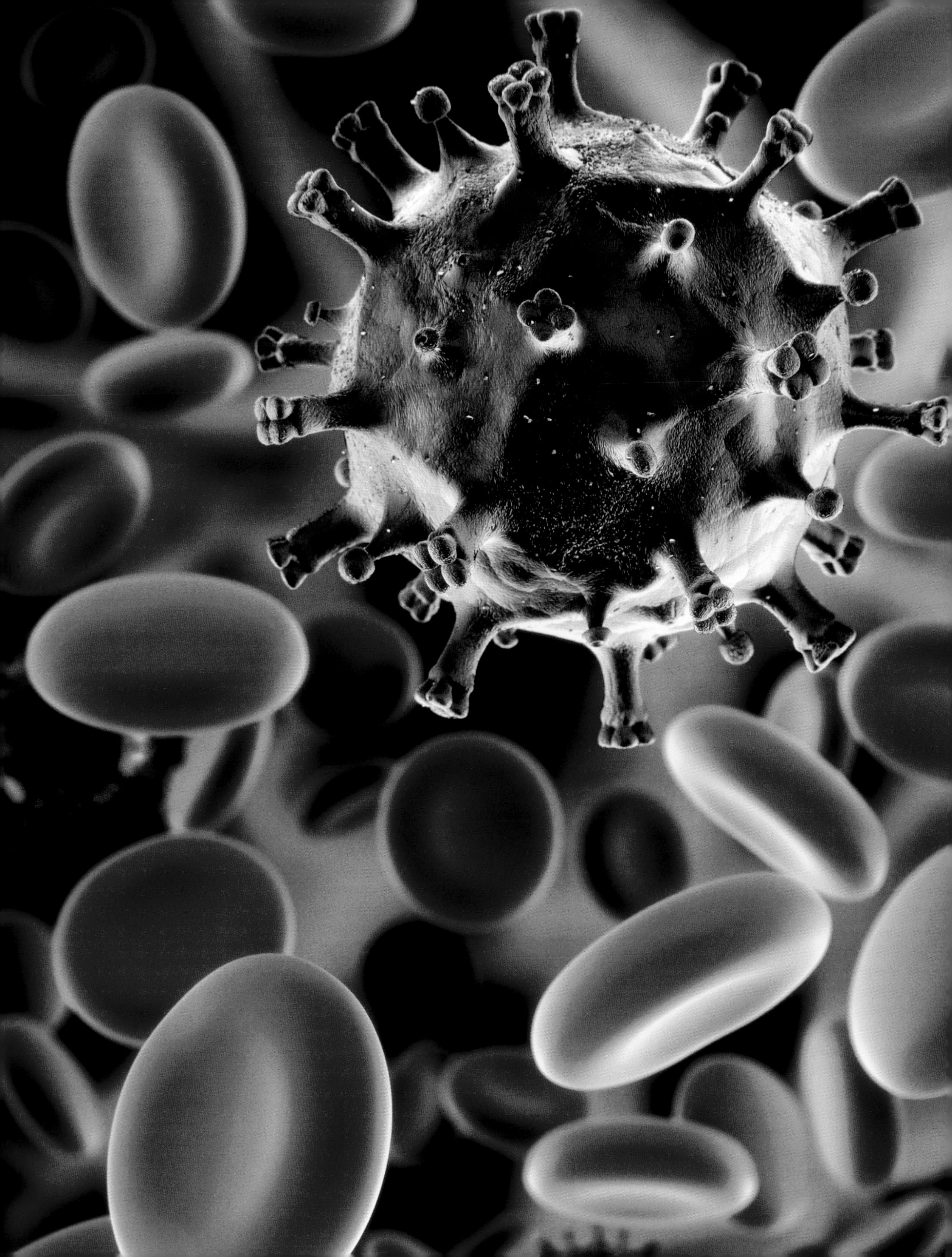

PRINCIPLE

RALLYING THE TROOPS

Immunobiology

The body knows what's good for it. It recognizes its own boundaries, the difference between itself and foreign bodies. It can mount a custom-designed defense against an array of invaders. And it seldom forgets an enemy.

The immune system governs these uncanny abilities. Without it, no human being would live long. Our environment is alive with bacteria, viruses, fungi, and parasites ready to take up residence in the warmth of human tissues. The immune system can sometimes recognize and destroy marauding cancer cells, too.

This intricate defense system is made up of an army of cells, all arising from stem cells in the bone marrow. Some immune cells release large proteins called antibodies into the bloodstream that rush to meet and surround a foreign invader. Others bind to the enemies and attack with lethal chemicals, and still others envelop and consume the offenders. Immune cells multiply in response to a threat and signal other immune cells to join the fight.

After successfully fighting off an infection—a bout of flu, for example—the immune system leaves behind a small contingent of memory cells that stand ready to obliterate that particular strain of influenza in case it comes back, even many years later.

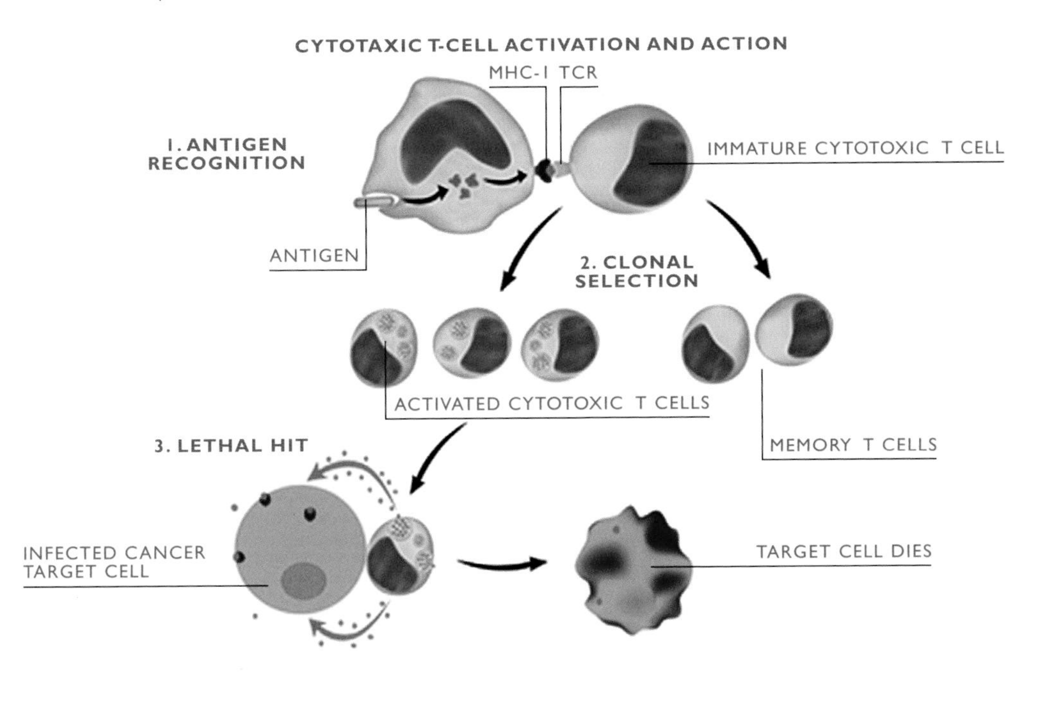

Testing the efficacy of antibiotics in the laboratory

Immunobiology

ANTIBIOTICS

Faced with a bacterial invader that's new and especially virulent or that is able to establish a beachhead by taking advantage of temporary weakness or injury, the immune system can be overwhelmed. That's where antibiotics come in. Like an air strike, these medicines reinforce the immune system's mission to kill invaders while sparing host cells.

"Thanks to PENICILLIN ... He Will Come Home!" proclaimed one World War II–era advertisement, featuring the image of a medic injecting a fallen soldier with the new wonder drug. It was an entirely reasonable claim for the world's first truly potent antibiotic, capable of clearing away diseases that had plagued humanity for countless generations, from pneumonia and gonorrhea to wound infections once treatable only with ghastly rounds of surgical drainage.

Penicillin was discovered in 1928, when a British bacteriologist first noted the antibacterial properties of the blue mold Penicillium notatum (now known as P. chrysogenum). It works by interfering with a bacterium's ability to form a cell wall, the outermost layer of a cell found in plants, bacteria, fungi, and algae that protects and strengthens it. That interference prevents the bacterium cell from dividing and multiplying; it eventually bursts. Human cells don't have cell walls, so they're unaffected. Other antibiotics interfere with bacteria's production of proteins that support cell functions. In this case, the antibiotic exploits differences between human and bacterial ribosomes, the cell structures that assemble amino acids into proteins.

Unfortunately, some bacteria can survive antibiotics. They multiply extremely rapidly and can swap genetic material with one another. This allows them to adapt to threats—including antibiotics. The result: drug-resistant superbugs that many common antibiotics can no longer kill.

»»See Also
RELATED PRINCIPLES: Ecology, 322 • Fermentation, 314
RELATED APPLICATIONS: Allergy, 362 • Vaccines, 360

Immunobiology

VACCINES

If antibiotics provide reinforcement for an immune system already besieged by bacterial infection, vaccines offer a kind of advance warning, arousing the immune cells to prepare defenses against specific pathogens.

They do this by mimicking a natural infection—getting the threatening organism into the body, but in a form the immune system can easily fight off. A vaccine may be filled with bacteria or viruses that have been weakened in the laboratory or killed outright. It may contain chemically defanged toxins released by the organism. Or it may contain just pieces of a pathogen—the antigens that cover the surface of a microbe and mark it as foreign to the immune system.

It's the unique antigens of a given organism that the immune system will file away for future reference. In the meantime, when you get the vaccine, your body mounts an immune response you don't even feel. Among the many antibody-producing immune cells (called B cells) contained in the blood are some whose antibodies match the antigen introduced by the vaccine. The antibodies coat the foreign material, marking it as an invader, while the immune cells multiply and secrete more matching antibody. Other immune cells eat the invaders and carry their antigens to show ally cells.

As the threat dies down, some of the immune cells turn into memory cells, capable of reproducing and, on short notice, churning out more matched antibody should the same threat return.

Vaccination campaigns are designed to protect not just individuals but whole populations as well. Once a critical proportion of a community has been vaccinated, even those too vulnerable to receive the vaccine (because of a compromised immune system, for example) are safe from the disease. They simply won't be exposed to it.

SUCCESS STORY. Highly effective worldwide vaccination efforts have eradicated or nearly eradicated many deadly childhood diseases.

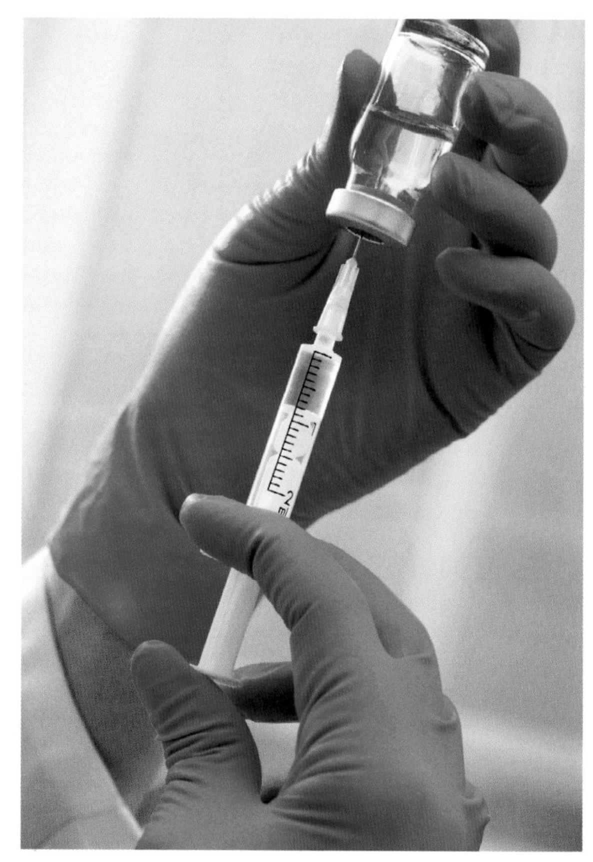

»»See Also

RELATED PRINCIPLE: DNA, 338
RELATED APPLICATIONS: Allergy, 362 • Antibiotic, 359

Immunobiology

ALLERGIES

Sometimes the immune system becomes primed to fight an imaginary foe—harmless grass pollen blowing in the breeze, for example, or cat hair on the sofa.

These false threats are called allergens, and the symptoms they cause are related to a particular pathway of the immune response and a class of antibodies called immunoglobulin E (IgE).

If you have allergies—you have an overactive immune system—then when you inhale grass pollen, for example, your immune cells identify the pollen as a foreign invader and produce IgE antibodies specific to this particular antigen. These antibodies bind to receptors on special immune cells called mast cells, found in connective tissue and mucous membranes, and basophils, a type of white blood cells, that circulate in blood. As a result, you become sensitized to the grass pollen.

The next time you encounter the pollen, its antigen attaches to matched antibodies already on the surfaces of your mast and basophil cells. Proteins in your blood are drawn to the bonded antibody-antigen complex and, step-by-step, assemble into a cell-puncturing cylinder. This process disrupts the outer membrane of the immune cell, which releases histamine, serotonin, heparin, and other chemicals. Depending on the area of the body affected, these chemicals can produce swelling, itching, sneezing, vomiting and diarrhea, airway spasm, and dilation of blood vessels. Meanwhile, the immune cells also give off chemical signals telling other immune cells to make more IgE.

Though people don't inherit an allergy directly, the tendency to develop allergies is highly hereditary. Most allergy drugs block the histamine that causes so much unnecessary drama when released from mast cells. That's why they're called antihistamines.

»»*See Also*
RELATED PRINCIPLE: DNA, 338
RELATED APPLICATIONS: Antibiotic, 359 • Vaccines, 360

ALLERGY OR INTOLERANCE

A food allergy involves an immune response to a normally harmless substance such as a peanut. Once established, the reaction tends to come on quickly and can be triggered by exposure to even a small amount of the offending substance. In severe responses, the release of histamine and other chemicals can trigger anaphylactic shock: the throat swells, blood pressure drops, and the sufferer is in serious peril.

A person suffering from a food intolerance, on the other hand, simply experiences some ill effect—unrelated to immunity—from an ingredient others handle readily. Lactose intolerance is a common example. People with this sensitivity lack sufficient quantities of the enzyme lactase in their small intestine, so they have trouble digesting the sugar in dairy products. The unpleasant effects of lactose intolerance can be avoided with a lactase supplement.

PRINCIPLE

SYNTHETIC SOLUTIONS

Bioengineering

People used to say that life can't be engineered. It evolves according to its own laws. In this worldview, the melding of technology with living things is relegated to science-fiction fantasies. But in recent decades, the field of bioengineering has produced results that are quite real. It turns out that we can apply the principles of engineering to organisms.

Living things can be studied and monitored using computers and quantitative techniques. Scientists can engineer tissue and change the characteristics of cells. They can tweak gene sequences to produce a soybean plant that resists herbicides or a lab mouse highly susceptible to developing cancer. At least experimentally, they can genetically engineer bacteria to deliver drugs to microscopic targets and convert plant matter into transportation fuel. They can make an artificial nose that detects diseases such as cancer by smell.

Bioengineering is a sprawling field. Its broadest definition encompasses any interface between technology and a living system. Perhaps the most familiar products of bioengineering are the electronic and mechanical devices that replace biological parts or functions in human beings. That concept is as old as the wooden leg, but its applications have become more sophisticated—and far more powerful in their ability to support and enhance human life.

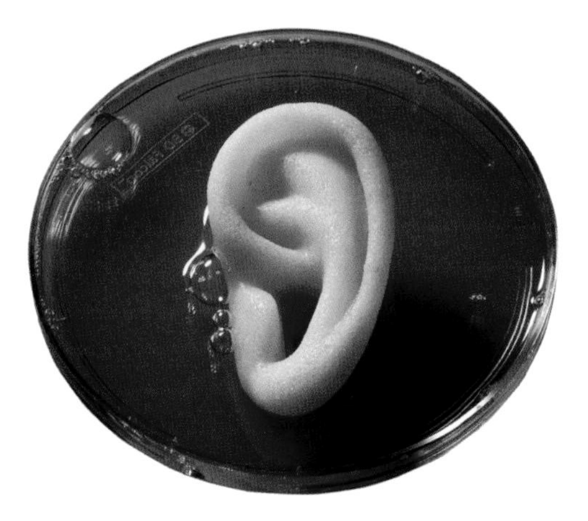

A synthetic ear bathed in cartilage-producing cells

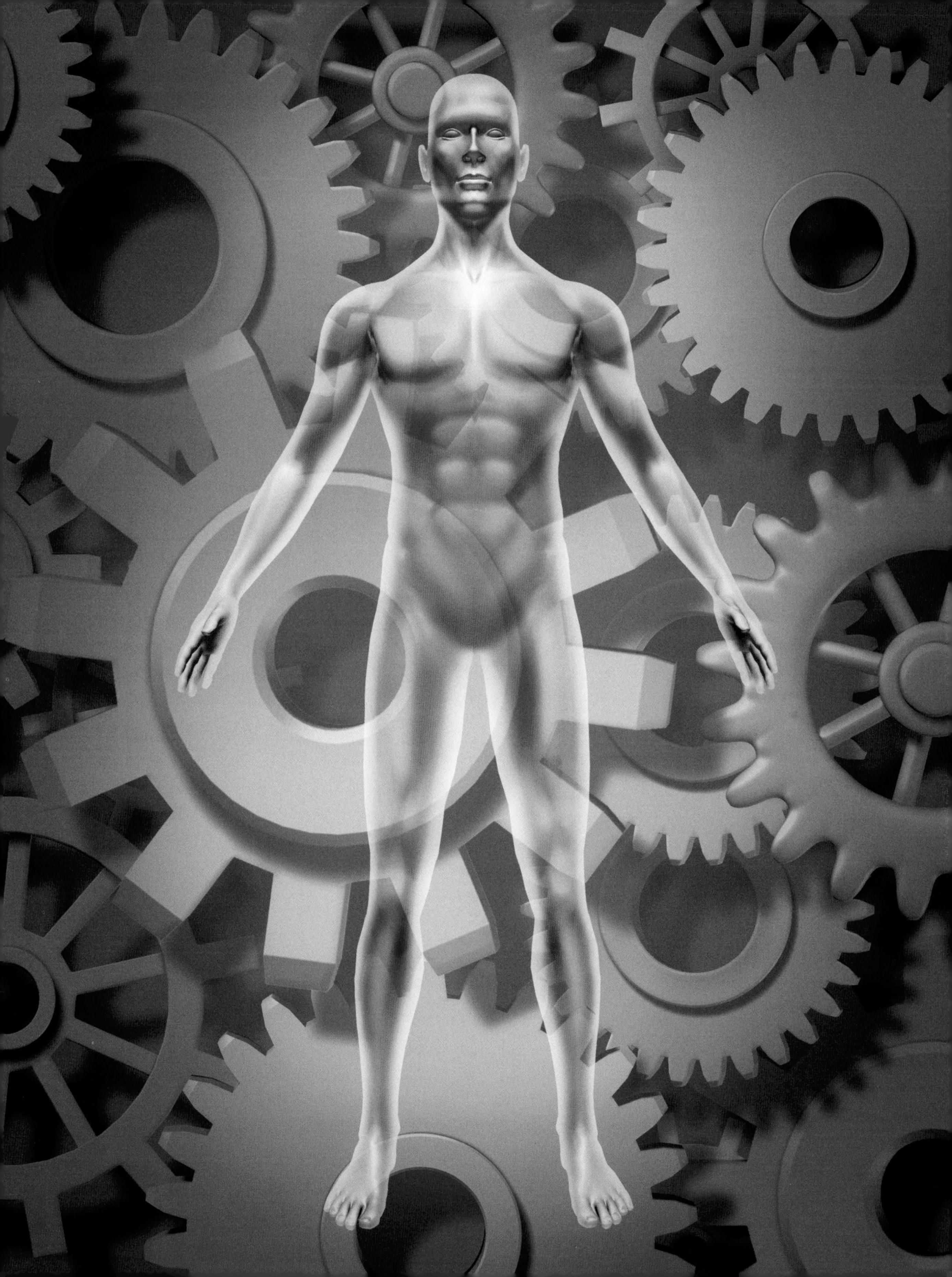

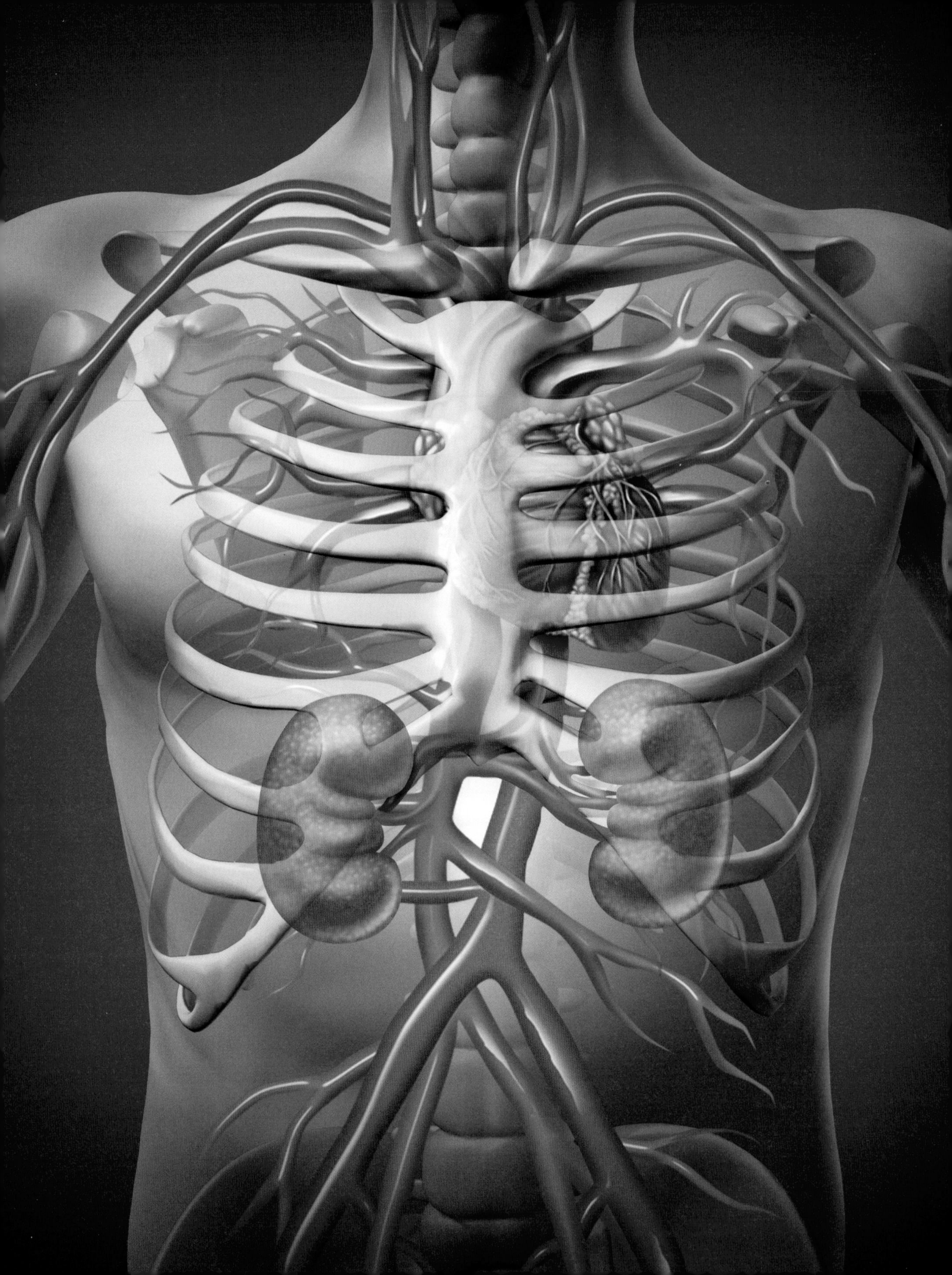

Bioengineering

KIDNEY DIALYSIS

Kidney function can be knocked out suddenly in the case of injury or overdose, or it can decline gradually as a result of diabetes or high blood pressure. Either way, the failure of these bean-shaped organs in the lower back to rid the blood of wastes and excess water can be deadly. The blood becomes overloaded with urea, a metabolic waste product, and minerals such as sodium (which can increase blood pressure), potassium (which can lead to irregular heartbeat), and phosphorus (which causes a loss of calcium in the bones).

In dire cases of kidney failure, patients can turn to a treatment called dialysis, which takes over the kidneys' most critical roles. It can be used during recovery from an acute problem or on an ongoing basis. The machine works by drawing the patient's blood through tubing into a bath of purifying liquid, then returning it to the body. The tubing is made of semipermeable membrane. When exposed to such a membrane, particles in solution always move from an area of higher concentration to one of lower concentration, according to the principle of osmosis. So in dialysis, the toxic waste particles in the blood cross the membrane into the dilute water; the larger particles, such as red blood cells, are too big to pass through the membrane's pores.

It was the anguish of watching a young man die slowly of kidney failure that moved the Dutch physician Willem Kolff to design the first dialysis machine in the early 1940s. He had to scrounge parts for its construction. His initial test of the prototype came in 1943; the first several patients died, but in 1945, the machine revived a comatose patient. By keeping patients with kidney failure alive, dialysis paved the way for the first successful kidney transplants. Although kidney dialysis is most often administered in a dialysis center, home dialysis has been growing as an option.

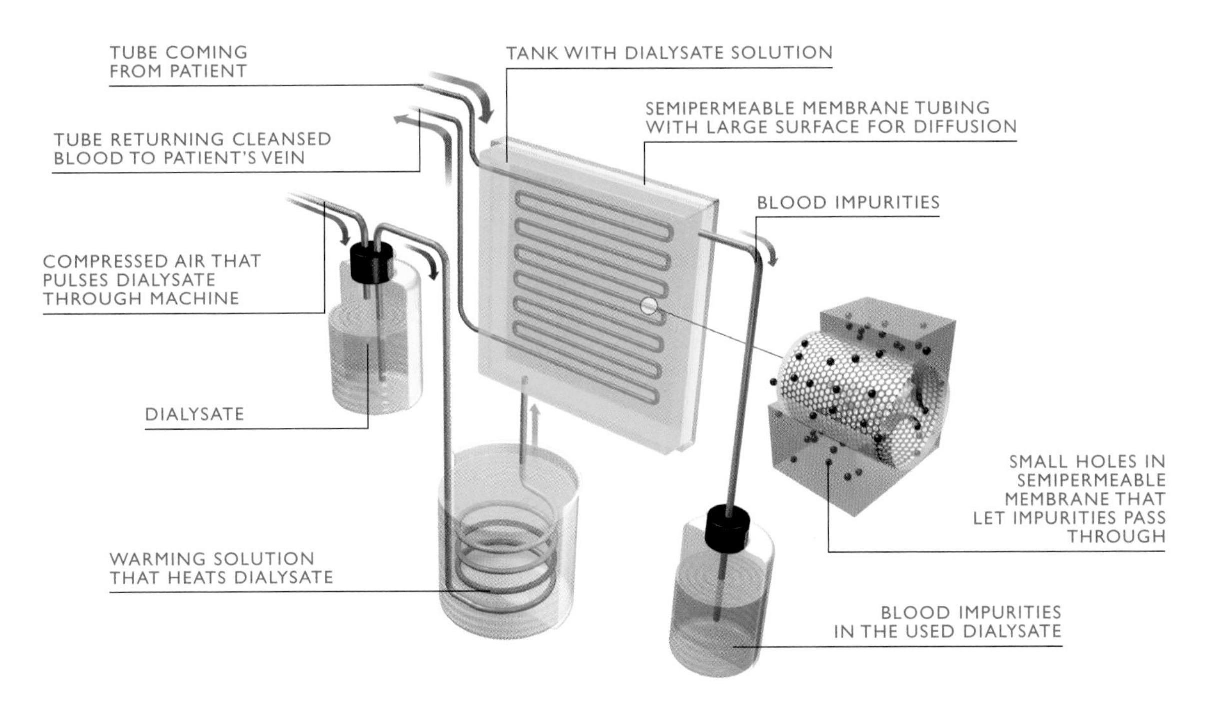

»»See Also

RELATED PRINCIPLES: Binary Code, 190 • Elasticity of Air, 78 • Electromagnetic Induction, 157 • Ohm's Law, 164 • Pascal's Law, 80
RELATED APPLICATIONS: Cochlear Implant, 373 • Pacemaker, 368 • Prosthetic Disc, 372 • Prosthetic Limb, 370

Bioengineering

PACEMAKER

A pacemaker is a battery-powered, computerized generator connected to sensor-tipped wires that are threaded through veins to the heart. Its job is to keep the heart beating steadily. Every year, some 300,000 of these devices are implanted in Americans who suffer from an inadequate heart rate or abnormal heart rhythm, whether due to faults in the heart's electrical conduction system or weakness of the muscle itself. The device is tucked under the skin just below the collarbone.

A pacemaker's electrodes sense the heart's electrical activity and relay this information through the wires to its miniature computer. If something is wrong—the heart is beating too slowly, for example—the generator sends an electrical impulse to one or more chambers of the heart, causing it to contract.

Doctors at Sweden's Karolinska Institute implanted the first pacemaker into a patient in 1958. It kept the man alive, but the bulky device had to be replaced many times. Today, most pacemakers are quite small—about the size of a dollar coin. Also, early pacemakers kept the heart beating at a constant rate, so patients often felt overstimulated when resting but tired during exertion. Newer models introduced in the 1980s are rate-responsive: they can sense changes in body movement or breathing and adjust the target heart rate accordingly.

Pacemakers have to be replaced when their lithium battery runs out, usually after about five years. For many, that's a small price to pay for feeling better and more energetic despite a heart that isn't up to its vital task without help. On a typical day, the heart pumps about 2,000 gallons of blood through the body.

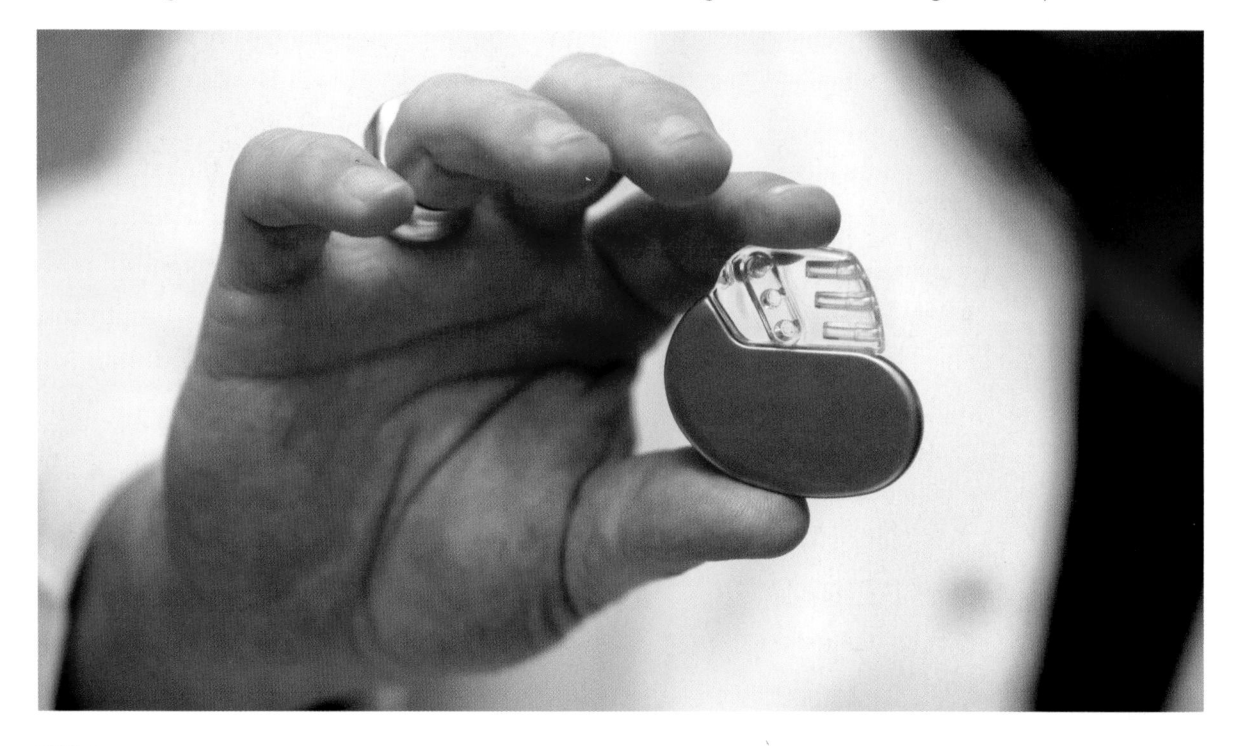

»»See Also

RELATED PRINCIPLES: Binary Code, 190 • Electrochemistry, 187 • Electromagnetic Induction, 157 • Electromagnetism, 147 • Ohm's Law, 164 • Pascal's Law, 80 • Robotic Surgery, 375

RELATED APPLICATIONS: Cochlear Implant, 373 • Kidney Dialysis, 367 • Prosthetic Disc, 372 • Prosthetic Limb, 370

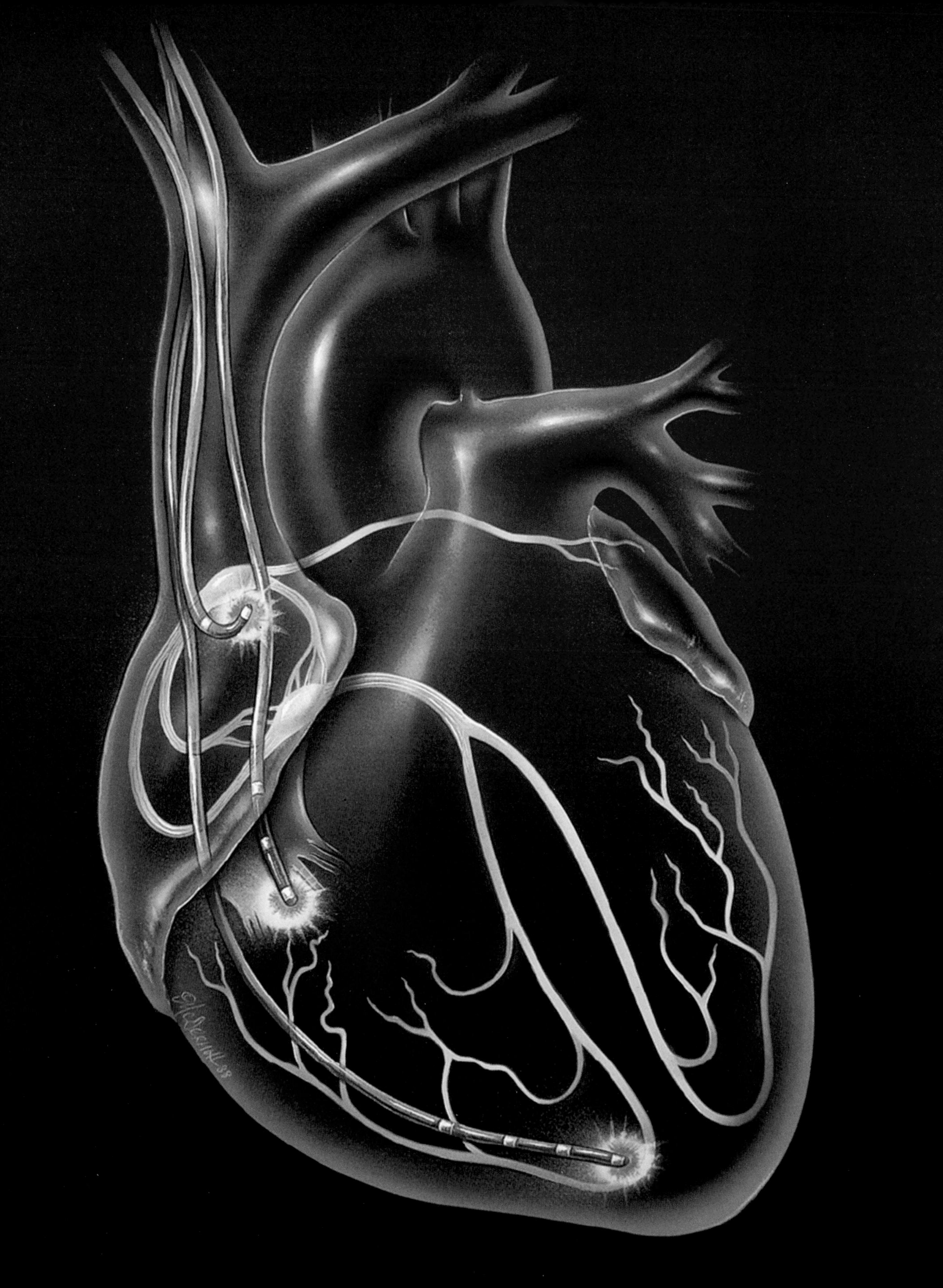

Bioengineering

PROSTHETIC LIMB

In ancient Rome and medieval Europe, soldiers who had lost an arm were sometimes fitted with an iron one so they could carry a shield into battle. Today, prosthetic limbs are far more flexible, designed to help their users grasp objects or walk with a natural gait.

Some of the newest prosthetics combine strong, lightweight materials with sensors and microprocessors that make them responsive to the user and conditions. A new aluminum knee, for example, uses a variety of sensors to measure angle, load, and movement in the joint. This information goes to a built-in controller that analyzes what the knee is doing and, through a motor, generates power and movement appropriate to the task.

Other prosthetic devices respond to muscle contractions in the limb. In a myoelectric arm, electrodes on the skin record the electric impulse created by a muscle contraction. A controller responds to the muscles' electric signals by turning on electrical motors that move the elbow, wrist, or hand with impressive dexterity.

An even more astonishing technique called targeted muscle reinnervation allows a person to move an artificial limb just by thinking about it. Surgeons relocate nerves that once controlled the lost arm or hand to the chest. When the user's brain sends the signal to use the arm, the chest muscle automatically contracts, and sensors on the muscle translate the signals into electric-powered movement of the prosthetic.

Carbon-fiber running blades may be the most well-known example of today's high-performance prosthetics. Their spring-like action lends speed and endurance. South African runner Oscar Pistorius, whose legs were amputated below the knee when he was 11 months old, competed in the 2012 London Olympics after successfully refuting arguments that his prosthetics gave him an unfair advantage—although his later notoriety for shooting his girlfriend eclipsed his athletic fame.

See Also

RELATED PRINCIPLES: Macromolecular Chemistry, 288 • Mechanical Advantage, 38 • Ohm's Law, 164 • Quantum Mechanics, 298 • Robotic Surgery, 375

RELATED APPLICATIONS: Cochlear Implant, 373 • Kidney Dialysis, 367 • Lever, 41 • Pacemaker, 368 • Prosthetic Disc, 372

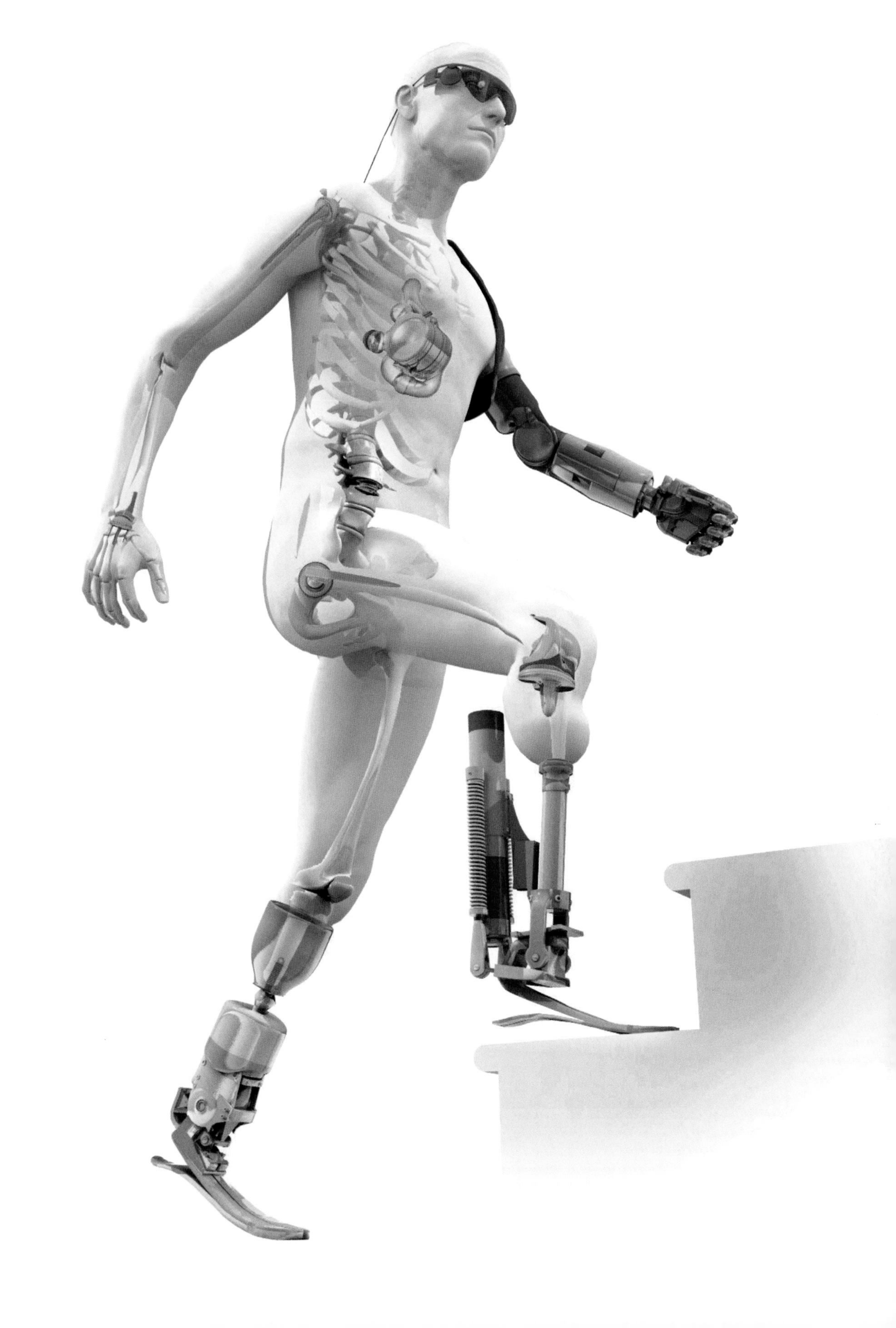

Bioengineering

PROSTHETIC DISC

The soft discs nestled between the bony vertebrae of the spine provide cushioning and flexibility. These discs sometimes flatten and degenerate with age, and that can hurt—a lot. If conservative treatments don't work, surgeons usually tackle the problem by fusing the two vertebrae together, reducing pain in many cases but greatly diminishing range of motion.

For decades, medical engineers labored to design a product that could allow surgeons to replace spinal discs the same way they do worn-out hips and knees. In 2004 the U.S. Food and Drug Administration approved the first prosthetic disc. Like a natural disc, which has a tough fibrous outside and jelly-like center, the artificial version consists of two metal plates with a movable, rubbery center.

Implanting a prosthetic disc requires great skill. Though manufacturers claim the discs will last around 40 years, that hasn't yet been tested by experience—and removing one in the case of malfunction is hazardous. But a study of more than 200 patients published in 2011 found those who'd undergone disc replacement were more satisfied and less likely to require follow-up surgery than people who had spinal fusion. Titanium is favored as a material in artificial discs because it conjoins well with the bone.

TITANIUM DISC. (Left) A doctor holds a titanium disc that can be implanted as a replacement in the treatment of a herniated vertebral disc of the cervical spine.

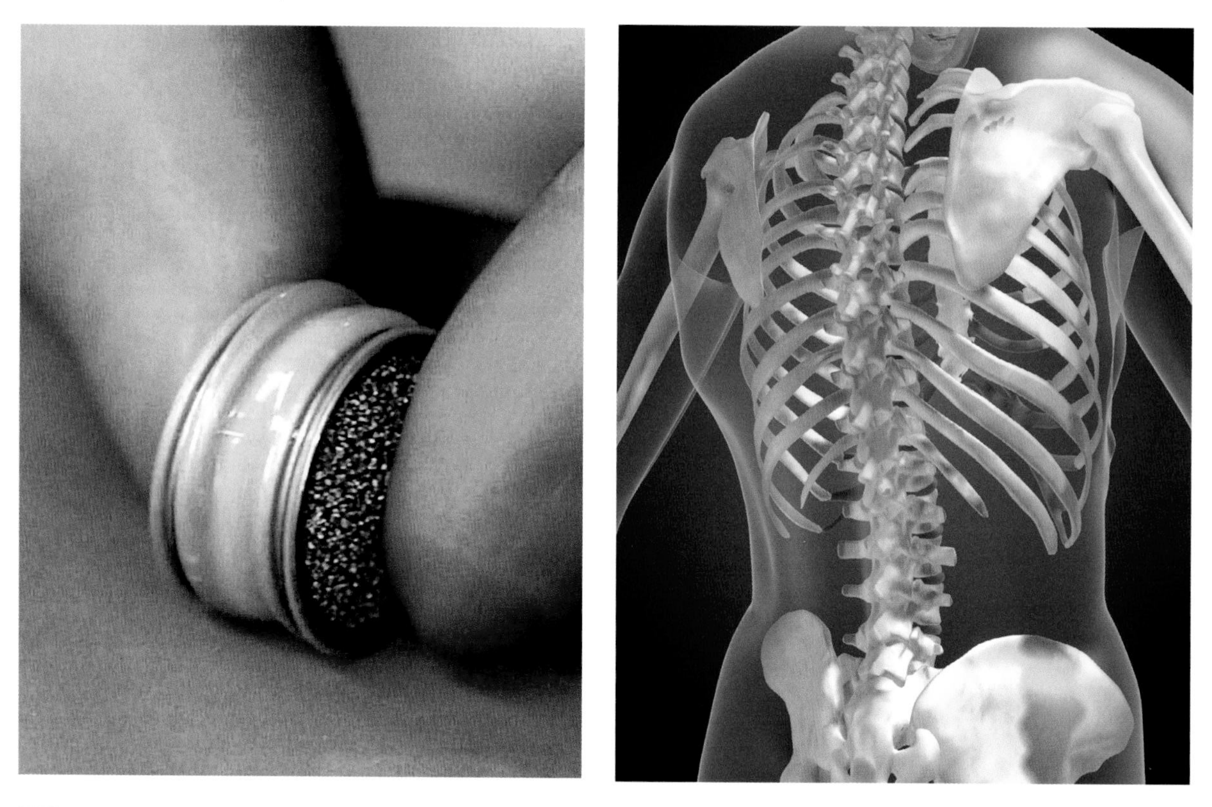

»»See Also

RELATED PRINCIPLES: Macromolecular Chemistry, 288 • Robotic Surgery, 375

RELATED APPLICATIONS: Cochlear Implant, 373 • Kidney Dialysis, 367 • Pacemaker, 368 • Prosthetic Limb, 370

Bioengineering

COCHLEAR IMPLANT

The sense organ of the ear, the organ of Corti, lies deep in the inner ear, at the center of a spiral-shaped cavity called the cochlea. The organ of Corti contains electrically active hair cells that, when agitated by the pressure waves of sound, release neurotransmitters that pass electrical signals along the auditory nerve to the brain.

Cochlear implants restore hearing by sending information straight to this critical sound translation center, bypassing the tiny vibrating bones of the middle ear. Approved in the United States for adults in 1985 and for children in 1990, the implants bring sound to those who are deaf or extremely hard of hearing and cannot benefit from an ordinary hearing aid. After getting an implant, the typical adult can understand most conversation, and young children who receive the implant early enough can develop language skills along with their peers.

A cochlear implant has several parts. A microphone behind the ear and processor worn externally pick up sound and turn it into a digital code. This digitized sound is sent to a transmitting coil, also behind the ear, which transmits the code as electrical impulses to a radio receiver just under the skin. The receiver passes the impulses along a wire to an array of electrodes placed within the cochlea, stimulating the auditory nerve. With practice, implant recipients learn to interpret this stimulation as sound.

»»See Also

RELATED PRINCIPLES: Binary Code, 190 • Electrochemistry, 187 • Electromagnetic Induction, 157 • Electromagnetism, 147 • Ohm's Law, 164 • Pascal's Law, 80 • Robotic Surgery, 375

RELATED APPLICATIONS: Electrical Wiring and Power Grid, 166 • Kidney Dialysis, 367 • Pacemaker, 368 • Personal Computer, 195 • Prosthetic Disc, 372 • Prosthetic Limb, 370

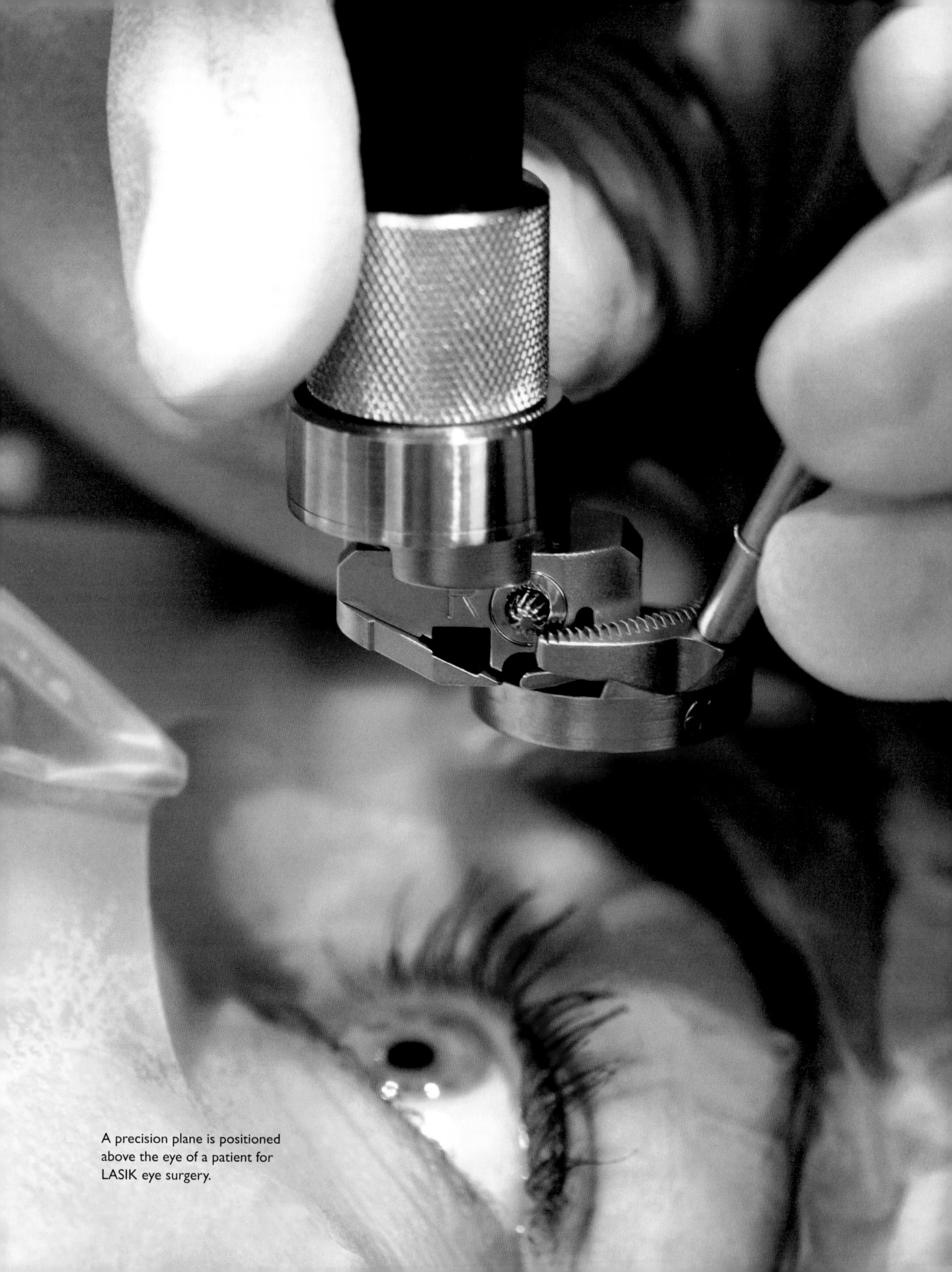

A precision plane is positioned above the eye of a patient for LASIK eye surgery.

PRINCIPLE

ROBO DOC

Robotic Surgery

The delicate, high-stakes job of performing surgery on a human being doesn't seem the most likely candidate for mechanization. But there are certain things robots can do that even the most skilled surgeon cannot. They can see inside the patient at high magnification and perform tiny, ergonomically awkward movements there with zero tremor. Plus, they never get tired.

Introduced in the 1980s, robot-assisted surgery is increasingly used today in prostate surgery, gynecologic surgeries such as hysterectomy, and heart surgeries, including coronary bypass and mitral valve repair. The surgeon sits at a computer console that operates a separate unit consisting of three robotic arms and a high-definition camera. These reach into the patient through keyhole incisions.

Robotic surgery is available at some of the most advanced surgery centers in the country. Many people claim that it allows accurate, minimal movements that mean a quicker and less painful recovery for the patient. But the approach is controversial. Some surgeons argue that while robotic assistance is a clear plus in cases where they'd otherwise have to make a large incision, there's little evidence that it's better for patients than ordinary keyhole surgery by an experienced surgeon.

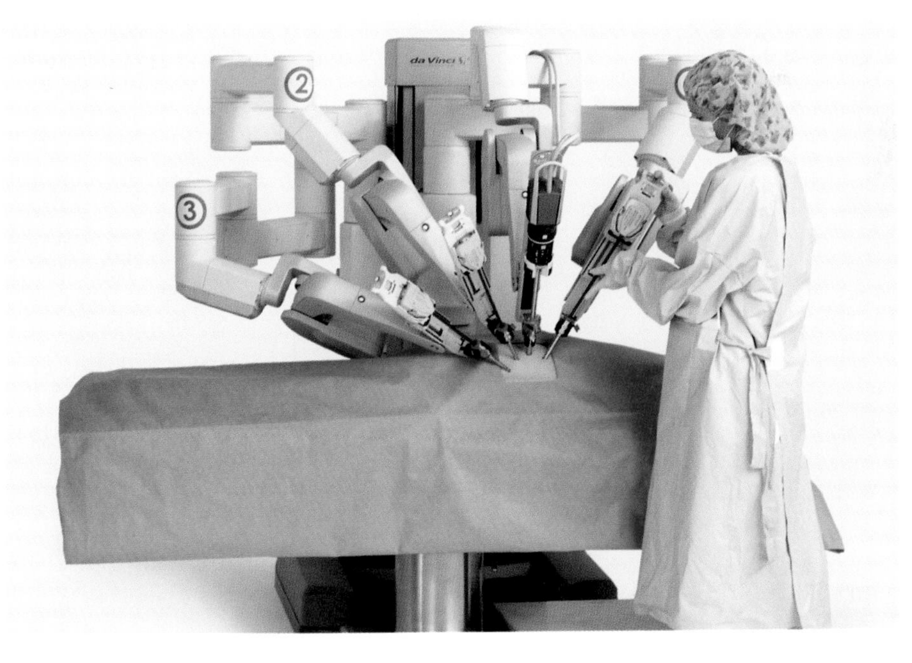

Robotic Surgery

ENDOSCOPY

An endoscope is a long, usually flexible tube fitted with a light and tiny camera that lets physicians examine the inner recesses of their patients' bodies on a video screen. It's a mainstay of medicine, used in a variety of diagnostic procedures—to examine the colon, for example, or the airways—and to guide minimally invasive (keyhole) surgery.

The idea of peering inside the human body without making a large surgical incision has been around since at least the early 19th century. In 1806 the German physician Philipp Bozzini invented a "light conductor," an internal imaging device consisting of a tube illuminated by an ordinary candle and a set of mirrors.

In 1910 the Swedish internist Hans Christian Jacobaeus published a report describing a diagnostic procedure he had performed on 17 patients with fluid in the abdomen. Using an endoscope featuring a lens system and a light, Jacobaeus examined his patients' abdominal and chest cavities in a procedure he dubbed laparothorakoskopie.

But such scoping devices weren't very useful until well into the 20th century, when major technological advances allowed doctors not only to see inside body cavities but also to operate there with accuracy. The 1950s saw the invention of a system, made up of flexible glass rods (fiber optics), for sending light into the body from an external source, coupled with a television camera for projecting the images onto a screen. The watershed technology was the charge-coupled device (CCD) system, essentially a miniature digital camera. By the mid-1980s, these cameras could provide surgeons with clear video from within the body.

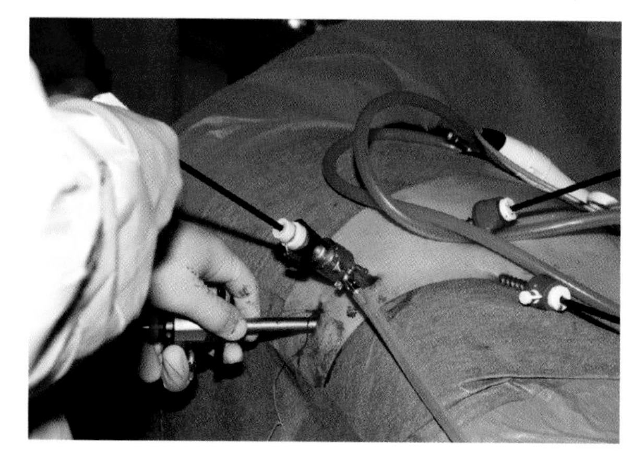

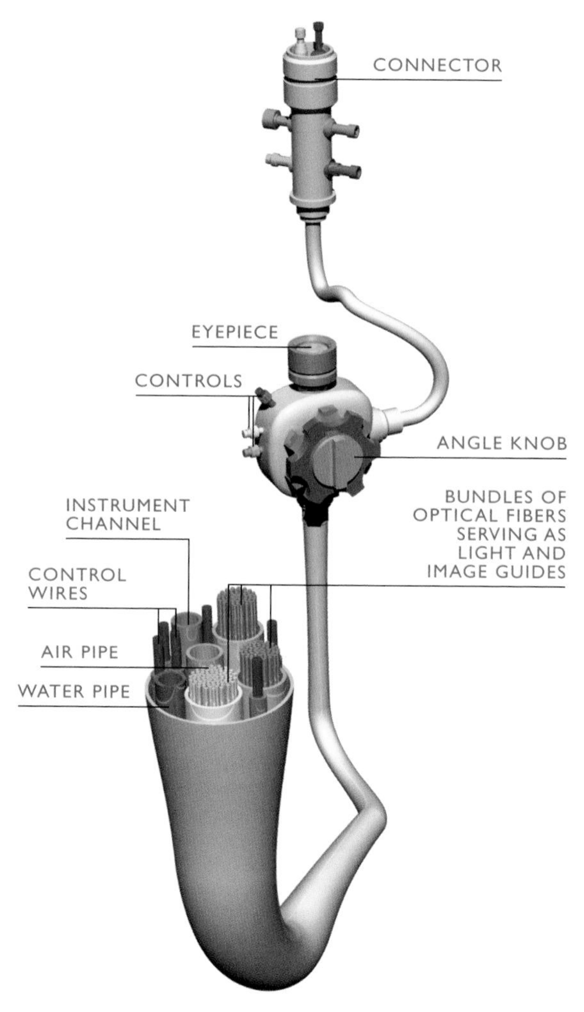

See Also

RELATED PRINCIPLES: Binary Code, 190 • Electromagnetic Induction, 157 • Electromagnetism, 147 • Fiber Optics, 229 • Ohm's Law, 164 • Optics, 225
RELATED APPLICATIONS: Color Television, 144 • Digital Video Camera, 192 • Electrical Wiring and Power Grid, 166 • Fiber Optic Cable, 229 • Integrated Circuit, 191 • Laser Surgery, 377 • Personal Computer, 195 • Plasma, LCD, and LED Screens, 145 • Radial Keratotomy, 378

Robotic Surgery

LASER SURGERY

Ordinary white light is made up of many wavelengths or colors. But a laser emits monochrome light in waves that are in phase—cresting together—and moving in one direction rather than spreading out.

Wielding these beams of focused light to cut or vaporize tissue offers big advantages over the scalpel. A laser can pinpoint a tiny portion of tissue. Its heat can sterilize and cauterize, or burn, the edge of an incision, reducing the risk of infection and bleeding. Using fiber optics, the light from a laser can be bent through a tube to reach inside the body and remove growths.

A laser's single-wavelength light makes it possible to target particular types of tissue because different colors absorb particular wavelengths. Hemoglobin, the pigment in red blood cells, absorbs the green light of an argon laser used to stop bleeding. A carbon dioide laser emits invisible infrared light, which the water present in tissues easily absorbs. It's used to remove thin layers of tissue, for example, in skin and certain gynecologic cancers, by vaporizing—not burning—the cells.

In photodynamic therapy currently used in esophageal and some lung cancers, doctors inject patients with a photosensitizing agent that, when exposed to a particular wavelength of light, produces toxins that kill nearby cells. The agent collects in cancer cells, and a laser is used to focus monochrome light on the tumor.

»»See Also
RELATED PRINCIPLES: Absorption of Light, 204 • Binary Code, 190 • Electromagnetic Induction, 157 • Electromagnetism, 147 • Ohm's Law, 164 • Stimulated Emission, 230
RELATED APPLICATIONS: Electrical Wiring and Power Grid, 166 • Endoscopy, 376 • Integrated Circuit, 191 • Laser, 232 • Personal Computer, 195 • Radial Keratotomy, 378

LASER TATTOO REMOVAL

A dark-colored, splashy tattoo that stands out against pale skin is actually easier to remove than one whose colors are closer to skin tones.

That's because modern laser tattoo removal applies a principal known as selective photothermolysis. Specific wavelengths of light are absorbed by specific colors (or, to be more precise, by objects of certain colors). And different media used to make lasers produce different wavelengths of focused light. So in zapping that unwanted tattoo, a dermatologist chooses a laser or changes settings on the laser to select colored ink particles embedded in the skin, sparing adjacent tissues. The laser heats the particles, breaking them into tinier pieces that will be absorbed by the body; the tattoo fades after treatment. The latest lasers emit a high-power, pulsed beam; they heat target particles very quickly, minimizing scarring due to heat damage in surrounding cells.

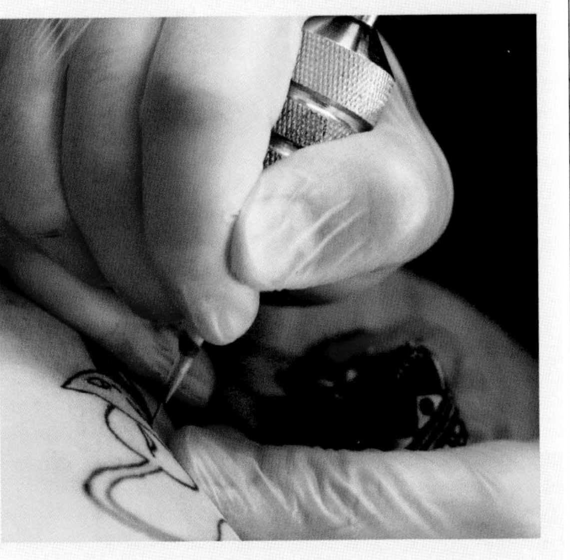

Robotic Surgery

RADIAL KERATOTOMY

You can't make out far-off street signs no matter how much you squint. Or perhaps you have difficulty reading the newspaper unless you hold it at arm's length. Such troubles are called refractive errors. They result from distortions in the cornea at the front of the eye. They're the most common vision problems.

For centuries we've solved such problems by wearing eyeglasses that compensate for the cornea's shape. But over the past two decades, more and more people have tossed their glasses after undergoing surgery to sculpt the corneas themselves.

The cornea is a clear, dome-shaped structure lying over the iris and pupil. It refracts, or bends, light rays to focus them on the retina at the back of the eyeball, creating an image that is carried by the optic nerve to the brain. If the cornea is overly curved, it bends light too sharply, focusing the image in front of the retina; things at a distance appear blurry. If the cornea is too flat or the eye too shallow, the light bends too little, and the image will focus on a point behind the retina; close-up objects blur.

In radial keratotomy, the surgeon makes spoke-like cuts around the center of the cornea, which causes it to flatten, reducing nearsightedness. But since the 1990s this procedure using a diamond knife has been largely replaced with laser-assisted in situ keratomileusis (LASIK) or photorefractive keratectomy (PRK). In LASIK the eye surgeon cuts and lifts a small flap in the top layer of the cornea and then, using a cold ultraviolet laser, removes material from the middle layer to make the cornea steeper or flatter. In PRK the surgeon applies the laser directly to the cornea to reshape it.

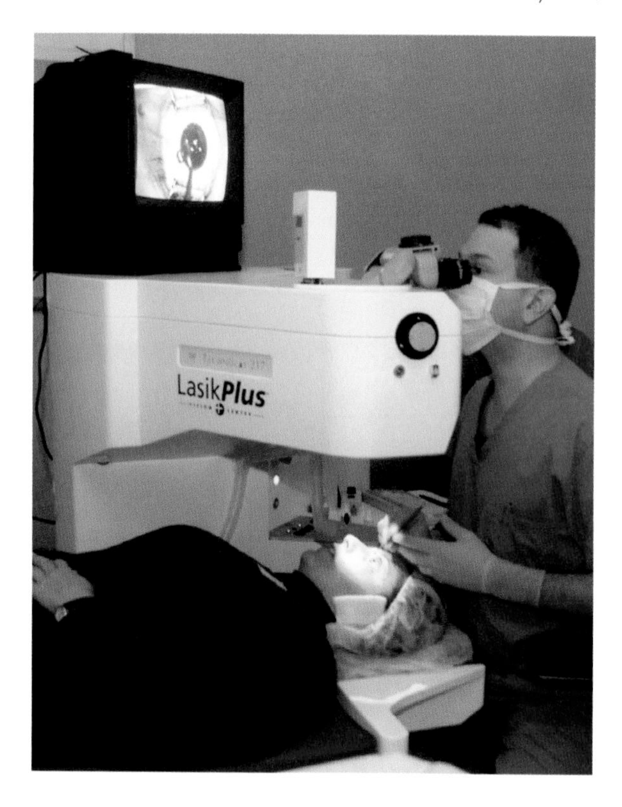

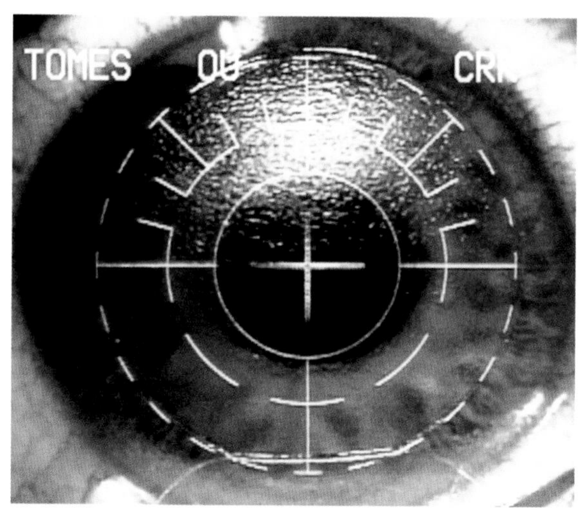

PRECISION VISION. (Left) An ophthalmologist performs laser LASIK eye surgery. (Top) The television monitor mounted above the laser machine shows the image from the operating microscope as the doctor precisely aligns the center of the pupil and the cold ultravioletexcimer laser.

»»See Also

RELATED PRINCIPLES: Binary Code, 190 • Stimulated Emission, 230 • Electromagnetic Induction, 157 • Electromagnetism, 147 • Ohm's Law, 164 • Optics, 225
RELATED APPLICATIONS: Age and Vision, 225 • Electrical Wiring and Power Grid, 166 • Eyeglasses and Contact Lenses, 226 • How We See, 227 • Laser, 232 • Personal Computer, 195

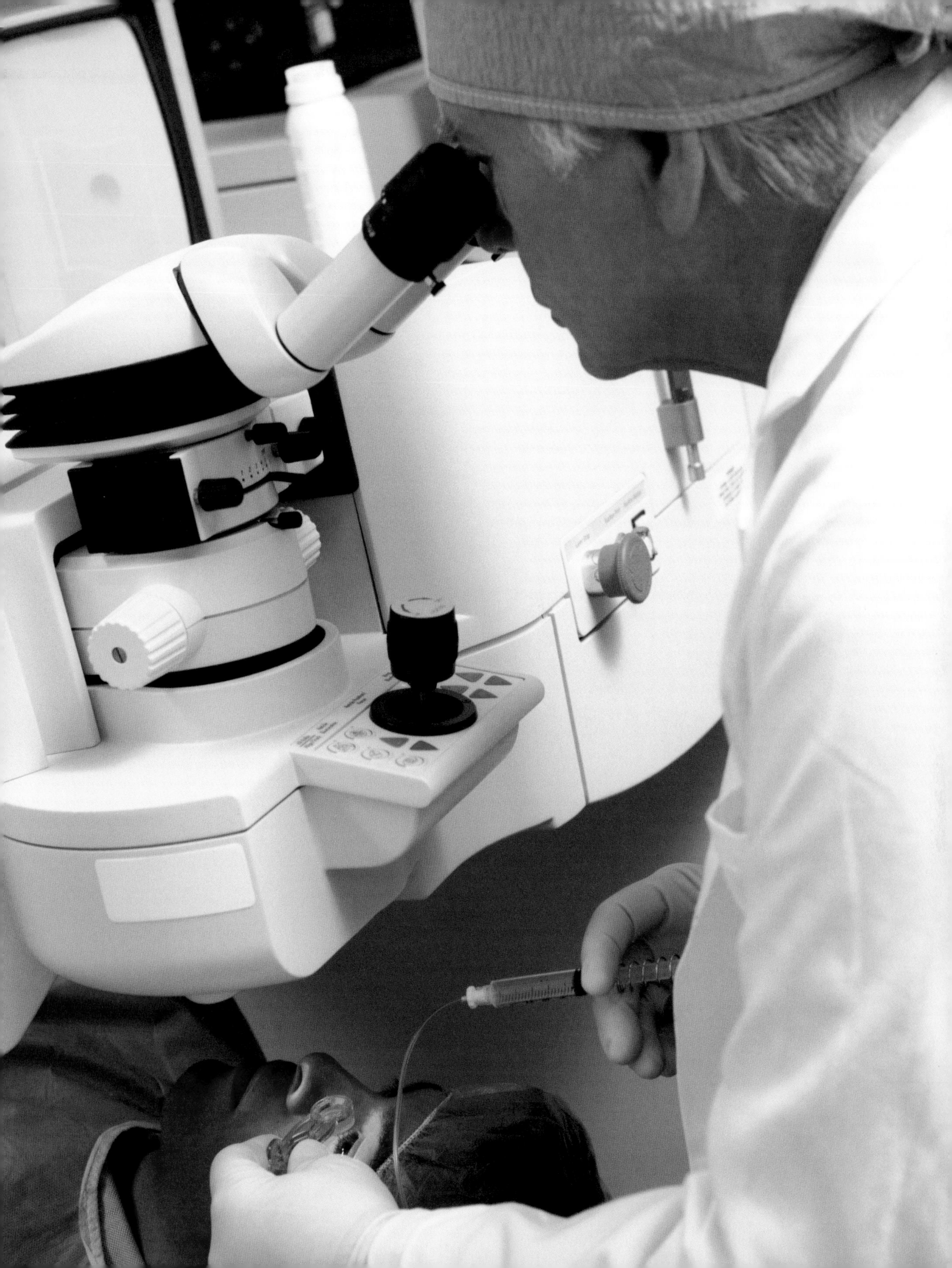

ABOUT THE AUTHORS

MECHANICS
BIOLOGY AND MEDICINE

KATHARINE GREIDER

Katharine Greider is a freelance writer living in New York City. Her work has appeared in dozens of national and local publications. She is the author of two books, most recently *The Archaeology of Home: An Epic Set on a Thousand Square Feet of the Lower East Side* (PublicAffairs, 2011).

NATURAL FORCES

LISA McCOY

Lisa McCoy has been a freelance writer and editor for more than thirteen years. Her work covers a wide range of subjects but is focused primarily on technology and medicine. She has authored six books and edited hundreds. A recent short story she wrote was awarded an Honorable Mention in the L. Ron Hubbard's Writers of the Future contest. When she isn't writing or editing, McCoy competes with her springer spaniels in obedience, agility, tracking, and hunt work.

MATERIALS AND CHEMISTRY

KELLY KAGAMAS TOMKIES

Kelly Kagamas Tomkies is an author, freelance writer, and editor. In addition to her contributions to two National Geographic books, she is the author of three Career Launcher books, two Career Guides for Vault.com, and has written articles for *USA Today*, *Business First*, and many others. She lives in Columbus, Ohio, with her husband, Kevin, and her son, Duncan.

ILLUSTRATIONS CREDITS

All artwork is from the National Geographic book ***How Things Work*** **by Pete Samek, Andy Christie, and Bryan Christie unless otherwise noted.**

2–3, Steve Goodwin/iStockphoto; 10–11, Richard McGuirk/iStockphoto; 12–13, TheCrimsonMonkey/iStockphoto; 4 (LE), jabiru/Shutterstock; 4 (CTR LE), jordache/Shutterstock; 4 (CTR), Mark Thiessen, NGS; 4 (CTR RT), Rich Carey/Shutterstock; 4 (RT), Stephen Mcsweeny/Shutterstock; 6, Sarah Leen/National Geographic Stock; 14, Sebastian Kaulitzki/Shutterstock; 16 (LE), EpicStockMedia/Shutterstock; 16 (CTR LE), Michael Coddington/Shutterstock; 16 (CTR RT), Vibrant Image Studio/Shutterstock; 16 (RT), Wittaya Leelachaisakul/Shutterstock; 18, Michel Hans/Vandystadt/Science Source; 20 (LE), Segway Inc.; 20 (RT), Segway Inc.; 21 (LE), Print Matters Inc.; 21 (RT), David Yu/Shutterstock; 22 (LO), Andrej Pol/Shutterstock; 23 (RT), Christopher Parypa/Shutterstock; 24, Mediagram/Shutterstock; 26 (UP), Luis Miguel Cortes/National Geographic Your Shot; 26 (LO), Ted Kinsman/Science Source; 27, David Liittschwager/National Geographic Stock; 28 (UP), Mikhail Starodubov/Shutterstock; 29, NASA; 30, nirijkedar/Shutterstock; 31, U.S. Department of Energy/www.fueleconomy.gov; 32–33, IM_photo/Shutterstock; 34–35, jabiru/Shutterstock; 35 (RT), Joy Fera/Shutterstock; 36 (LE), AP Photo/Mike Egerton, PA; 36–37, Vit Kovalcik/Shutterstock; 38, Christian Lagerek/Shutterstock; 40, Christian Lagerek/Shutterstock; 41, pryzmat/Shutterstock; 42–43, Rena Schild/Shutterstock; 44, Shutterstock; 45, Thomas Barrat/Shutterstock; 46, Darren Baker/Shutterstock; 47 (LO), Studio Wood Ronsaville Harlin, Inc.; 47 (UP), lebanmax/Shutterstock; 48 (LO), Michael Coddington/Shutterstock; 50, yuyangc/Shutterstock; 51 (UP), Borislav Bajkic/Shutterstock; 52, Andrey Eremin/Shutterstock; 53, Dimitry Kalinovsky/Shutterstock; 54, Stokkete/Shutterstock; 55 (LO), AP Photo/Kathy Willens; 55 (UP), Maxisport/Shutterstock; 56, Zacarias Pereira da Mata/Shutterstock; 58, Claus Lunau/Science Source; 59, EpicStockMedia/Shutterstock; 60, AP Photo/Sadatsugu Tomisawa; 61, Robert W. Tope, Natural Science Illustrations; 62, Lamainix/Shutterstock; 63, Andy Dean Photography/Shutterstock; 65 (UP), Simon Pedersen/Shutterstock; 65 (LO), Tyler Olson/Shutterstock; 66, Serg Shalimoff/Shutterstock; 68, Web Picture Blog/Shutterstock; 69, stocker1970/Shutterstock; 70 (UP), Sally J. Bensusen, Visual Science Studio; 70 (LO), Vibrant Image Studio/Shutterstock; 71 (RT), AP Photo/Michael Sohn; 72 (LO), Sally J. Bensusen, Visual Science Studio; 73, Kellis/Shutterstock; 75, Iscatel/Shutterstock; 76, Jannis Tobias Werner/Shutterstock; 77 (UP), Inga Ivanova/Shutterstock; 78, R-O-M-A/Shutterstock; 79 (LE), ambrozinio/Shutterstock; 80, thewada1976 /Shutterstock; 81, totophotos/Shutterstock; 83 (LO), Rechitan Sorin/Shutterstock; 83 (UP), Morphart Creation/Shutterstock; 85, Aleksey Stemmer/Shutterstock; 86–87, ded pixto/Shutterstock; 88, Igor Leonov/Shutterstock; 89, AP Photo/Daniel Ochoa De Olza; 90, OlegD/Shutterstock; 91 (LO), AP Photo/Tatro; 91 (UP), Stephen VanHorn/Shutterstock; 92, Robert W. Tope, Natural Science Illustrations; 93, Volodymyr Goinyk/Shutterstock; 94 (UP), Aleksey Stemmer/Shutterstock; 95 (UP), AP Photo/Department of Defense; 96, Mechanik/Shutterstock; 98, North Wind Picture Archives via AP Images; 99, Kinetic Imagery/Shutterstock; 101 (LE), Anna Hoychuk/Shutterstock; 102, Creative Travel Projects/Shutterstock; 103, ssuaphotos/Shutterstock; 104 (UP), F.Schmidt /Shutterstock; 105, Wittaya Leelachaisakul/Shutterstock; 106–107, Anettphoto/Shutterstock; 108, Mattia Menestrina/Shutterstock; 109, AP Photo/NOAA; 110, mikeledray/Shutterstock; 111, Tamara Kushch/Shutterstock; 112, trekandshoot/Shutterstock; 113, javi_indy/Shutterstock; 114, Andraž Cerar/Shutterstock; 115, Vit Kovalcik/Shutterstock; 116–117, Lawrence Manning/Corbis; 118 (LE), Federico Rostagno/Shutterstock; 118 (CTR LE), Tatiana Popova/Shutterstock; 118 (CTR RT), Balefire/Shutterstock; 118 (RT), Bizroug/Shutterstock; 120–121, Christophe Michot/Shutterstock; 122, Stephen Mcsweeny/Shutterstock; 123, Ted Kinsman/Science Source; 124–125, European Space Agency; 126 (UP), hxdbzxy/Shutterstock; 127 (LE), iurii/Shutterstock; 128–129, David Persson/Shutterstock; 130 (LO), fluidworkshop/Shutterstock; 130 (UP), Anna Morgan/Shutterstock; 131, Alex Mit/Shutterstock; 132, Albert Siegel/Shutterstock; 133, Jose Antonio Sanchez/Shutterstock; 134, Tek Image/Science Source; 136, kuppa/Shutterstock; 137 (UP), NASA; 137 (LO), Milagli/Shutterstock; 138, Rich Carey/Shutterstock; 139, Diego Cervo/Shutterstock; 140, Konstantin L /Shutterstock; 141 (UP), Pressmaster/Shutterstock; 141 (LO), stocksolutions/Shutterstock; 143, Mark Harmel/Science Source; 144 (UP), Stocksnapper/Shutterstock; 145, Shanghai Daily - Imaginechina via AP Images; 146, roadk/Shutterstock; 147 (LO), Friedrich Saurer/Science Source; 148 (LO), Amra Pasic/Shutterstock; 148 (UP), Alexander Demyanenko /Shutterstock; 149 (UP), Vladnik/Shutterstock; 150 (UP), CERN; 150 (LO), CERN; 151, NASA; 152 (RT), Scott Camazine/Science Source; 152 (LE), Pavel L Photo and Video/Shutterstock; 155 (UP), Nikada/iStockphoto; 156, KROMKRATHOG/Shutterstock; 157, Lisa F. Young/Shutterstock; 160 (LE), Blaz Kure/Shutterstock; 160 (RT), Rob Hyrons/Shutterstock; 161 (LO), Viappy/Shutterstock; 162–163, michaket/Shutterstock; 164, Rick Parsons/Shutterstock; 165 (LE), Library of Congress Prints & Photographs Division, LC-B2-1026-9; 165 (RT), Leo Blanchette/Shutterstock; 166 (LO), sydeen/Shutterstock; 168 (LO), timy/Shutterstock; 168 (UP), Maksym Bondarchuk/Shutterstock; 170 (UP), pryzmat/Shutterstock; 171 (UP), Khomulo Anna/Shutterstock; 171 (LO), Boyan Dimitrov/Shutterstock; 173, Liu Jiao -Imaginechina via AP Images; 174, Rob Wilson/Shutterstock; 175, Steve Mann/Shutterstock; 176–177, Brian A. Jackson/Shutterstock; 176 (LE), Teun van den Dries /Shutterstock; 178, NASA; 179, Markus Gann/Shutterstock; 180 (UP), Craig Kiefer; 180 (LO), Vaclav Volr/Shutterstock; 182, Federico Rostagno/Shutterstock; 183, neijia /Shutterstock; 184–185, David Nunuk/Science Source; 186, Studiotouch/Shutterstock; 187, Coprid/Shutterstock; 188 (UP), GIPhotostock/Science Source; 188 (LO), Mona Makela/Shutterstock; 189 (RT), PRNewsFoto/DuPont/AP Images; 189 (LE), Hywit Dimyadi/Shutterstock; 190, S.Borisov/Shutterstock; 191 (UP), Nils Petersen/Shutterstock; 192 (UP), Oleksiy Mark/Shutterstock; 192 (LO), Maxim Blinkov/Shutterstock; 193, chaoss/Shutterstock; 194, Morgan Lane

Photography/Shutterstock; 195, Africa Studio/Shutterstock; 196 (UP), Sashkin/Shutterstock; 197, Adam Gregor/Shutterstock; 198 (UP), kubais/Shutterstock; 199, ArchMan/Shutterstock; 200 (UP), Tatiana Popova/Shutterstock; 200 (LO), Quka/Shutterstock; 201, AP Images/Associated Press; 202, FloridaStock/Shutterstock; 204, Teri J. McDermott, M.A./Phototake; 205, Vinicius Tupinamba/Shutterstock; 206–207, GoodMood Photo/Shutterstock; 208 (LO), Press Association via AP Images; 208 (UP), Nils Z/Shutterstock; 209, Grande-Duc/Shutterstock; 210, Chubykin Arkady/Shutterstock; 211 (UP), Dja65/Shutterstock; 212, NASA; 213, NASA; 214, Geoffrey Jones/Shutterstock; 215, Chrisferra/Shutterstock; 216, Sandia National Laboratories; 217 (UP), Yellowj/Shutterstock; 217 (LO), AlikeYou/Shutterstock; 218, James Doss/Shutterstock; 219, NOAA; 220 (UP), Katarzyna Mazurowska/Shutterstock; 221, AP Photo/Itsuo Inouye; 222, Jim Barber/Shutterstock; 223 (LO), Balefire/Shutterstock; 223 (UP), basel101658/Shutterstock; 224, (NYC Street Scene), composite image created using Shutterstock images: Stuart Monk/Shutterstock; (hands holding glasses), rangizzz/Shutterstock; 225, Jorg Hackemann/Shutterstock; 226, Gang Liu/Shutterstock; 227 (RT), Daniel Rajszczak/Shutterstock; 227 (LE), Alila Medical Images/Shutterstock; 228, asharkyu/Shutterstock; 231, Kyodo via AP Images; 232–233, GIPhotoStock/Science Source; 234 (UP), Human Media Lab/Rex Features via AP Images; 234 (LO), A.Punpleng/Shutterstock; 235, AP Photo/Jersey Government; 237, Miles Boyer/Shutterstock; 238–239, Rebecca Pitt, for the exhibit "Discovering Particles: Fundamental Building Blocks of the Universe" (University of Birmingham and University of Cambridge); 240, Gines Romero/Shutterstock; 241, Ensuper/Shutterstock; 242, REDAV/Shutterstock; 243, Odua Images/Shutterstock; 244 (LO), Pincasso/Shutterstock; 245, European Space Agency; 246 (LO), arka38/Shutterstock; 246 (UP), DeiMosz /Shutterstock; 247 (LO), Alexander Kirch/Shutterstock; 247 (UP), Isma Riza/Shutterstock; 248, fluidworkshop/Shutterstock; 249 (LE), Kevin Tietz/Shutterstock; 250 (UP), trailexplorers/Shutterstock; 251 (UP), VladKol/Shutterstock; 251 (LO), pedrosala/Shutterstock; 252, Mikael Damkier/Shutterstock; 253, Burben/Shutterstock; 254, Bizroug/Shutterstock; 255, Paul Hakimata Photography/Shutterstock; 256 (UP), Andrea Danti/Shutterstock; 256 (LO), Monkey Business Images/Shutterstock; 257, Paul Faith/PA Wire URN:14461022 (Press Association via AP Images); 258 (LE), Rido/Shutterstock; 258 (CTR LE), moneymaker11/Shutterstock; 258 (CTR RT), M. Shcherbyna/Shutterstock; 258 (RT), Rod Ferris/Shutterstock; 260, isak55/Shutterstock; 262, Imagewell/Shutterstock; 263, Mitar Vidakovic/Shutterstock; 264 (LO), Ken Schulze/Shutterstock; 264 (UP), BKingFoto/Shutterstock; 265 (UP), AP Photo/Juan Karita; 265 (LO), Dennis Steen/Shutterstock; 266, Lisovskaya Natalia/Shutterstock; 267 (UP), coprid/Shutterstock; 267 (LO), Dimec/Shutterstock; 268 (UP), MarcelClemens /Shutterstock; 268 (LO), moneymaker11/Shutterstock; 269 (LO), David Huntley Creative/Shutterstock; 269 (UP), Fribus Ekaterina/Shutterstock; 270 (LO), Rod Ferris/Shutterstock; 270 (UP), hddigital/Shutterstock; 271 (LO), Dennis Burdin/Shutterstock; 271 (UP), Iamnao/Shutterstock; 272 (UP), bornholm/Shutterstock; 272 (LO), jordache/Shutterstock; 273, Barna Tanko/Shutterstock; 274–275, Nikita G. Sidorov/Shutterstock; 276, kingero/Shutterstock; 277, amybbb/Shutterstock; 278, Natalia7/Shutterstock; 279, Lisa S./Shutterstock; 280 (LO), AP Photo; 280 (UP), Dmitry Kalinovsky/Shutterstock; 281, Beto Chagas/Shutterstock; 282, anaken2012/Shutterstock; 283, ene/Shutterstock; 284 (UP), Thomas M. Perkins/Shutterstock; 284 (LO), Neil Bradfield/Shutterstock; 285, FloridaStock/Shutterstock; 286, NitroCephal/Shutterstock; 287 (UP), DGF72/Shutterstock; 287 (LO), bikeriderlondon/Shutterstock; 288, keantian/Shutterstock; 289, videnko/Shutterstock; 290, nobeastsofierce/Shutterstock; 291, Nadezda Cruzova/Shutterstock; 292, Evlakhov Valeriy/Shutterstock; 293, Rex Features via AP Images; 294, Ruggiero Scardigno/Shutterstock; 295, AP Photo/Jae C. Hong; 296, ogwen/Shutterstock; 298, Andris Torms/Shutterstock; 299, Jan-Peter Kasper/picture-alliance/dpa/AP Images; 300–301, Africa Studio/Shutterstock; 302, M. Shcherbyna/Shutterstock; 303, simonalvinge/Shutterstock; 304 (UP), Peter Sobolev/Shutterstock; 304 (LO), Peter Sobolev/Shutterstock; 305, AP Photo/Mike Groll; 306 (UP), AP Photo/U.S. Army Dental and Trauma Research Detachment; 306 (LO), STILLFX/Shutterstock; 307, gosphotodesign/Shutterstock; 308 (LO), Pilkington Deutschland AG; 308 (UP), MANDY GODBEHEAR/Shutterstock; 309 (RT), Rido/Shutterstock; 309 (LE), Adrov Andriy/Shutterstock; 310 (LE), Christopher Meade/Shutterstock; 310 (CTR LE), Asianet-Pakistan/Shutterstock; 310 (CTR RT), AP Photo/John Chadwick; 310 (RT), Fotokostic/Shutterstock; 312–313, Bowen Clausen Photography/Shutterstock; 314, Ehpoint/Shutterstock; 315, Marcin Balcerzak/Shutterstock; 316 (RT), Gtranquillity/Shutterstock; 316 (LE), jeka84/Shutterstock; 317 (LO), Subbotina Anna/Shutterstock; 317 (UP), SeDmi/Shutterstock; 318 (LE), Elena Elisseeva/Shutterstock; 318 (RT), Jesus Cervantes/Shutterstock; 319 (LE), rebvt/Shutterstock; 319 (RT), Quayside/Shutterstock; 320, Sebastian Kaulitzki/Shutterstock; 321 (UP), Tomislav Pinter/Shutterstock; 321 (LO), Alexey Lysenko/Shutterstock; 322, hinnamsaisuy/Shutterstock; 323, Fotokostic/Shutterstock; 324, Michael Hero/Shutterstock; 325, fotogenicstudio/Shutterstock; 326, Vladislav Gajic/Shutterstock; 328, donatas1205/Shutterstock; 329, ggw1962/Shutterstock; 330, saddako/Shutterstock; 331, Jiri Hera/Shutterstock; 332–333, Markus Mainka/Shutterstock; 334, AP Photo/Michael Stravato; 335, NASA; 336–337, Kjpargeter/Shutterstock; 338, Vasilius/Shutterstock; 339, Kirk Moldoff; 340–341, Yu Lan/Shutterstock; 342, AP Photo/John Chadwick; 343, Reuters/Francois Lenoir/Landov; 344, Monkey Business Images/Shutterstock; 345, Monkey Business Images/Shutterstock; 346, Lightspring/Shutterstock; 347, Lightspring/Shutterstock; 348–349, Christopher Meade/Shutterstock; 350, Jens Goepfert/Shutterstock; 351, Spectral-Design/Shutterstock; 352, Teri J. McDermott, M.A./Phototake; 353, AP Photo/Dr. Carl June; 354, design36/Shutterstock; 356, PLRANG/Shutterstock; 357, Alila Medical Images/Shutterstock; 358–359, ggw1962/Shutterstock; 360 (RT), Alexander Raths/Shutterstock; 360 (LE), Christian Delbert /Shutterstock; 361, Asianet-Pakistan/Shutterstock; 362, Anelina/Shutterstock; 363, Elena Elisseeva /Shutterstock; 364, Rebecca Hale/NGS; 365, Lightspring/Shutterstock; 366, Lightspring/Shutterstock; 368, Picsfive/Shutterstock; 369, Teri J. McDermott, M.A./Phototake; 370, Mark Thiessen/NGS; 371, Bryan Christie Design/National Geographic Stock; 372 (LE), dpa Universitîtsklinik Magdeburg/picture-alliance/dpa/AP Images; 372 (RT), Sebastian Kaulitzki/Shutterstock; 373, AP Photo/St. Joseph News-Press, Ival Lawhon Jr.; 374, Rolf Vennenbernd/picture-alliance/dpa/AP Images; 375, Reuters/Courtesy of Intuitive Surgical/Landov; 376 (LE), kurhan/Shutterstock; 377, Alexander Mozymov/Shutterstock; 378 (LE), AP Photo/Al Behrman; 378 (RT), AP Photo/Charles Rex Arbogast; 379, Monkey Business Images/Shutterstock.

INDEX

C

D

F

J

K

L

Q

R

S

NATIONAL GEOGRAPHIC
SCIENCE OF EVERYTHING

PUBLISHED BY THE NATIONAL GEOGRAPHIC SOCIETY

John M. Fahey, Chairman of the Board and Chief Executive Officer

Declan Moore, Executive Vice President; President, Publishing and Travel

Melina Gerosa Bellows, Executive Vice President; Chief Creative Officer, Books, Kids, and Family

PREPARED BY THE BOOK DIVISION

Hector Sierra, Senior Vice President and General Manager

Janet Goldstein, Senior Vice President and Editorial Director

Jonathan Halling, Design Director, Books and Children's Publishing

Marianne R. Koszorus, Design Director, Books

R. Gary Colbert, Production Director

Jennifer A. Thornton, Director of Managing Editorial

Susan S. Blair, Director of Photography

Meredith C. Wilcox, Director, Administration and Rights Clearance

STAFF FOR THIS BOOK

Susan Hitchcock, Senior Editor

Gail Spilsbury, Project Editor

Elisa Gibson, Art Director

Marshall Kiker, Associate Managing Editor

Judith Klein, Production Editor

Galen Young, Rights Clearance Specialist

Katie Olsen and Linda Makarov, Production Designers

James S. Trefil, Chief Science Consultant

Developed and produced by Print Matters, Inc.
(www.printmattersinc.com)

PRODUCTION SERVICES

Phillip L. Schlosser, Senior Vice President

Chris Brown, Vice President, NG Book Manufacturing

George Bounelis, Vice President, Production Services

Nicole Elliott, Manager

Rachel Faulise, Manager

Robert L. Barr, Manager

Art Hondros, Imaging Technician

The National Geographic Society is one of the world's largest nonprofit scientific and educational organizations. Founded in 1888 to "increase and diffuse geographic knowledge," the Society's mission is to inspire people to care about the planet. It reaches more than 400 million people worldwide each month through its official journal, *National Geographic*, and other magazines; National Geographic Channel; television documentaries; music; radio; films; books; DVDs; maps; exhibitions; live events; school publishing programs; interactive media; and merchandise. National Geographic has funded more than 10,000 scientific research, conservation, and exploration projects and supports an education program promoting geographic literacy. For more information, visit www.nationalgeographic.com.

For more information, please call 1-800-NGS LINE (647-5463) or write to the following address:

National Geographic Society
1145 17th Street N.W.
Washington, D.C. 20036-4688 U.S.A.

For information about special discounts for bulk purchases, please contact National Geographic Books Special Sales: ngspecsales@ngs.org

For rights or permissions inquiries, please contact National Geographic Books Subsidiary Rights: ngbookrights@ngs.org

Library of Congress Cataloging-in-Publication Data

National Geographic science of everything : how things work in our world / foreword by David Pogue.
p. cm.
ISBN 978-1-4262-1168-3 (hardback)
1. Science--Miscellanea. I. National Geographic Society (U.S.) II. Title: Science of everything.
Q173.N275 2013
500--dc23

2013012981

Copyright © 2013 National Geographic Society
All rights reserved. Reproduction of the whole or any part of the contents without written permission from the publisher is prohibited.

ISBN: 978-1-4262-1168-3 (regular)
ISBN: 978-1-4262-1320-5 (deluxe)
Printed in the United States of America
13/RRDW-CML/1